Handbook
of
Genetics

Handbook of Genetics

Classical Genetics, Molecular Genetics, Gytogenetics, Plant Breeding, Biotechnology, & Evolution

Dr. A.V.S.S. Sambamurty
Reader in Botany
Sri Venkateshwara College
University of Delhi
New Delhi - 110021

Shri Sai Printographers
New Delhi - 110002

Handbook of Genetics
Author - Dr. A.V.S.S. Sambamurty

1st Indian Edition

ISBN - 978-81-87798-03-3

Published by
Shri Sai Printographers
102, 4787/23, Ansari Road, Daryaganj,
New Delhi - 110002
Phone - 011-23246222, 9312220441
e-mail : shrisaiprintographers@gmail.com

Typsetting by
Shri Sai, Daryaganj, Delhi - 110002

Cover Design by
Kamini Bijoria, Daryganj New Delhi - 110002

Printed in India by
Milan Entreprises, Daryaganj, New Delhi - 110002

Pages: 382/Paper - 80gsm
Size of Book: 5.5X8.5 | Price - Rs. 1995.00

DEDICATED TO

SHIRDI
KE
SHRI SAI BABA

PREFACE

The scope of Genetics is enormous in the present day. The present book '**Handbook of Genetics**' has been designed as a dictionary giving as many genetic terms as possible and defining them, sometimes giving elaborate meanings in the context of evolution, genetics, breeding, molecular genetics, eugenics, biotechnology, genetic engineering or recombinant DNA technology from a classical to modern sense of Genetics. It is a synthesis of genetics and evolution through technical terms which will be found at one place in one text here. For want of space diagrams are avoided; moreover this is not a text book for formal genetics students. Certain terms used in this book have far reaching conclusions in evolution closely related to genetics and eugenics. A formal knowledge of chemistry and biology is needed to understand the intricacies of the meanings.

The book consists of four parts, viz., Part I, deals with genetic terms closely associated with evolution; Part II deals with terms from a molecular and cytogenetic point of view; and Parts III & IV deal with biotechnology and cloning which occupy a pivotal position in modern genetics.

This book can be used as a reference and as an adjunct for the study of Genetics by all students of Botany, Zoology, Chemistry, Biochemistry, Microbiology, Medicine, Genetics, Pharmacy, Biotechnology, Immunology, Instrumentation, Biophysics and Biomedical Engineering.

The author expresses his sincere thanks to M/s. SHRI SAI PRINTOGRAPHERS for bringing out the book neatly and in a short time. He also expresses indebtedness to his family members for their patience and interest in the book.

New Delhi **_A. V.S.S.Sambamurty_**

CONTENTS

Part - 1

Genetic terms closely associated with evolution

A

A chromosomes, the normal chromosomes in a eukaryote. Occasionally extra chromosomes occur, which are known as>B chromosomes.

ABO blood group system, one of several genetic systems of antigens found on the surface of red blood cells in humans. It is medically important, because in a blood transfusion the donor's ABO blood group must be compatible with the recipient's. There are three alleles involved: G^A, G^B, and G^o; G^o is recessive to both the others, while G^A and G^B are codominant. The blood groups of the six possible genotypes are therefore as follows:

G^AG^A = group A

G^AG^B = group AB

G^AG^O = group A

G^BG^B = group B

G^BG^O = group B

G^OG^O = group O

The alleles G^A and G^B code for making an antigen, which is what is known as A or B; G^o makes no antigen. People with blood group A have in their bloodstream antibodies to B (known as anti-B), and people with blood group B have antibodies to A (anti-A); people with blood group O have antibodies to both A and B; and people with blood group AB have neither of the antibodies. In a blood transfusion, the person's circulating antibody would attack any incoming blood that had the corresponding antigen, setting off an adverse reaction within the bloodstream that could be fatal. Thus it would be dangerous to give group A blood to a person with blood group B (and therefore with anti-A antibodies). It follows that group O blood can be given in transfusion to a person of any type, as there are no antigens (hence O group is termed 'universal donor'), and conversely a person who is AB can receive blood of any group, as there are no antibodies (hence 'universal recipient').

The G^o allele is much the commonest in humans, being at an average global frequency of 62 per cent. European populations are about 68 percent G^o, 26 per cent G^A and 6 per cent G^B. The place with the highest frequency of G^B is north India, with about 30 per cent; this allele is virtually absent from Amerindian populations.

Abortion, the expulsion of the foetus from the uterus. Spontaneous abortion (commonly known as 'miscarriage') is quite common, occurring in perhaps as many as 20 per cent of human pregnancies, particularly within the first two months. About half these foetuses have chromosome abnormalities. Some diseases cause abortion, e.g. brucellosis in cattle, and many species will spontaneously abort if disturbed: for example, the smell of a male mouse other than her mate causes a female mouse to lose her litter. Induced abortion of normal foetuses is very widely used as a method of birth control legally in about half the countries of the world and illegally else where. Therapeutic abortion is practised in advanced countries to prevent the birth of deformed or diseased foetuses. (The word 'therapeutic' is curious in the context, as it means curative: the mother cannot be the person cured, as she was not ill, and neither can it be the foetus, who ends up dead.) In 1988, nearly one in five pregnancies in Britain were terminated, and of these 180,000 abortions, some 5 per cent were performed because of the risk that the foetus might be abnormal.

Acceleration, the appearance of a structure in >embryogenesis earlier than in ancestral forms. The opposite is >retardation.

Acceleration, rule of, the rule that in an organism's development the structures that are formed first are the ones most important to the organism's overall function.

Acentric chromosome, a chromosome with no >centromere. This may be merely a large part of a chromosome that has broken off from another one. Acentrics tend to get lost because they cannot be manoeuvred during the normal cell-division process.

Acetabularia, a single-celled green alga; a eukaryote (unlike the blue-green algae), with a diploid chromosome number of about 20. Because it is so large (about 1 cm long) it is useful for experiments in which a cell's nucleus is removed and implanted into a denucleated specimen of a different strain or species, thus clarifying the role of chromosomal versus >cytoplasmic inheritance.

Achiasmate meiosis, a >meiosis in which there are no chiasmata (see chiasma), i.e. no >crossing-over takes place. This happens in some species, including *Drosophila melanogaster*, where only the homogametic sex has crossing-over.

Achondroplasia, a human disease caused by a dominant allele, Easily recognizable at birth, the features of the disease are very short limbs and a large head (but usually a small nose); adults grow to a height of 4 feet (1.2 m) or less. Achondroplastics are perfectly healthy and have normal intelligence, fertility and so on. About one baby in 50,000 is born with achondroplasia. Some 80 per cent of them have normal parents, which means that the achondroplasia gene must have arisen by a new> mutation in these cases, which seems to happen slightly more often when the father is over the age of 30. People with achondroplasia tend to marry others with the same condition, i.e. the marriage is between heterozygotes (Aa, where A is the dominant allele for achondroplasia and a the corresponding normal one). Their offspring are therefore expected to be ¼ aa (normal), ½ Aa (achondroplastic) and ¼ AA. These last are not in fact the same phenotyically as their Aa parents. The A allele is strictly speaking semi-lethal, with the homozygote more severely affected than the heterozygote; AA babies die shortly after birth.

Acquired characteristics, inheritance of, a mistaken idea that has dogged genetics for years and still persists;. It seems to stand to reason that if an animal acquires at some point during its lifetime a trait that is useful or that helps it to survive, it will pass it on to its offspring (and that in this way evolutionary progress will be made). It is plausible, but it does not happen. Wallace, co-proposer of the theory of evolution by natural selection, wrote an excellent refutation:

'Neither did the giraffe acquire its long neck by desiring to reach the foliage of more lofty shrubs, and constantly stretching its neck for the purpose, but because any varieties which occurred among its antetype with a longer neck than usual at once secured a fresh range of pasture over the same ground as their shorter-necked companions, and on the first scarcity of food were thereby enabled to outlive them.'

Let us look at Wallace's example in genetic terms. Even if a giraffe could make its neck longer by stretching, its offspring would have a normal-length neck, and each would have to do its own stretching to get a longer one. If there was a gene n for neck length, all giraffes were of genotype nn, and all their offspring would be nn, normal-length neck. But suppose a mutation arises, N, which is dominant over n, and gives a longer neck; then the giraffes that carry this gene (genotype Nn or NN) are longer-necked than the normal type, with the result that Wallace describes.

The point is that there is no way in which anything that happens to the individual giraffe's neck can be incorporated into what it passes on to its descendants. The only genetic route from one generation to the next is via the DNA in the chromosomes, and this DNA is always (barring the very rare occurrence of mutation) an exactly faithful copy of the DNA in the preceding generation.

The key figure in disproving the inheritance of acquired characteristics was the German zoologist August > Weismann. He wrote several impeccably argued treatises (bases on theory rather than on experimental results), including Das Keimplasma (1892, trans. The Germ-Plasm: A Theory of Heredity). His insistence that heredity must reside in units unaffected by what happens to the individual's own body was soon supported by Mendel's findings, which came to light in 1900. Since that time the idea of the inheritance of acquired characteristics has been rejected by scientists.

Acrocentric chromosome, a chromosome whose >centromere is towards one end. If the centromere is right at the end, the chromosome is called >telocentric.

Acromegaly. See growth hormone.

Active immunity, immunity to a disease that has been acquired through previous exposure to it. See immune system; passive immunity.

Adaptation, the acquisition of characteristics that increase an organism's reproductive success in a particular environment. Adaptation can happen to an individual organism if it carries a favourable mutation, but in evolution it is a property of populations, not individuals. From the evolutionary point of view it is only reproductive success that counts, not the efficiency or otherwise of the animal's interaction with its environment. See also acquired characteristics, inheritance of.

Adaptive radiation, the evolutionary diversification of a number of species from a common ancestral species, made possible by the appearance of vacant ecological niches into which new species can move. This can happen after a catastrophe or a colonization.

Adaptive value, the amount of increase or decrease in reproductive > fitness (relative to other genotypes and in a particular environment) that a given genotype or mutation confers.

Additive variance, the component of >genetic variance that is due to the additive effect of two or more genes on the trait in question, i.e. excluding any dominance or interaction.

Adenine (abbreviation: A), one of the four nitrogenous bases which form the core of both DNA and RNA. It pairs with thymine (T) in DNA and uracil (U) in RNA. See also replication.

Adenosine deaminase (ADA), an enzyme involved in the immune system. Lack of it is one of the causes of >severe combined immune deficiency.

Adenoviruses, a group of viruses which cause certain respiratory diseases (e.g. laryngitis and some kinds of pneumonia) in humans and other mammals. Their genome consists of double-stranded DNA but contains only about 12 genes.

Adoption, the raising of a baby by a person or persons not its biological parents. It is frequently found in non-human species; orphaned infants are not usually left to die among animals that live in social groups. A macaque monkey mother who has lost her own infant will even kidnap another as a replacement. Legal adoption was formerly a common means of providing a childless couple with a child, or an important man with an heir (such as Augustus Caesar, adopted by Julius), but it now occurs far less frequently in developed countries. The availability of abortion has drastically reduced the number of candidates for adoption (causing some couples to turn to the Third World as a source of babies), and the new reproductive technologies have given other possibilities to childless people. Adopted children have sometimes been used in the investigation of questions to do with > nature versus nurture, such as the heritability of intelligence (see IQ) or obesity; theory predicts that, if environmental influence is more important, they will resemble their adoptive parents more than their biological ones, and vice versa if genotype is more important.

Agamospermy (from the Greek a, 'without', gamos, 'marriage', sperma, 'seed'.) a process of seed formation in plants in which fertilization does not take place. See also apomixis.

Ageing, the phenomenon of progressive failure of self-repair of an organism. The process is not understood. The Various theories include: build-up of >somatic mutations, age as a non-specific >auto-immune disease, and increasing inefficiency of DNA repair leading to >error catastrophe. These are all physiological explanations, but the >disposable soma theory is based on the population geneticists' view of a species' total biological investment. See also tissue culture.

Aggregation chimaera, an organism artificially created in vitro by mixing cells from two genetically different early embryos. The embryo is then reimplanted into a surrogate mother. The individual is a >chimaera, as can easily be seen if two very different strains are used, e.g. one black and one white mouse. See tetraparental mice.

Aggression, any behaviour intended to frighten or harm another. There has been heated discussion (decorated with such cliches as 'nature red in tooth and claw') as to whether aggression is genetically determined. Darwin felt that aggression within and between species must be part of the struggle for existence and hence a necessary feature of evolution.

In his book On Aggression (English edition 1966), Konrad >Lorenz proposed that it is natural, occurs in all species and is genetically determined. He also suggested that in most species there are (genetic) mechanisms that make the individual direct its aggressive behaviour outside its own group, and that in species which have the most dangerous natural weapons (fangs, claws, etc), there are highly developed inhibitory mechanisms to cause individuals to use bluffing, submissive gestures, etc., rather than actually fight and harm one another. He went on to say that the problem with humans was that they were not naturally armed and therefore did not have these inhibitory mechanisms; so once they had armed themselves with artificial weapons, there was nothing to stop them killing one another.

Lorenz's facts were challenged by the sociobiologist E.O. Wilson, who noted that in species that have been observed for long periods in the wild aggression leading to serious injury or death is often seen: hyenas, lions, macaque monkeys and chimpanzees all kill members of their own species, and sometimes eat them too. Nothing that these incidents only become apparent when a species has been observed for at least 1000 hours, Wilson remarked that one murder per 1000 hours is a great deal of violence by human standards. Human aggression is therefore neither unique nor extreme. Sociobiologists believe that aggressive behaviour, including both fighting to the death and ritualized conflicts, is genetically determined, and using >game theory analysis, they calculate the evolutionary advantage to the individual of handling aggression by fighting or by submissive behaviour. See also sociobiology.

Agouti, a coat colour in mammals. The coat appears dull brown, but actually each hair has bands of black and light brown. It is the wild-type colour of mice, rabbits, etc., and is controlled by a dominant allele, A, in a series with numerous other alleles. Variation at the agouti locus is responsible for many of the fancy colours that hobby breeders concentrate on. The word itself is from Guarani, a Brazilian language.

AIDS, acquired immune deficiency syndrome, a viral disease affecting humans and closely related species, in which the body's immune system becomes inoperative so that the patient ultimately dies of infection by some other disease. The agent that causes it was identified in 1983 as the human immunodeficiency virus (HIV), which can only be passed from one person to another when blood, semen or vaginal fluids pass directly into the bloodstream. This can happen during sexual intercourse or as a result of sharing needless for drug injection; and if a donor of blood, semen, breast milk or an organ for transplantation carries the virus, the recipient may become infected. (However, donated blood and semen are now both rigorously treated to avoid this possibility.) Not everyone who has the virus in their bloodstream will develop AIDS, and it is not true that homosexual men are particularly susceptible to either the virus or the disease. The virus is passed on in exactly the same way by AIDS patients and by HIV-positive people (those who have the virus but no symptoms).

AIDS has no genetic component, but there is confusion about this, as a mother who has the virus can transmit it to her baby during pregnancy or birth, or in her breast milk. See also severe combined immune deficiency.

Albinism, a genetically determined condition in which there is no pigmentation in the skin, hair, eyes, etc. The cause is a recessive allele; the corresponding normal allele codes for the enzyme tyrosinase which is one step in the production of > melanin, the pigment that usually gives these tissues their colour. The albino gene in humans is the same as the one in mice, rabbits and other animals, and was one of the very first traits to be identified as showing Mendelian inheritance in humans (reported in 1903). Albino humans are healthy and not at a disadvantage except that they must avoid bright sunlight which damages their skin and eyes. There is an interesting group of North American Indians, the Hopi, who have a relatively high incidence of albinism; this is believed to be due to the fact that albino men and women spend more time inside (and consequently reproductively busy) than their normal comrades. Very occasionally, a different mutant causing albinism arises which is dominant.

Algeny, >genetic engineering. A word formed to be like 'alchemy', algeny includes any sort of manipulation of genetic material in vivo or in vitro.

Alkaptonuria, a human disease caused by recessive allele. The actual mutation is a failure to produce an enzyme that converts homogentisic acid into urea to be excreted in the urine; thus the urine contains homogentisic acid, which turns black. There is sometimes also degeneration of cartilage (especially in the spinal discs). This was one of the >inborn errors of metabolism identified by Sir Archibald >Garrod.

Allele (from the Greek for 'each other') one of the variant forms of a gene. An individual has two alleles for each gene, one on each of the two homologous chromosomes, at the appropriate locus for that gene. If the two alleles are identical, the individual is a >homozygote (in respect of that locus), and if they are different, the individual is a >heterozygote. Each gene may have a very large number of alternative alleles, since an allele need only differ from the original form by one base-pair out of thousands (see mutation). The allele that is the original form is sometimes known as the >wild type. See also dominant; recessive.

Allele-specific oligonucleotide, an artificially produced sequence of nucleotides that is known to be exactly complementary to a particular sequence of >DNA. These sequences, usually about 18 >base-pairs long but sometimes as long as 80 base-pairs, can be used to detect the presence or absence of the sequence in question (perhaps a disease-producing mutant) in a sample of natural DNA: when a quantity of the nucleotide is added to the natural sample, it will bind if it can find its exact complementary sequence, but not otherwise, being accurate to a single base. In a case such as sickle-cell anaemia, where the exact sequence of the mutation site is known, an allele-specific oligonucleotide is an efficient way of detecting the presence of the mutation, and is used in prenatal diagnosis.

Allelic exclusion, the fact that the lymphocytes of an individual who is heterozygous at one of the >immunoglobulin loci only make one of the immunoglobulin types. The mechanism by which the other allele is prevented from being active is not known; it is not a simple case of dominance.

Allelic series, the set of alleles available for a given locus. They can be listed in order of >dominance, though this order might not be a simple hierarchy and there might be some >co-dominance. An example known to mouse breeders is the >agouti series, with five alleles in use **in the fancy** out of some 20 known ones.

Allergy, the body's response to a substance to which it has been previously hypersensitized, by contact with a particular antigen (known as the allergen). If the allergen enters the gastrointestinal tract, the symptoms are vomiting or diarrhoea; if it enters the respiratory tract, the result is sneezing, rusty eyes and inflammation of the membranes; and if it makes contact with the skin, there are rashes or swellings. Almost any molecule can act as an allergen, although some - such as pollen and animal fur-are particularly common. There may be a genetic basis to an individual's liability to become allergic to particular substances. Some patients respond to immunotherapy (repeated small doses of the allergen to alter the hypersensivity response); others are treated with antihistamine.

Allometry, the differential growth of one part of the body compared to another or to the body as a whole. It is one of the most frequent phenomena in evolution (e.g. the positive allometric growth of the human cranium compared with that of the immediately ancestral form), and is perhaps under very simple genetic control: all that is needed is to keep a given gene switched on for longer than was normal.

Allomones, a group of chemicals that act as chemical messengers between members of different species, in the same way as >pheromones do within species. An interesting example is the substance secreted by aggressive species of ants to frighten and subdue other species of ants that they intend to capture as slaves.

Allopatry, the occupation by populations of geographically separate ranges. The formation of new species can occur if two groups of the same species become geographically separated and develop divergent changes in their gene pools: this is allopatric speciation. See also adaptive radiation.

Allophene, an organism that has received a transplant of foreign cells. The transplanted cells have their own genotype different from the host's, and they make their own gene products. An example is a patient who receives a bone-marrow transplant to replace their own defective marrow tissue. The word is also used for >tetraparental mice.

Allopolyploid, a >polyploid organism in which the chromosome sets have come from more than one species. See *Triticum aestivum*; autopolyploid.

Allozyme, one of the variant forms of an >enzyme coded for by alleles at a given locus. See isoenzyme.

Alternation of generations, the system by which plants have a life cycle consisting of a haploid phase and a diploid phase. The diploid stage, which is the 'plant' itself as visible, makes male microspores in the anthers; these divide by meiosis and each produces four microspores. These are the male gametophytes, and are haploid. At the same time the plant also produces female megasporocytes (diploid) which undergo meiosis to give rise to megaspores (haploid): these are the female gametophytes, and they go through some mitotic divisions (see mitosis) but still remain haploid. Fertilization unites a megaspore and a microspore to form a diploid zygote, and the sporophyte phase begins again at the point. In higher plants the sporophyte is much the longest phase, the gametophyte being a transient phase involving a few microscopically small cells.

Altruism, behaviour by one organism that benefits not itself but another. More specifically, from the genetical point of view, it is behaviour that increases the reproductive performance of another. It has been a puzzle to evolutionists to explain how altruism could have evolved, since it appears to be contrary to the way natural selection works (i.e. the genotypes that are 'selected' are those that leave offspring), but it is undeniably common in nature.

There are three categories of altruism:

- straight altruism - e.g. lower-ranking wild turkeys who assist their brothers in inter-group fights over females with which they themselves will not be allowed to mate.
- compulsory altruism - e.g. the killing of one offspring to feed it to the others.
- reciprocal altruism - 'one good turn deserves another', a much rarer phenomenon but sometimes found in bands of baboons which beg food on one occasion and repay it on another.

It must be emphasized that there is no suggestion that altruistic behaviour is in any way conscious; indeed, it is a tenet of sociobiology that altruistic behaviour (like all other sorts of behaviour) is directly dictated by the genes.

Altruism was supposed by some to have evolved because such behaviour was of the benefit to the group (see group selection), but it has been shown that group selection does not work mathematically. The current thinking in sociobiology is that altruism evolved through >kin selection because by sacrificing its own reproductive performance the altruist increases the number of its genes (i.e. identical copies) transmitted by its relatives. Even this has been strongly criticised by geneticists because of the closeness of the relationship needed to make the mathematics work out. One particular case in which kin selection could explain altruism is in the >social insects. In humans it is possible to imagine how kin selection could explain such altruism as a father going without food to fed his children, but it is nowhere near explaining the mass altruism seen in world wars and other disasters. This anomaly is admitted by those sociobiologists who agree that culture makes humans a special case (which is by no means all of them).

Alu sequence, a sequence of highly repetitive DNA in the human genome. It is about 300 base-pairs long, and is repeated 300,000 times, the copies being scattered around the chromosomes. The Alu sequence copies make up some 3 per cent of the total human DNA. The sequence does not seem to have any genetic meaning, but as it is somewhat related to the sequence of one type of RNA, it is possible that the Alu sequence arose by the action of >reverse transcriptase and hence could have originated from >retroviruses. The copies exist in slightly varying forms, and it is the combination of these variants that gives the individuality analysed in >genetic fingerprinting. The Alu sequence got its name from that of the restriction endonuclease, Alu I, by which it was isolated.

Alzheimer's disease, a human disease existing in several forms, one of which may be genetically determined. The symptoms are mental confusion and loss of memory, often with loss of speech and lack of muscular control. Brain scans show deposits of plaques on the nerve cells, which are made of the protein beta-amyloid, and it is an abnormality in this protein that is suspected of being the cause of Alzheimer's disease. Most people who get Alzheimer's disease do not develop it until they are over the age of 70. The early-onset form (developing at about 50)is the one that has a genetic component, the gene responsible having been located on chromosome 21. The late-onset form, which develops when the patient is over 70, is linked to two markers on chromosome 19. The exact mode of inheritance has not been worked out, owing to uncertainties in the interpretation of family histories when most of the people who may have been affected by the disease are already dead and may not have been correctly diagnosed one way or the other.

Amber codon, one of the three >nonsense codons in DNA that code for 'stop' and therefor signal the termination of the protein being synthesized. Amber is UAG. See protein synthesis.

Ames test, a technique for testing whether a substance is a >mutagen. A culture is made consisting of a strain of *Salmonella* that cannot synthesize the amino acid histidine but must get it from its food source. This strain is set in a nutrient medium lacking histidine, and the substance to be tested is added; if the latter is mutagenic, there will be some mutations of the *Salmonella* enabling it to make its own histidine and therefore to survive; if the substance is not mutagenic, nothing will happen.

Amino acids, the sub-units out of which proteins are made. There are 20 that commonly occur in animals, and about 100 more that are rare and found only in plants. Most amino acids can be synthesized in the body, but some, known as the 'essential amino acids', must be got from the diet. The essential amino acids for humans are: isoleucine, leucine, lysine, methionine, phenylalanine, threonine, tryptophan and valine. The list is a little different for other species. See genetic code; protein synthesis.

Amish, the, or the Old Order Amish, a religious community living in three areas of the United States. Descended from a small group of eighteenth century German emigrants, the Amish have kept themselves apart from the rest of the American population in order to preserve their puritanical beliefs. They refuse to use any machinery, and they work their farms with hand or horse-drawn implements.

The Amish are an ideal subject for genetical research: the population size is small (about 50,000) and marriage with outsiders rare, so that there is much inbreeding, which can be traced through the excellent family records that the Amish have always kept. Several genes have been found at high frequency (due presumably to genetic drift), one being a recessive causing the formation of six digits on the hands. It was also among the Amish that the first evidence was found for a gene causing >manicdepression.

Amitosis, a form of cell division in which two new cells are formed without >mitosis. The nuclear membrane simply pinches itself in two, with equal numbers of chromosomes in each half; the genetic composition of the two new cells is therefore not the same. This happens in species that are highly >polyploid.

Amniocentesis, the drawing off of some of the amniotic fluid (the fluid in the sac surrounding the foetus). This fluid contains some cells from the foetus, and these can be cultured for >prenatal diagnosis of genetic defects. The procedure involves putting a needle through the skin of the mother's abdomen and through the wall of the uterus. It cannot be carried out until at least the twelfth week of pregnancy, and then several weeks are needed for tissue culture and tests, so that the results are not known until about the sixteenth week at the earliest. There is a 0.5-1.0 per cent risk of losing the foetus from the procedure itself. See also chorionic villus sampling; polymerase chain reaction.

Amphimixis, sexual reproduction in plants, including both self-fertilization and cross-fertilization.

Anaemias, a group of diseases of the blood, several of which are inherited. See glucose-6-phosphate dehydrogenase deficiency; sickle-cell anaemia; thalassaemia.

Anaphase, one of the phases of the two types of cell division, >meiosis and >mitosis.

Aneuploidy, the condition in which the number of chromosomes in a cell is not an exact multiple of the >haploid number. This can be either because one or more chromosomes have been lost or because of the presence of extra chromosomes. One form of aneuploidy is >trisomy, in which there are three copies of a chromosome: a well-known example is Down's syndrome, also called trisomy 21, in which there are copies of chromosome 21.

Angiosperm, (from the Greek angeio, 'vessel', and sperma, 'seed') a plant that has an ovary to contain its seeds. Angiosperms comprise the >flowering plants, and are contrasted with the >gymnosperms.

Animalia, one of the five kingdoms of living things. See classification.

Anisogamy, the production of two unequal-sized types of >gametes ina species. It is the system used byall higher eukaryotes, in which the ovum is much larger than the sperm. The opposite is >isogamy.

Ankylosing spondylitis, a human disease probably caused by virus but with an element of genetic susceptibility. The symptoms are the fusing together of the vertebrae of the spine, starting from the lowest, but even though there may be almost no movement in the back, the patient is usually well enough to lead a normal life. Men are affected four times more often than women, and there is often more than one case in a family. The factor that makes someone liable to develop ankylosing spondylitis seems to be linked with a gene in the >HLA complex, part of the immune system: almost all cases of the condition in people with HLA type B27, though this type occurs in only 3 or 4 per cent in the population. The HLA genes may be directly involved in the disease, in causing an attack on the body's own tissue, or they may only be genetic markers linked very closely to the real genetic factor.

Annealing, the reassociation of complementary strands of DNA after they have been denatured (usually by melting at a high temperature); see denaturation. The annealed strands can either be original complementary strands or they can be any two strands that have enough base-pairs matching for reassociation to occur. This hybridization of DNA is one technique used in >genetic engineering. Annealing can also occur between a strand of DNA and a strand of RNA.

Anthropomorphism, the attribution of human characteristics to animals, things or deities. It is an idea that, if put into reverse, is useful in thinking about the evolution of humankind. It is not that animals are like us; we are like animals. Humans share 97.5 per cent of their genes with their nearest living non-human relatives, chimpanzees. Rabbits and pigs do not really wear clothes as they do in Beatrix Potter's, but their mammalian physiology is near enough to our own for us to think it worth while to use these animals for testing whether various substances may be harmful to ourselves.

Antibiotics, resistance to. See resistance.

Antibody, a protein, part of the >immune system in vertebrates. Antibodies are >immunoglobulins (abbreviation: Ig), of which there are several types). Each Ig molecule consists of four chains, each with a constant (C) region, a joining (J) region and a variable (V) region, the last occupying from a quarter to half of each chain's length. The function of an antibody is to recognize and combine with an invading >antigen. It is the variable region of the antibody molecules are produced by the >B Lymphocytes in the spleen and lymph nodes, the appropriate cell line that makes a particular antibody being selected when the antigen specific to it enters the body (see clonal selection theory). See also monoclonal antibodies.

Anticoding strands. See sense strand.

Anticodon, the message part of a molecule of >transfer RNA. It matches on to the corresponding codon in messenger RNA (see protein synthesis). See also genetic code.

Antigen, any substance recognized by the body as foreign, against which >antibodies are made. Antigens include viruses, bacteria, fungi, parasites and non-living particles. See clonal selection theory; immune system.

Antiparallel alignment, the arrangement in which the two strands of DNA or RNA lie. They are each oriented from the 5' to the 3' end, and in antiparallel alignment the 5' of one is opposite the 3' of the other.

Antiserum, blood serum containing antibodies against a particular antigen. >Passive immunity is given to patients by injecting into them antiserum made by another animal so that the antibodies will enter their bloodstream immediately, rather than waiting for their own >immune system to produce them. A common example is administering or injecting antiserum against tetanus, the serum being derived from horses. Treatment with antiserum can produce an unpleasant side-effect (serum sickness) if the patient has an adverse reaction to other proteins present in the serum.

Apo-activator. See apo-inducer.

Apo-inducer, a protein that combines with the inducer to initiate gene transcription. See gene regulation.

Apomixis, reproduction without a sexual process, in plants. It can be either >vegetative propagation or >agamospermy, both resulting in groups of offspring that form a >clone. Sexual reproduction in plants is called >amphimixis.

Apo-repressor, a protein that becomes a repressor when combined with another molecule (co-repressor). See gene regulation.

Arboviruses, a group of viruses that can attack both insects and humans. Yellow fever and one type of encephalitis (inflammation of the brain) are caused by arboviruses. The genome is single-stranded RNA.

Aristotle (384-322 BC), Greek philosopher. Plato was his teacher for ten years, and he became tutor to Alexander the Great. He founded a school at the Lyceum in Athens, and most of his writings are notes for lectures that he delivered there. They cover an enormously wide range - all of the range of learning at that time. His system of logic was the basis for all the range of learning at that time. His system of logic was the basis for all the rest of his thought, and his views on politics and ethics have had an unparalleled influence on Western thought. So also have his thoughts on biology, which is less fortunate. Aristotle seems to have found biology, a practical subject with which his pupils would not have had to concern themselves, far less interesting than the important matters such as politics and virtue. He was an uncritical collector of biological 'facts', some quite bizarre, such as the notion that a female mouse can become pregnant before it is born. He (c. 335 BC) classifies animals into those that arise spontaneously from mud, etc. and those that reproduce sexually, the latter group being further divided into those that lay eggs and those that have live young. At least three of his ideas about genetics were totally wrong: (1) that some animals reproduce by >spontaneous generation
(2) that the father is responsible for the 'form' of the offspring and (3) that acquired characteristics can be inherited. It is fair to say that these were not Aristotle's own ideas; he was simply repeating the general beliefs and superstition of the time; once they had appeared in his writings, they acquired the prestige of his name and became virtually irremovable.

Artificial insemination, making a female pregnant by manually introducing sperm. It was first done to a spaniel bitch, by the Italian scientist >Spallanzani in the eighteenth century. It has been practised in humans since befor World War II. However, donor insemination (formerly called AID, artificial insemination by donor, an acronym that was abandoned after the spread of AIDS), as distinct from artificial insemination by husband (AIH) has been extremely controversial on moral grounds. Jewish belief is against it, considering it to be 'bringing orphans into the world'. In Britain in 1948 the Archbishop of Canterbury stated that donor insemination should be a criminal offence, and a Royal Commission in 1960 reported that 'both society and the medical profession are opposed to AID [i.e. donor insemination]'. It was not until 1973, following the report by the Peel Committee, that donor insemination was officially approved as a treatment, and even then it was thought suitable to restrict its use as much as possible. Artificial insemination is widely used in cattle breeding. However, it is not permitted in racehorse breeding, in which the fee charged for the services of a stud horse is kept high because of the limit to the number of mares that he can mate during the season.

Artificial selection, selection deliberately carried out by humans, as opposed to natural selection. All examples of >domestication involve selectively breeding from organisms whose phenotype suits some purpose, and the changes so produced can be enormous. Artificial selection was understood by both Lamarck and Darwin, and the search for an equivalent force in natural conditions led the latter to his theory of natural selection. See selection limit.

Ascospore. See ascus.

Ascus, a sort of pod containing the four cells produced at >meiosis by certain fungi, e.g. >***Neurospora*** and >***Saccharomyces*** (yeast). Each of the four undergoes a mitosis division so that the ascus finally contains eight cells, known as ascospores. See tetrad analysis.

Asexual reproduction, reproduction by simple cell division. This commonly occurs in prokaryotes, and also fungi, protozoa and some plants. Bacteria normally reproduce asexually, but are also capable of a form of sexual reproduction known as >conjugation.

Aspergillus, a genus of fungi used in genetic experiments. The species most often used is ***A. nidulans***. Another species. ***A. fumigatus***, causes a lung disease, but the majority are among the many fungi found on decaying food. **Aspergillus** is a eukaryote and may be either haploid or diploid (16 chromosomes).

Assortative mating, any system of mating in which mate choice is influenced by phenotype. In positive assortative mating, mates are more like each other than they would be by chance. This is quite commonly found, and has been noticed in human populations: in advanced countries, there is positive assortative mating both for height and for socioeconomic status (which traits are positively correlated with each other). In negative assortative mating, mates are less like each other than they would be by chance. From the point of view of the genetics of a population, positive assortative mating tends to increase homozygosity and negative assortative mating to decrease it.

Atavism, the reappearance of a form that used to occur in an ancestral stock, due to recessive alleles meeting each other or to recombination. A respectable term in modern ecology, it has a bad history in that it was used in the late nineteenth century to describe various human disorders as being reappearances of traits found in pre-humans (another example of the perversion of Darwin's theory of the descent of man). The popular term is 'throwback'.

Attenuation. Ther are two definitions of this word in genetics. First, it is the name given to a form of >gene regulation which operates when the gene product is an enzyme (or more usually a group of enzymes) that synthesizes an >amino acid. There is a leader sequence before the structural genes, and this includes a few repeats of the codon for the amino acid in question; if this sequence can be encoded - i.e. if there is already a sufficient quantity of that amino acid in the cell - the messenger RNA forms a terminator loop and transcriptin stops. If there is an insufficient amount of the amino acid - that is, not enough molecules of its transfer RNA - transcription of the repeat codons cannot take place, the terminator loop is not formed, and the gene(s) are transcribed through to the end. Therefore shortage of the amino acid turns the gene on, and when the supply is sufficient it is turned off.

Second, attentuation is the word used for loss of virulence in a strain of viruses or bacteria. Viruses can be attenuated by treating them with ultraviolet radiation; attenuated strains of bacteria are obtained by selection in vitro. Both types of attenuated micro-organisms are used in >vaccination.

Attenuator, a sequence in a messenger RNA molecule where transcrption stops, in the system where the gene is regulated by >attenuation. It comes immediately before the coding sequencs.

Australopithecus. See human evolution.

Autogamy (from the Greek auto, 'self, and gamos, 'marriage'), self-fertilization. Usually the term refers to self-fertilization in plants, but it can also describe the process in which a >*Paramecium* divides its macronucleus to form a new micronucleus and macronucleus after cell division.

Auto-immune diseases, a group of diseases in which the symptoms are caused by the >immune system attacking the body's own tissues. In many of these diseases there seems to be a genetic element, in that the disease occurs among relatives; an example is systemic lupus erythematosus, in which many different organs, including the skin, heart , spleen and kidneys, are attacked as 'foreign'. A few auto-immune diseases have been linked to specific types in the >HLA complex, part of the genetic system controlling the immune response. >Rheumatoid arthritis is known to be four times more frequent in people with HLA type B27. None the less, it is not possible to say that merely having this or that HLA type is enough to cause an auto-immune disease; there must be some trigger factor, such as infection by a virus, which is more dangerous to one HLA type or another.

Autogenous regulation. See autoregulation.

Autopolyploid, a >polyploid organism in which all the sets of chromosomes come from the same species. See allopolyploid.

Autoregulation, the system in which a gene's activity is controlled by its own gene product. See gene regulation.

Autosexing, a technique for identifying the sex of organisms before the age at which the sexual organs are differentiated. The classic example is chicks: in certain strains the males and females are a different colour, distinguishable at one day old.

Autosome, a chromosome that is not a sex chromosome. The autosomes always outnumber the sex chromosome (of which there are only two).

Avery, Oswald Theodore (1877-1955), American doctor and bacteriologist, who discovered that the genetic material in chromosomes is >DNA. He repeated F. Griffith's experiment in which non-virulent viruses are transformed into virulent ones by something contained in an extract of heat-killed virulent virus. In 1944 Avery showed that the transforming factor was DNA, a molecule that no biologist was very interested in. Avery did not make the claim that genes consisted of nothing but DNA, but he did state that DNA was an important element of genetic material. This finding made very little impact at the time.

B

B chromosome, one that is an extra to the normal diploid set (the A chromosomes). B chromosomes are more often found in plants than in animals (though grasshoppers specialize in them). Because they do not seem to have any genetic effect it has been suggested that they are 'selfish chromosomes' (see selfish DNA).

B Lymphocyte, a type of white blood cell, part of the immune system. The B lymphocytes' function is to produce the antibody molecules that bind antigens in humoral immunity (see immune system). When an antigen arrives in the body, it is recognized by a virgin B Lymphocyte in the bone marrow (see clonal selection theory); this cell is stimulated by the antigen arrives in the body, it is recognized by a virgin B Lymphocyte in the bone marrow (see clonal selection theory); this cell is stimulated by the antigen into proliferating and establishing a clone of descendant cells (plasma cells) in the spleen and lymph nodes which produce large amounts of antibodies. B Lymphocytes (so called because they mature in and are released from the bone marrow) recognize antigens by their three-dimensional shape (unlike T Lymphocytes).

B_1, B_2, etc., designations for the generations involved in a >backcross. The cross between the first hybrid generation (>F_1) and the parental type is the first backcross, and the offspring of this are the B_1 generation. If individuals from the B1 generation are again backcrossed to the parental type, the resulting offspring are the B_2 generation, and so on. B_2 is also sometimes used to designate the offspring of the backcross of F_1 to the other of the two original parental types.

Back mutation, a >mutation that happens to an already mutant allele, correcting the mutation and restoring the >wild type. It may be an exact reversal of the first mutation, or it may really be a second >forward mutation that happens to have the effect of cancelling out the effect of the first.

Backcross, a cross between >heterozygous individuals of the >F_1 generation and individuals of one of the parental genotypes. The result is a generation in which the hybrid and parental types are present in the ratio of 1:1 - for example, Aa (the F_1) x AA (the parental type) gives ½ AA and ½ Aa. The term 'backcross' is also used for the cross between the F_2 generation and one of the parental types. A >testcross is a particularly useful form of backcross.

Background genotype, the part of the genotype that is concerned with traits not normally discernible in the >phenotype.

Bacteria, a large group of single-celled organisms. They can occur either free-living in air, water, soil, etc., or they may be parasitic upon a host plant or animal. Only a minority cause any disease, and more are actually useful to their hosts, e.g. in breaking down food as part of the digestive process (see *Escherichia coli*). The three basic shapes are bacillus (rod), spirillum (spiral) and coccus (spherical), but some bacteria may link up to form chains or clumps. Much genetical research is done on bacteria because, being >prokaryotes, they have simple genetic machinery. Bacteria normally reproduce asexually (i.e. by simple binary cell division), but occasionally 'conjugate' to reproduce sexually (as discovered by Joshua >Lederberg and E.L. >Tatum). The word itself comes from the Greek for 'walking stick'.

Bacteriophage, a virus that infects bacteria. A 'phage' (as it is usually called) consists of an outer capsule made of protein and a core of nucleic acid, which can be either DNA or RNA. The phage attaches itself to the cell wall of the bacterium, sometimes by means of a spiky 'tail', and delivers its DNA or RNA into the host cell. The host then replicates the virus, either by treating the DNA as one of its own genes or by using the RNA as mRNA: in either case, the phage's entire genetic constitution is expressed by the host cell's protein synthesis apparatus. Thus the host cell begins to contain a number of viruses that the cell itself has made (this process was discovered by >Hershey). Sometimes the mature viruses leave the host cell by being extruded through the cell wall, but more usually the result is >lysis, when the viruses produce an enzyme that burst the cell wall. **Phages** (especially the >lambda and >T4 varieties) have been intensively studied because the way they get bacteria to utilize their DNA may be comparable to the way in which genes are switched on in normal cell metabolism (see gene regulation).

Von Baer's law, the rule that, in their earlier stages of development, organisms resemble each other more than they do at maturity. It was formulated by the Estonian embryologist Karl von Baer (1792-1876) who also discovered the mammalian ovum, and published in his Uber der Entwickelungs Geschichte der Thier (part I 1828, part II 1837, History of the Development of Animals). See also Haeckel's law.

Balanced Lethal, a system in which a population is consistently heterozygous for two (or more) recessive >mutations that are lethal when homozygous and are not linked. Let us take a two-locus example. All individuals have genotype m+n+ (where m and n are the mutant alleles and + are the wild-type alleles). This is a >dihybrid cross, so the result is a 9:3:3:1 ratio, in which the first nine are all viable because they are double heterozygotes, but the remaining seven are non-viable, being homozygotes for one or other of the lethal recessives. The survivors appear to breed true, but only in the sense that all their surviving descendants are genotypically the same as themselves; they have had other offspring that have died, so this is not the same thing as >true breeding.

Balbiani rings, See puff.

Baldness, a classic example of a sex-limited trait. Premature balding only occurs in men, but this does not mean that the gene for it is on the Y chromosome. It can be shown in a pedigree that it is an autosomal dominant, which can be transmitted to either daughters or sons. A female who has the allele will not herself go bald but will pass on the baldness allele to her offspring; thus half her sons will go bald. The gene is only expressed in men, because it requires the presence of testosterone (see sex hormones).

Baltimore, David (1938-), American molecular biologist; professor of microbiology at the Massachusetts Institute of Technology 1968-90; president of Rockefeller University since 1990; discoverer of >reverse transcriptase. After important work on how the polio virus replicates, Baltimore found that in some tumour viruses there is an enzyme that can make a DNA molecule by reading off from RNA. This the exact opposite of what the central dogma states, i.e. DNA RNA protein. Baltimore's discovery of reverse transcriptase won him the 1975 Nobel prize for physiology or medicine (jointly with H. Temin who had independently made the same discovery).

Banana. See triploidy.

Banding, the pattern of light and dark regions in chromosomes revealed by various staining techniques. The bands can be used to identify individual chromosomes within the karyotype. In the giant >polytene chromosomes in **Drosophila** it is believed that the light interbands mark the areas where >housekeeping genes are located and the dark bands mark the >luxury genes. The different appearance is due to the fact that where genes are being transcribed the DNA has to be loosely wound, whereas the DNA for inactive genes is highly condensed and shows up darker (see heterochromatin).

Barr body, a dark-staining particle seen in the nucleus of any cell (when it is not dividing) of a female animal. The Barr body is actually one of the two chromosomes that has become inactivated (see Lyon hypothesis). As the Barr body can be seen in any non-dividing cell, it is extremely easy to distinguish the sex of an individual from a small sample (e.g. in prenatal sex testing), though the Barr body cannot be seen in red blood cells because they have no nuclei. A Barr body is also seen in the cells of an individual whose complement of sex chromosome is XXY (see Klinefelter's syndrome) even though they are phenotypically male, but an individual with >Turner's syndrome (sex chromosomes XO), phenotypically female, has no Barr body, A person with three X chromosomes will have two Barr bodies in each cell, as there is always one Barr body fewer than the number of X chromosomes. In insects and others where it is the male that is the >homogametic sex Barr bodies are seen in the cells of males. See also heterochromatin.

Barrier, in genetic terms, any impediment to the exchange of genes between one population and another. Barriers are usually geographical features, but might be other factor, e.g. different behaviour patterns. See also isolating mechanism.

Base-pair (abbreviation: 'bp'), one of the pairs of nitrogenous bases that form the core part of a molecule of >DNA or >RNA. They are either >purines (guanine and adenine) or >pyrimidines (cytosine, thymine and uracil), and in each pair there is one of each, guanine always pairing with cytosine, and adenosine always pairing with thymine (in DNA) or uracil (in RNA), Genes are often measured in terms of how many base-pairs they consist of (see kilobase).

Bates, Henry Walter (1825-92), British naturalist. Accompanied by >Wallace, Bates went to South America where between 1848 and 1859 he collected a vast number of specimens, including some 8000 species (mostly insects) that had never before been described. His theory of >mimicry was one of the pieces of evidence that Darwin used to support his idea of evolution through natural selection.

Bateson, William (1861-1926), British geneticist; professor of biology at Cambridge and director of the John Innes Institute; inventor of the term 'genetics'. During the 1890s Bateson's interest in evolution led him to the conclusion that species evolve by 'jumps' rather than gradually, and in trying to prove this experimentally he came up with ideas very similar to Mendel's. When Mendel's work was rediscovered in 1990, Bateson translated it into English and became its leading advocate. He found the first Mendelian genes in animals, and correctly interpreted Garrod's findings as showing Mendelian genes in humans. He obtained some results that did not match Mendel's law of independent assortment, and this led to T.H. Morgan's discovery of linkage.

Bayes' theorem, a way of calculating the relative probability of an event. It is much used in medical genetics (see genetic counselling) to predict the probability of a person being a carrier of a deleterious recessive gene in a family where such a gene is known to be present. It combines a prior probability, based on the occurrence of the gene in the pedigree, with a conditional probability, based on the phenotype of the individual, which together lead to the posterior (or final) probability.

Beadle, George Wells (1903-), American geneticist; professor of biology at the California Institute of Technology for 14 years before going to the University of Chicago in 1961. With his microbiologist colleague Edward >Tatum, in the 1940s Beadle investigated mutant genes in the pink bread fungus >*Neurospora crassa*, a good subject for genetical research since it is haploid (i.e. had only one set of chromosomes) so that recessive mutants are expressed. Beadle and Tatum gave the mutant strains of bread fungus different mixtures of nutrients to grow on, and found that specific mutations led to an inability to digest one nutrient. Their conclusion was that each mutation caused the loss of a particular enzyme, hence the statement 'one gene, one enzyme'. They did not know that this idea had been put forward by Garrod some 40 years earlier. Beadle and Tatum shared the 1958 Nobel prize for physiology or medicine with Joshua >Lederberg.

Beneden, Edouard van (1846-1910), Belgian zoologist. Beneden, a pioneer in the study of chromosomes, made the discovery that the number of chromosomes is the same in all the cells of an animal's body, except for those found in the sex cells, which have half. He also discovered that the chromosome number is not the same for every species. Beneden did not publish an adequate interpretation of his findings; but they were taken up by others, notably Boveri, de Vries and Weismann.

Berg, Paul (1926-), American geneticist; professor of biochemistry at Stanford University since 1959, Berg took up Francis >Crick's 1955 suggestion that there must be an intermediate molecule between the RNA template and the protein being formed during protein synthesis; within a year, he had found the first example of tRNA, or transfer RNA, which is this intermediate. Berg was also a pioneer of techniques for introducing specific genes into host bacteria (see recombinant DNA), an important part of genetic engineering. He was one of the first to see the risks involved in recombinant DNA research, and urged extreme caution in pursuing it. He shared the Nobel prize for physiology or medicine in 1980.

Bidirectional replication, DNA replication in which, from a single starting point, replication proceeds in both directions along the molecule.

Biffen, Sir Rowland Harry (1874-1949), British geneticist. Biffen was among the first to put Mendel's laws to practical use when they were rediscovered in 1900. He was then teaching agricultural botany at Cambridge University, and he at once showed that physiological traits, as well as the visible ones that Mendel had studied, could be controlled by single genes. He was therefore able to breed strains with characteristics such as resistance to fungal infection. His wheat 'Little Joss' was the outstanding variety until after World War II.

Bimodal distribution, a statistical distribution in which the individuals cluster round two different values. On a graph, this would show as a curve with two peaks separated by a dip. Continuous variation due to a single pair of alleles and with environmental influence shows a bimodal pattern, if there is complete dominance.

Binary character, a genetically determined trait that exists in two states, which can be recorded as + or -, 0 or 1, etc. An example is the presence or absence of horns (in some breeds of cattle and sheep; some are more complicated, having multiple horns). See also multistate character.

Binomial nomenclature, the system, devised by >Linnaeus in the eighteenth century, under which all species have been given a two-part name. The first part of the name is the genus and is a noun; the second is the adjective that designates the particular species. If the species is the type-species of the genus, the two parts may be the same, e.g. Gorilla gorilla. Both parts of the name are in Latin (which is useful in that it is to some extent still a lingua franca), and printed in italic. The specific name always begins with a small letter, even when it is based on a proper name, e.g. *Fregata andrewsi* (one of the frigate birds). See also classification.

Biological warfare, the use of pathogens (disease-causing agents such as viruses, bacteria, fungi, parasites) as weapons. The target may be humans, or it may be the enemy's agricultural animals and plants. Biological warfare was banned under the 1925 Geneva Convention, but only for 'first strike' use; this proviso enabled governments to develop biological weapons as long as they could be described as 'defensive'. The 1975 Biological Weapon Convention (signed by over 100 countries) goes further, banning all development and production of biological weaponry; unfortunately it allows research for defence and does not define the amount of stock that can be held for research as distinct from operational purposes. It is believed that at least ten countries (including Britain and the United States) have developed a biological warfare technology; the United States spends $60 million per year on its research programme. This research probably consists to a large extent of genetic engineering, manipulating the virulence of common pathogens, and also of immunization procedures for one's own troops or population. Deliberate ecological disturbance, e.g. the use of defoliants, is in a sense biological warfare too.

Biometrics, the science of applying statistical methods to biological problems. It is much used in plant and animal breeding for agriculture, where the data typically is concerned with >continuous variation.

Biotechnology. In its narrowest sense, this term means the use of micro-organisms to make an end-product. This includes such things as the ancient technique of using yeast to make wine from grape juice, as well as such recent ones as the production of quantities of once-scarce human hormones from genes >cloned into >*Escherichia coli*. (See genetic engineering.) Biotechnology can also mean anything that improves human utilization of animals and plants, such as >artificial insemination and >embryo transfer for cattle , or the improvement of crops by selection or genetic manipulation. Bioengineering is the word used for the development of such things as artificial limbs, kidney machines and so on. Biotechnology can make an immense contribution to 'green' environmental policies. One key example is the production of fuel for road vehicles by using yeasts to make alcohol from vegetable matter; this means that vehicle fuel can be a renewable resource, in contrast to the finite stocks of oil and gas that are fast running out (of course, vehicles themselves would have to be changed). Another example is the use of ecological manipulation (e.g. the introduction of sterility genes into wild populations of insects) rater than chemical pesticides to protect crops.

Bivalent, a pair of homologous chromosomes at the time when they lie alongside each other during the first meiotic division (see meiosis). Each chromosome in a bivalent consists of two >chromatids.

Blastocyst, a stage in the >embryogenesis of mammals, prior to implantation. It is the first stage at which there is any differentiation among the ball of cells, which then number about 64. The blastocyst consists of an outer layer of protective cells (the trophectoderm) surrounding a fluid-filled cavity containing the inner cell mass that goes on to develop as the embryo. The human blastocyst is about 0.15 mm in diameter.

Blastomere, one of the undifferentiated cells in the very early life of the embryo. It makes up the >blastula. See embryogenesis.

Blastula, a stage in the >embryogenesis of animals that consist of a ball of cells which have not yet undergone any differentiation. It is a hollow ball of cells (called blastomeres) either one or several layers thick. The next stage is the >blastocyst.

Blending inheritance, the mistaken idea that heredity involves an irrevocable mixing, as of fluids, between the two parental contributions.

Blindness, total or partial loss of sight. It is usually caused by disease or accident, but there are a few forms of blindness that are genetic. One is early-onset blindness, which starts by the age of five and affects one child in 10,000; it is inherited as an autosomal recessive. Another form is caused by a dominant gene, and affects about the same number of children, though the blindness does not begin until they are about ten. The inherited cancer >retinoblastoma also causes blindess, and other genetic diseases have blindness as a secondary effect. See also retinitis pigmentosa.

Blood group systems, the various sets of >antigens found on human red blood cells. Over 30 blood group systems have been discovered since the first - the >ABO - in 1990. Only the ABO and the >Rhesus systems are normally of medical importance. See also Duffy blood group system; MN blood group system.

Blunt end, the end of a DNA molecule where there are strands of DNA right down to the last base-pair, as opposed to one where the last few bases on one side are unpaired (see sticky end). See also ligation.

Bombyx mori. The silk moth. See silk.

Bottleneck, in genetic terms, a decrease in population size that suddenly reduces the amount of genetic diversity in the population, especially if the survivors are not genetically a random sample of the former population.

Boveri, Theodor Heinrich (1862-1915), German zoologist. Boveri started from Beneden's observations that the chromosome number is constant in any species and that a sperm and an egg each contain a half-set of chromosomes that unite at fertilization to give the complete number found in all other cells. He went on to observe (in sea urchins) that an individual that has a deficient number of chromosomes will develop abnormally; this was an added piece of evidence that chromosomes are indeed the 'packages' in which the hereditary material is carried.

Boyer, Herbert Wayne (1936-), American biochemist; professor of biochemistry at the University of California at San Franciso since 1976, and co-founder of Genentech, one of the earliest commercial genetic engineering firms. In 1973 Boyer developed the technique of splicing together DNA from two different plasmids in >*Escherichia coli* to produce a hybrid functional 'gene', known as a >chimaera.

Bp, abbreviation for >base-pair, used as a unit of length to describe a sequence of DNA.

Breakthrough, an individual that is >homozygous for a lethal recessive mutant gene but none the less manages to survive (though not necessarily to reproduce).

Brenner, Sydney (1927-), South African-born British molecular biologist; director of the Medical Research Council Laboratory of Molecular Biology 1979-86, and since 1986, director of the Molecular Genetics Unit at the same site; joint discoverer of the genetic code. In 1957 Brenner joined the MRC's molecular biology laboratory at Cambridge, where the structure of >DNA had been discovered by Francis >Crick and James >Watson four years earlier. His work showed that the bases in DNA form a code if read in threes, and that these >triplets (or >codons) are not overlapping; GCATAGATC is GCA-TAG-ATC, not GCA-CAT-ATA and so on. Brenner and Crick found in 1961 that the molecule that carries the code from the DNA in the chromosome to the >ribosomes is messenger RNA (m RNA). Brenner was also the prime mover of the British project to map the human genome (>Human Genome Project.

Buffon, Georges Louis Leclerc, Comte de (1707-88), French scientist, one of the first to suggest the evolution of species. He was keeper of the Jardin du Roi (now Jardin des Plantes) in Paris, and of the French royal museum of natural history; his 44-volume catalogue of the latter (Histoire naturelle, 1749-1804) grew into the first serious attempt to describe the whole natural world. On the evidence of certain vestigial features such as the side toes on pigs, Buffon suggested that similar animals are descended from 'common ancestors'. This idea, and Buffon's realization that species in nature are variable not uniform, were influenced upon Darwin.

Burkitt's Lymphoma, a tumour of the lymph glands, common in some parts of West Africa. It appears to be triggered by a virus infection, and is one of the commonest cancers found in people with AIDS.

Burt, Sir Cyril Lodowic (1883-1971), British psychologist, remembered for his falsificaton of results in studies on the heritability of intelligence (see IQ). Burt was professor of psychology at University College, London, and a passionate believer in selective education; his advice to the British government at the time of the Education Act 1944 was crucial in setting up the two-tiered system of grammar and secondary modern schools. Burt claimed to have data derived from studies of identical twins separated at birth that proved intelligence to be almost entirely genetically determined; but in 1976, five years after Burt's death. the Sunday Times journalist Oliver Gillies exposed the data as fraudulent. It seems that Burt did originally have some rather unsystematic data on a few sets of twins but that this information had been lost in the war. Therefore, when he came to write his most polemical papers, after his retirement, he made up data to suit his argument, and invented a research assistant who never existed. For some of his data, see twin studies.

C

C region, or constant region, part of the heavy and light chains of an >immunoglobulin molecule. See antibody.

C value, the amount of DNA in a genome (in its haploid condition). Very generally, the C value is higher in organisms that are evolutionarily more advanced, but sometimes closely related species have widely different C values, and there are sporadic cases of very high C values with no obvious explanation (this is known as the C-value paradox); for example, the C value of humans is 3.0 picograms per cell, of yeast 0.024, of broad beans 24.0, of geese 1.4. In any case, there seems to be far more DNA per cell in higher animals than can be necessary for the genes in that animal; much must be redundant or repetitious.

Camerarius, Rudolf Jakob (1665-1721), German botanist; simultaneously professor of medicine and director of the botanic garden at Tubingen; discoverer of sexual reproduction in plants. Camerarius was not the first to notice that pollen causes fertilization in plants (it was known, in some plants at least, to the ancients; see Theophrastus), but in 1694 he proved experimentally that the male and female flowers of plants are exactly analogous to the sexual parts of animals, with stamens the male organs and pistils the female.

Canalization, the hypothetical tendency of an organism during development (particularly >embryogenesis) to follow the right pathways, irrespective of any adverse influence from the environment.

Cancer, a blanket term for a group of diseases with the common factor that they involve the uncontrolled proliferation of cells. Humans, animals, birds and plants can all be afflicted with cancer. Virtually every part of the body is susceptible, though some are much more commonly affected than others (cancers of the lung, stomach, bowel, breast, and prostate are among the frequent ones in European populations; incidences vary enormously between populations). The cancer sometimes remains localized in tne tissue, and sometimes spreads via the bloodstream or Lymphatic system to other parts of the body (metastasis). The many different cancers have different causes, and only in a few can a single cause be identified. There is an environmental factor in some - for example, the high frequency of cancer of the scrotum in chimney sweeps was due to their contact with soot, and radiation and chemical pollution are today strongly implicated in some cancers. Others are caused by a virus, for instance, Burkitt's Lymphoma. Genetic analysis of the family histories of cancer patients shows that, while most of the common cancers show no genetic pattern, a few particular cancers do run in families.

Some cancers are caused by a recessive gene. One of these is >retinoblastoma, tumour of the eye. Children who are homozygous for the abnormal allele develop eye tumours early in life. In other cases, there is a later onset, and it seems that these patients are heterozygous in whom the protection given by the normal allele is lost when the allele itself is lost in one cell (possibly by the deletion of a small segment of a chromosome or non-disjunction during cell division). Other cancers have an inheritance pattern that suggests a dominant allele, but with variable penetrance (i.e. not everyone who carries the gene will get the disease); an example is the type of breast cancer that develops in both breasts before the menopause. There are also cancer-related conditions that are inherited; one of these is >polyposis coli, which is often the precursor of cancer of the colon and is inherited as an autosomal dominant. One of the most puzzling findings is of 'cancer families', in which the members of the family get different kinds of cancer, but the getting or not getting of the disease is inherited as a single gene; these families are very rare.

Cap site, a modification to the 5 end (the 'front') of a messenger RNA molecule in eukaryotes. It seems to assist the molecule in attaching itself to a ribosome. The other end has a >poly-A tail.

Carcinogen, any environmental agent that causes cancer. A carcinogen may be a chemical substance, such as tar (associated with lung cancer) or aniline dyes (bladder cancer), or it may be a form of radiation, such as >ultraviolet light (skin cancer) or the radiation leakage from nuclear power stations that is now suspected of causing leukaemia in the children of male power-station workers.

Carrier, an individual who is heterozygous for a recessive gene. Such individuals do not show any of the effects of the gene but transmit it to their offspring. The term is also used for a person who has become infected with a bacterium or virus, and although he or she does not have the symptoms of the disease in question, he or she can pass it on to others. See also X linkage.

Carrying capacity, the maximum number of organisms that can inhabit a given area. See density-dependent factors.

Caste, in ecology, a group of individuals within a population who share some morphological or behavioural characteristic. It is a useful term for describing the different groups within a colony of social insects. In human society, especially in the Indian subcontinent, castes ar hierarchical, closed social units, each with prescribed occupations and with obligatory endogamous marriage. Although there is some gene flow between castes, endogamy has been strictly enforced for long enough for there to be differenes in gene frequencies.

Catalyst. See enzyme.

Catastrophism, the idea that the geological features of the Earth were formed in a series of violent and catastrophic events. Baron Cuvier believed that the succession of fossils was accounted for by a series of catastrophes, after each of which life was created anew. It was essentially an anti-evolutionary point of view, and, as far as geology went, was replaced by the theory of uniformitarianism advanced by the British geologist James Hutton in the late eighteenth century. Catastrophism is now taken to mean that the evolutionary process does not go at a uniform speed but is fairly slow until enormously speeded up in the aftermath of catastrophes. See also gradualism.

cDNA. See complementary DNA.

Cell cycle, the periodic phases in a cell's life. The two major phases are >mitosis (cell division) and >interphase, the period between divisions when the cell is growing and then doing whatever it is specialized to do. Interphase is itself subdivided into three parts; G1 (gap 1), in which each chromosome is represented by only one >chromatid; S (synthesis), during which DNA is replicated so that each chromosome again has two chromatids; and G2 (gap 2), leading to mitosis (also called 'M phase'). Interphase takes up about 90 per cent of the time of a >eukaryotic cell, with the duration of G1 being anything from a few hours to several years long.

Cell line, a population of cells maintained in tissue cultre. If the original cells were differentiated, i.e. had become specialized for some function, the cells will continue to show that differentiation when kept > in vitro. Over a long period of time, cumulative chromosome changes will become apparent and the cell line may finally become unviable. See HeLa cells.

Centimorgan, a unit (named in honour of T.H. > Morgan) used to measure the frequency of >crossing-over and therefore the distance apart of linked genes: 1 centimorgan is equal to a 1 per cent frequency of cross over, and is therefore equivalent to a >map unit. The original unit, the morgan, is equal to 100 per cent cross-over.

Central dogma, the basic principles of molecular biology as stated by Francis >Crick. They are:

- DNA is replicated to produce more DNA.

DNA
DNA [replication]
DNA

• DNA is transcribed into RNA which is translated into protein.

DNA [transcription] RNA [translation] protein

Common to both parts is the idea that DNA is the source of information, not the recipient. However, an exception happens with >reverse transcriptase, an enzyme that can make a DNA molecule from an RNA template.

Centriole, one of a pair of organelles in the cell nucleus, which make the fibres for the >spindle and also form the poles from which the spindle fibres radiate in >anaphase.

Centromere, the point at which the two arms of a chromosome join together end to end. The centromere is also the point at which the chromosome is attached to the spindle fibres during cell division (see meiosis; mitosis). Chromosomes may have the centromere right at the end (>telocentric), at or near the middle (>metacentric) or towards one end (>acrocentric). The DNA at the centromere consists of repeated short sequences that carry no genetic information. see also acentric chromosome.

Chargaff, Erwin (1905-), Czechoslovakian-born American biochemist; professor of biochemistry at Columbia University, New York, from 1952 until his retirement; famous for his basic work on the properties of DNA. Working at a time (the late 1940s) when few biologists were interested in the nucleic cids, Chargaff discovered two very important facts about DNA. The first was that each species has its own characteristic DNA, differing from other species in the relative amounts of four nitrogenous bases; adeine (A), cytosine (C), guanine (G) and thymine (T). The second was that, in any sample of DNA, the amount of A is equal to the amount of T, and C is equal to G. These two principles became known as 'Chargaffs rules', and were used by Francis >Crick and James >Watson in their search for the structur of DNA; the significance of the rules was immediately apparent once that structure was established.

Chiasma (pl. chiasmata), the point at which non-sister chromatids of homologous chromosomes join during >crossing-over. As seen microscopically, each chiasma does indicate a crossing-over event in genetic terms, but the site of the chiasma does not correspond exactly to the molecular point of exchange.

Chimaera (or chimera; from the name of a creature in Greek mythology, with a lion's head, goat's body and snake's tail), an organism which has cells of two or more different genotypes, derived from differnt zygotes. One example is the >aggregation chimaera of two different strains of mice. A chimaera created by adding a single genetically marked cell to an

8-cell embryo can be used to study embryonic development.

Chimaeras can also be created out of two different species, e.g. sheep and goat. A plant chimaera is a plant in which part of one individual has been grafted on to another and the resulting plant contains cell lines derived from both; a flowering plant could thus bear flowers of two different strains or species. See also mosaic.

Chi-square test, a statistical technique that tests whether or not the observed occurrence of something fits with its expected occurrence. It can only be used with either/or variables, not continuous ones. An example would be the number of patients who died or who survived under (a) the traditional treatment or (b) treatment with a new wonder drug. Testing the >null hypothesis - in this case, the proposition that the new drug is not improvement - one compares the proportion of survivors to see whether it is or is not significantly different from what was expected (i.e. from what it had been with traditional treatment).

Chlamydomonas reinhardtii, a single-celled green alga, often used in genetical research. For most of its life cycle it is haploid, only becoming diploid when two individuals fuse in order to reproduce; the resulting >zygote immediately divides by meiosis to form four new individuals.

Chlorella, a genus of green alga. It is studied for the genetic control of photosynthesis.

Cholesterol. See hypercholesterolaemia.

Chorionic villus sampling (CVS), a technique for sampling the genetic make-up of the foetus, using cells from the chorion, a layer derived from part of the blastocyst (see embryogenesis) and therefore genetically the same as the foetus but attached to the wall of the of the uterus. Enough of these cells can be gathered for their DNA to be directly analysed, or else the cells can be carried out a 8-10 weeks of pregnancy and involves much less stress for the mother than amniocentesis, as it is done with a catheter through the cervix. There is probably a higher risk to the foetus than with amniocentesis (of the order of 3.5 per cent casualties) but this is disputed because there is a significant loss through spontaneous abortion at about that time of a pregnancy. See prenatal diagnosis; polymerase chain reaction.

Chromatid, half of a duplicated chromosome. When a chromosome is about to divide, it doubles itself longitudinally into two sister chromatids joined at the >centromere; each chromatid is therefore equivalent in content to a single chromosome. At cell division (see meiosis; mitosis), the chromatids part and each then becomes a new chromosome.

Chromatin, the substance of which chromosomes are made. It consists of >DNA and the protein >histone in roughly equal proportions, arranged in bead-like units called >nucleosomes. It gets its name from the Greek chroma, 'colour', because it appears coloured after biological staining. Regions of chromosomes that are not genetically active have chromatin that is highly condensed (see heterochromatin), whereas active regions have a much more open configuration (see euchromatin).

Chromomere, a very small concentrated region of >heterochromatin within a chromosome, seen only during meiosis. The chromomers are landmarks by which different chromosomes can be identified. In >lampbrush chromosomes they show up as loops, and in >polytene chromosomes they look like bands.

Chromosome, the structure which carries the genes. The chromosomes lie within the cell nucleus, and divide when the cell divides: this process (called >meiosis) is the key to heredity. It is useful to think of a chromosome as a string of beads, each gene being a bead (and its place on the string being its >locus). The genes are made of >DNA, and the chromosome also contains a protein called >histone; this mixture is known as >chromatin. The chromosomes are not visible within the cell (because they unwind themselves into a structure too loose to be resolved by a microscope) except when cell division is about to take place. Then they become condensed, and it can be seen that they consist of two arms attached end-to end at a point called the >centromere. At this stage the chromosomes also split themselves into two identical strands called >chromatids (see mitosis). The complete set of chromosomes is called the >karyotype. This includes the two >sex chromosomes plus the general purpose chromosomes known as the >autosomes. The number of chromosomes varies between species from three pairs to over a hundred, but bacteria each have only a single chromosome in the form of a ring.

Chromosome abnormalities, any irregularity in the number or gross structure of the chromosomes. The great majority of these abnormalities are lethal, the organism being unable to sustain the loss of so much genetic material or the surplus genetic dose, as the case may be. It is estimated that about 10 per cent of human conceptions are spontaneously aborted (miscarried) because of chromosome abnormalities.

Over 60 disorders in humans caused by chromosome abnormalities are known, and the most common involve the sex chromosomes. A female may have either only one X chromosome (see Turner's syndrome) or three, instead of two; a male may be XXY (an extra X; see Klinefelter's sysdrome) or XYZ (see XYY syndrome), instead of XY. The overall incidence of the sex-chromosome abnormalities is about 2 per 1000 live births. Abnormalities involving one extra autosome (i.e. >trisomy) are also quite common, totalling about 1.7 per 1000 live births, and they include the well-known trisomy 21, commonly known as >Down's syndrome. There are also abnormalities in which a chromosome is missing (see monosomy). All these arise from non-disjunction of the chromosomes when ova and sperm are being made. Structural abnormalities such as major translocations and inversions occur in about 1.3 per 1000 live births; they may be the result of radiation damage to chromosomes, but little is known about what causes them in normal circumstances.

Cis configuration, in an individual that is heterozygous one of the two possible arrangements for the alleles for mutations at two linked loci. Cis (from the Latin, 'on this side of) means that the two mutant alleles are both on one chromosome and the two wild-type alleles are on the homologous chromosome. This is also known as 'coupling'. The opposite is >trans.

Cistron, a sequence of DNA that codes for a >polypeptide, i.e. that functions as a gene. See intron.

Clade, a group of species or other taxa that share a common ancestry and are more closely related to each other than to any other taxa. See phylogenetic classification.

Cladistic distance, a measure of how closely two species (or other >taxa) are to each other, in terms of the number of branching points between them in the tree-diagram of their ancestry. See also phenetic distance.

Cladogenic evolution, evolutionary change in which there is a repeated pattern of one species splitting off from another. The resulting 'tree' (see dendrogram) has many branches, with numerous species co-existing at the same time. An example is the natural populations of the genus *Drosophila*, with nearly 1500 species known. See also phyletic evolution.

Classification, the systematic ordering of all living things. There have been many different attempts at this (see Aristotle for an early one), and the version in use today is doubtless not definitive. For a long time it was accepted that there were two kingdoms, Animalia (animals) and Plantae (plants), but the rise of microbiology has caused a rethink about the simpler life forms, so that the usual scheme now is to have five kingdoms:

(1) Monera: all the >prokaryotes, including >bacteria and blue-green algae.

(2) Protoctista: single-celled >eukaryotes, including amoebae, higher algae, slime moulds, Protozoa, etc.

(3) Fungi: fungi of all sorts from yeast (>***Saccharomyces***) to mushroom like forms. The genetically important >***Aspergillus*** and >***Neurospora*** are in this kingdom.

(4) Plantae: all plants from the simplest (mosses and liverworts) to the trees and >flowering plants.

(5) Animalia: including sponges, sea anemones, worms, molluscs, spiders, millipedes, insects, fish, amphibians, reptiles, birds and mammals.

Note that viruses are not included, since although they are made of the same stuff as living things, they are not capable of reproducing themselves except via a host organism. See also origin of life; taxonomy.

Cleavage, the division of the fertilized egg into, first, a two-celled stage, then the four-celled, etc. (see embryogenesis). The word is also used for the cutting of a DNA molecule by an >endonuclease.

Cleft palate, a condition associated with more than one congenital disorder. The most usual from is inherited via a combination of genes, but a rare form is caused by a dominant allele. There is much variability as to how much babies are affected: in some, only the lip is cleft (hare lip); in others, the palate is cleft right to the back of the mouth. Surgery can repair the damage to a great extent, surgical techniques in this area having greatly improved. Boys are more commonly affected than girls; about 1-2 per 1000 white babies have cleft lip/palate, but the frequency is lower in blacks.

Cline, a gradient of variation in a continuously varying character, over a geographical range. It can also be a gradient in the frequency of a phenotype or gene.

Clonal selection theory, an idea put forward in 1951 by the Australian immunologist Sir Macfarlane Burnet to explain how the >immune system recognizes antigens and produces the appropriate antibodies. He proposed that, during foetal life, the body's lymphoid tissue produces numerous clones of cells, each specialized for making a particular antibody. The clones of cells, each specialized for making a particular antibody. The clones remain small and inactive until 'their' antigen arrives in the body, whereupon the clone becomes active and multiplies to produce large quantities of the antibody (see B Lymphocyte). Having been stimulated once, the clone persists at a more numerous and active level than clones that have never been activated; this explans how immunity against a disease can be acquird and retained. Burnet also proposed that, during early foetal life, the cell lines that would have made antibody against the body's own tissues are selectively destroyed; if this is not correctly done, an auto-immune disease occurs.

Clone, a group of organisms that are genetically identical because they are derived from a single cell. Clones occur in nature, as the result of >vegetative propagation. Experimentally, clones have been made of toads; they are especially suitable because the nucleus of the single-cell embryo (one egg in the spawn) is easily removed and replaced by one from any cell in the body of the toad to be cloned. The embryo develops as it normally would, except that its entire genetic apparatus is identical to that of the toad that donated the nucleus. The number of identical toads that can be produced in this way is limited only by the patience of the experimenter. The same system could work in any animal - humans included - provided there is access to the single-celled fertilized egg and there is the technical capability to enucleate it.

A simpler method that has been used in sheep and mice is to take the four-cell stage of the embryo, break it up into four individual cells (each genetically identical, and still with the full potential to develop into a complete individual), keep them viable and allow them to complete their development (mammalian embryos have to be reimplanted, whereas those of toadsand other creatures do not). This could also be done with humans, and is in fact similar to the process that happens in nature when the two-celled embryo splits in two and develops as identical twins, who are technically a clone. (Identical quadruplets are possible too, if the embryo splits up at the four-celled stage.)

The word 'clone' can also be applied to a group of cells which are all derived from a single parent-cell; this could be either in the body of an organism or in tissue culture (see monoclonal antibodies). See also gene cloning.

Coadaptation, the simultaneous selection of two or more genes which when present together give increased fitness. The resulting gene complexes become progressively established in the gene pool of the population or species. Coadaption can also mean simultaneous evolution of two different species who interact ecologically to their mutual benefit.

Co-dominant, term to describe two (or more) alleles which express their effect whether or not the other is present. The genes for A and B blood groups are co-dominant. See also semi-dominant.

Codon, a sequence of three bases in DNA or messenger RNA, forming a 'word' in the >genetic code. See also protein synthesis.

Colchicine, a substance derived from the autumn crocus and used in genetic research. If it is added to a culture of cells, it blocks the process of >mitosis; the chromosomes all become arrested at metaphase, the stage at which they are thickened and easily visible.

Colour blindness, inability to distinguish between certain colours, a genetically determined condition. There are several types of colour blindness in humans. The most common are the two forms of red-green colour blindness, both of which are due to recessive mutant alleles on the X chromosome. About 8 per cent of men in European populations have one or other of these genes and are therefore unable to distinguish red from green, while about 6 out of every 1000 women have this kind of colour blindness. Other populations have lower frequencies; e.g. 2.3 per cent of Mexican and 1.7 per cent of Congolese men are colour blind. A much rarer form of colour blindness is the total inability to see colours, which is inherited as an autosomal recessive.

Competitive - exclusion principle, the idea that two species that inhabit exactly the same ecological niche cannot co-exist indefinitely. Though they may do so for a time eventually one will drive the other out by being a more successful competitor for the available resources. The few examples found in ecological genetics that appeared to contravene this principle have turned out to involve species whose ecological niches were not after all exactly the same.

Complement, a group of proteins involved in the >immune response. After antibody molecules become bound to the invading antigens, complement is produced and facilitates the dissolving of the antigenic cells. There are nine separate steps (each involving an enzyme) in what is known as the 'classical pathway of complement activation'. After complement has reacted with the antibody-antigen complex, it is said to be 'fixed'.

Complement, a group of proteins involved in the >immune response. After antibody molecules become bound to the invading antigens, complement is produced and facilitates the dissolving of the antigenic cells. There are nine separate steps (each involving an enzyme) in what is known as the 'classical pathway of complement activation'. After complement has reacted with the antibody-antigen complex, it is said to be 'fixed'.

Complementary DNA (cDNA), a DNA molecule that has been artificially produced, not by replication, but by reverse transcription of a strand of RNA. See reverse transcriptase.

Complementary genes, two or more genes that are not alleles (i.e. they are at different loci, and may or may not be linked) that must work together to produce the effect in the phenotype. An example would be genes for two different enzymes, both of which are needed to synthesize a particular protein. Complementary genes can be either dominant or recessive.

Complementation test, a technique which tests whether two mutants are alleles of each other or are at separate loci. For example, let strain A be homozygous for the recessive mutation a (all individuals have genotype aa), and strain B for b (all individuals bb). If they are crossed and the genes are not alleles (i.e. they are at separate loci), the result will be a normal phenotype, because at each of the two loci, the mutant is opposite the >wild-type allele, i.e. the genotype is a+/+b. If a and b are alleles, the result will be a mutant phenotype, because the genotype will be ab++, i.e. at that one locus there is no wild-type allele, only a and b which are both mutats. The phenotype of these hybrids is not necessarily the same as that of either of the strains A and B, because although a and b are both recessive to the wild type one may be semi-dominant over the other.

Congenital defects, any of the various things that go wrong with a baby's development that are due not to genes but to the environment in the uterus during pregnancy. These include the results of infection (e.g. the deafness and blindness caused by >rubella), deficiency in the mother's diet, or contact with harmful chemicals. Such defects are not inherited if the affected child grows up to have children of its own (they are an example of >acquired characteristics). In some cases, there is an interaction between an environmental cause and a genetic predisposition to the defect; this seems to be true of >spina bifida, which is related to maternal diet but also has a genetic element. In Britain congenital defects account for some 18 per cent of all admissions to hospital for paediatric treatment.

Conidia, asexual spores produced by the haploid-phase >hyphae of some fungi. Conidia from type A hyphae can fertilize type a hyphae, and vice versa, but conidia cannot fertilize hyphae of their own mating type.

Conjugation, the form of sexual reproduction that single-celled organisms, including bacteria, sometimes use (as discovered by >Lederberg and >Tatum). Two cells form a bridge between their cytoplasm and exchange genetic material; this is carried on plasmids (extrachromosomal DNA rings) rather than on the chromosome. In >***Escherichia coli***, the transfer of DNA is one-way only, but in other genera, it is reciprocal. Conjugation can also occur between different genera (e.g. ***E. coli*** and ***Salmonella***).

Consanguinity, relatedness. The amount of genetic relatedness between individuals is measured by the coefficient of >relationship. The word is from Latin, meaning 'shared blood'. see p. 9 for the persistence of this idea. See also genetic distance.

Consensus sequence, a sequence of DNA that is repeated in a number of different places in the genome of an organism or in many different organisms, presumably because it serves a similar fuction each time. An example is the >TATA box.

Conserved sequence, a sequence of DNA that has varied little during evolutionary time. One example is the H3 and H4 types of histone. See molecular evolution.

Constant region. See C region.

Constitutive, a term used to describe an >operon or a gene that is more or less constantly transcribed, as opposed to an operon or gene that is under some form of >gene regulation. Also used for a constitutive enzyme, one that is constantly produced, as opposed to an inducible or repressible enzyme.

Containment, a term that embraces all the techniques for making sure that potentially dangerous viruses, bacteria, etc. are safely handled within the laboratory. Isolated areas, controlled air pressure, radioactive decontamination, and so on are used.

Continuous variation, variation between one individual and another such that the various phenotypes do not fall into distinct classes but are a continuum and can only be distinguished by some sort of measurement.

Much variation in biology is of this type, and it was originally believed that Mendelian inheritance could not explain it. In 1908 >Nilson-Ehle carried out classic experiments with a cross of wheat that had seeds continuously varying from dark red to white, and showed that only two gene pairs were involved. Later Sir Ronald > Fisher proved mathematically that data on continuous variation can only be explained as being due to separate Mendelian genes interacting with each other and with the environment. Current thinking is that even complex continuously varying traits such as body weight are probably controlled by a small number of genes. See also polygenes.

Contraception, Prevention of unwanted pregnancy. It is not an issue in genetics, unless it can be shown-as early proponents of eugenics believed - that contraception is only practised by persons of superior genotypes, leaving the inferior types to swamp the human race with their genes. If contraception is practised by large numbers of people irrespective of their genotype (as of course is the case) it is of no significance genetically.

Convergence, evolutionary change that brings two unrelated species to resemble each other, either in their overall >phenotype or else in one particular trait. An example is the black swan of Australia, which is remarkably similar to the white swans of the Northern Hemisphere, which have a completely different ancestry. See also parallelism.

Conversion, the phenomenon in which one allele is converted into another. It can happen during either mitosis or meiosis but is more easily detected in the latter. If the two alleles are A and a, the products of meiosis should be two A gametes and two as. If gene conversion occurs, the product is three As to one a (or the other way round). It may be that DNA polymerase molecule making the new copy switches temporarily from one homologous chromosome to the other.

Copy error, a >mutation that results from an error of copying during the replication of one strand of DNA from another.

Correlation, coefficient of, a measurement of how closely associated two paired sets of variables are. In genetical applications, the variables might be the same trait measured in pairs of related individuals or two traits measured in a sample of individuals. It is defined as the >covariance of two variables divided by the product of their own >variances. The values range from + 1 (total positive correlation: when one variable goes up, the other always goes up too) to 0 (no correlation: variation in one is random with respect to the other) and then to -1 (totally negative correlation: when one goes up, the other goes down), with all points in between being possible. These correspond to what one would see if the two variables were plotted against each other on a graph: +1 would be seen as points lying around a line rising diagonally up from lower left to upper right; 0 would be a scatter of points at random; -1 would be points around a line falling diagonally from upper left to lower right; and other values would be lines with slopes other than the perfect diagonal. The method of calculating correlations was devised by >Galton.

Correns, Karl Erich (1864-1933), German botanist, one of the three rediscoverers of Mendel's work. In 1899 Correns was working at the University of Tubingen on plant hybridization when his experiments with peas produced the same ratios in the hybridization as Mendel's original paper, and when de >Vries's paper proving the same theory came out in 1900 Correns had the best grasp of the theory of heredity. He was the first to prove that segregation of the alleles does not depend on one being dominant and the other recessive, and in 1903 he predicted that sex would be found to be inherited as a Mendelian factor - as of course it is.

Coupling. See cis configuration.

Covariance, as statistical technique to measure the association between two variables (Continuously varying). These might be two traits in the same individual (e.g. intelligence and shoe size) or the same trait in related (or otherwise paired) individuals. the covariance is the Product of the deviations from the two means, averaged over the number of observations. With no association, the individual values will vary from the mean in opposite directions or not at all as often as they vary together, giving a covariance value of about 0 (e.g. if very intelligent people have large feet as often as they have small); but if there is association, the covariance value increases. See also correlation, coefficient of.

Creationism, the belief that the Earth and the life forms on it were made by a creator. This could have happened all at once, as in the biblical account, or on more than one occasion, as Cuvier thought. A modernized version is that the first living cell was creator using >

natural selection as its instrument. As recently as 1981 the state of Arkansas passed a law requiring 'balanced treatment for creation-science and evolution-science', meaning that fundamentalist creationism had to be taught in schools alongside Darwinism; the law was overturned the following year after appeals by teachers.

Crick, Francis Harry Compton (1916-), British molecular biologist, joint discoverer of the structure of DNA. Crick trained as a physicist and served in World War II in naval weapons research, but after the war he became interested in biology and joined the Medical Research Council's Molecular Biology Research Unit then located in the Cavendish Laboratory at Cambridge. His research programme was in X-ray diffraction studies of protein structure, but when the geneticist James >Watson arrived, with his enthusiasm for finding out what genes were really made of, Crick turned his imaginative scientific mind to the problem. They used what clues were available - >Avery's identification of DNA as the genetic material; >Chargaffs rules on the ratios of the bases in DNA; >Franklin's X-ray diffraction photographs - but their final success in finding the double helix came when they built a scale model out of pieces of cardboard and wire. Crick and Watson's Paper in the scientific journal Nature in 1953 described their finding as one of the classics of scientific literature, and is only one page long. Crick's brilliant originality led him to further fundamental insights in to molecular biology, most notably the >central dogma, which states the following progression:

DNA [transcription] RNA [translation] protein

With Sydney > Brenner, Crick also did much of the work in establishing that the >genetic code is in triplets (see also wobble). He continued to make contributions to the theory of molecular biology, including proposing the idea of > selfish DNA. Crick shared the 1962 Nobel Prize for physiology or medicine with Watson and Maurice Wilkins. He continued to work at the MRC laboratory until 1977, when he went to the Salk Institute at La Jolla California.

• **J. D. Watson**, The Double Helix (1968): a good read, although Watson's Portrait of Crick is ungenerous.

Cro-Magnon Man. See human evolution.

Cross-over value, the number of individuals who show recombinant types (as the result of >crossing-over) expressed as the total number of individual offspring in a cross. See also map unit.

Crossing-over, the reciprocal exchange of segments between non-sister chromatids during > meiosis. This happens during the pachytene stage; the homologous pairs of chromosome lie alongside each other, and each chromosome is doubled longwise into two identical chromatids. One chromatid of one chromosome may at any point at random join up with the corresponding point on a chromatid of the homologous chromosome, and exchange all alleles from that point to the end of the chromosome (or to the next point of crossing-over, because it can happen more than once to each chromosome pair at each meiosis). this process results in > recombination of alleles between the maternal and paternal chromosomes. For instance, if the alleles on one chromosome were H L and on the other were h l, and there was a cross-over between two loci, the result would be one chromatid with H l and the other with h L. Note that there are also the two chromatids which were not involved in the cross-over, these retain the original arrangements, H L and h l. The point at which the chromatids cross is called a chiasma.

The farther apart on a chromosome two genes are, the greater is the probability that a cross-over will occur between them, and this fact is used to make >gene maps. See also intrachromosomal cross-over.

Cultivar, a plant variety that does not occur in nature. A cultivar can be produced either by selection within a wild derived strain, or by crossing either within a sepecies or between species or genera.

Cultural evolution. See exogenetic inheritance.

Cystic fibrosis, a human disease inherited as a recessive. The clinical effects are lung damage (always) and swelling and malfunction of the spleen, liver or pancreas. If untreated, 75 percent of babies die within their first year, of lung complications. If they are given treatement to prevent lung damage, sufferers live on average until their early 20s. Cystic fibrosis is the commonest genetic disease in Europeans, occurring in about 1 in 2000 births; this implies that about one person in 23 is a heterozygote. The frequency is much lower China, where the disease occurs at a rate of 1 per 100,000 births.

The gene responsible was known by the early 1980s to be situated on chromosome 7, and was found, isolated and cloned in 1989. It differs from the equivalent normal gene in having a deletion of three base-pairs (i.e. one >codon) So that one> amino acid is missing from the gene product; the biological implications of this have not yet been worked out. the fact that the gene has been identified means that in most carriers of the gene can now be identified, and prenatal diagnosis carried out.

Cystinuria, a genetically determined disorder in which there is a defect in the membranes in the kidney so that the amino acid cystine is not reabsorbed from the urine as normal. Stones made of cystine form within the kidneys and can ultimately destroy them. The reabsorption defect is due to a recessive gene, and about 1 in 12,000 babies is born with the disorder. Cystinuria was one of the first four Mendelian traits in humans that Sir Archibald >Garrod discovered in 1903.

Cytogenetics, the study of the number, behaviour and morphology of chromosomes as a means of understanding the genetics of an organism.

Cytoplasmic inheritance, genetic inheritance via the DNA-carrying >organelles in the cytoplasm, i.e. the chloroplasts (plants only) and mitochondria. In mammals and many other groups, the sperm contributes virtually no cytoplasm, whereas the ovum is a large cell, so that cytoplasmic inheritance is exclusively maternal. It is not, however, an important feature in mammalian genetics. See also maternal inheritance.

Cytosine (abbreviation: C), one of the four nitrogenous bases which form the core of DNA and RNA. It pairs with guanine (G). See also replication.

D

Darwin, a unit of evolutionary change. A darwin is equal to a change (increase or decrease) in any one character by a factor of 2.7 per million years. As this is a very fast rate of change, most evolutionary change is expressed in millidarwins.

Darwin, Charles Robert, (1809-82), British naturalist, co-discoverer of the principle of evolution by natural selection. Born into a distinguished family (his father's side were outstanding scientists, including his grandfather Erasmus Darwin, and his mother was one of the Wedgwoods of pottery fame), Darwin was himself an unpromising pupil who studied medicine at Edinburgh and theology at Cambridge but Graduated without honours. He had spent his whole time at university following his passion for the study of nature, and he was a formidably accomplished collector. When he heard that there was a place for an unpaid naturalist on board a naval exploration ship, he was determined to go, in spite of his family's opposition. Darwin boarded HMS Beagle in December 1831; the voyage lasted until October 1836, and included visits to South America, Australia, Tasmania, New Zealand, Tahiti, Mauritius, and many other places. Darwin was as interested in geology as in anything else, and in his Structure and Distribution of Coral Reefs (1842) he was the first to put forward a theory(still held) of reef formation. For an account of his arrival at the theory of evolution by natural selection.

After the momentous publication of the Origin of Species in 1859, Darwin wrote many books extending his basic theory, though his biological thought was always hampered by the fact that the mechanism of heredity was unknown (although it had, in fact, already been discovered by Mendel). The Variation of Animals and plants under Domestication (1868), The Descent of Man (1871), The Expression of the Emotions in Man and Other Animals (1872) and The Effects of Cross and Self Fertilization in the Vegetable Kingdom (1876) were the most important of his later works. His collected Letters to his many scientific friends contain many of his most interesting thoughts, including his remarkable speculation on the origin of life.

Darwin never held a paid post, being one of the last of the hitherto important 'gentleman-scientists'. his beliefs are summed up by his statement (1870) that'I cannot look at the universe as the result of blind chance, yet I can see no evidence of benefits design, or indeed of design of any kind, in the details. See also Wallace; Huxley; Lyell.

Deafness, the inability to hear because of (a) damage or malformation in the structure of the ear or (b) a neurological defect. Deafness has many different causes, but some forms are inherited. In one, the baby is born deaf; this is caused by a recessive gene and occurs in about 1 per 10,000 births. In another, inherited as a dominant, deafness sets in at about the age of one year; this is about twice as common. More frequent still is the form of deafness that sets in during early adult life, which is caused by a dominant allele and affects about one person per 1000. Persons deaf from birth or infancy are also mute.

The genetics of inherited deafness has been well studied because of the frequency with which deaf adults intermarry. This has led to the conclusion that there are at least two different recessive genes that cause total deafness, have children who are normal, which could not happen if the parents were both homozygotes for the same gene.

Degenerate code, See redundant code.

Degrees of freedom, the number of independent variables that can vary within a statistic or total. If one has a sample of individuals to assign to four classes, there are three degrees of freedom: there might be any number in the first class (first degree used up), and any number in the second (second degree used up), but once the number in the third class has been assigned (third degree used up), there is no freedom for the last class to vary - it must simply by the remainder (even if zero). The general formula is n-1.

Deletion, the loss of a part of a chromosome. The lost portion may be as small as a single base-pair or as large as a whole arm of the chromosome. Deletions can occur at any point along the chromosome, though those involving loss of an end are most common. A deletion is one form of >mutation, and is likely to be lethal when homozygous (i.e. it occurs on both chromosomes) but viable when heterozygous. The latter is true because the genes on the normal chromosome can do the work of the ones deleted on the opposite number; recessive genes will be expressed, and are said to be 'uncovered' in such a situation. Several syndromes in humans are caused by deletions; see chromosome abnormalities. See also deletion mapping.

Deletion mapping, a technique for mapping the position of genes on a chromosome by using variants with >deletions on that chromosome. The technique depends on the deletions being visible under the microscope, and has therefore been used most extensively with *Drosophila* (because of its large chromosomes). A recessive mutation can be located because when a strain carrying this mutant is crossed with a strain with a deletion, the recessive will be expressed if it is uncovered, i.e. if it is on the stretch of chromosome missing by deletion. The accuracy of the mapping increases if there are strains available with different overlapping deletions or if deletions of short sequences have been identified. See also genetic map.

Deme, a local interbreeding unit within which there is unrestricted random mating; part of a population.

Demography, the study of the dynamics of populations. Among the relevant variables are birth and death rates, are structure, immigration and emigration rates, some of which are genetically influenced and\or have an impact on the genetic composition of the population (which can be human or any other species).

Denaturation, generally the destruction of a macromolecule's three dimensional structure. With reference to DNA, it means the breaking down of the double-stranded form into single strands or fragments, usually by heating. The process is reversible, the two strands reassociating themselves as they cool (see annealing). Proteins can also be denatured, but they can very seldom regain their original structure.

Dendrogram (also known as a 'phylogenetic tree'), a diagram in the shape of a tree which shows the ancestry of different species (or other taxa). The trunk of the tree is the original ancestor, and each time a new species is formed, it splits off as a branch. The lower down a branch occurs, the earlier the species. See also family tree.

Density-dependent factors, a number of mechanisms by which a population's density in relation to its environment is self-regulating. When density is too great, organisms may become less fertile or even cease to breed, mortality may increase or there may be emigration, until the population declines to a level where its density is right. Conversely, at low density fertility tends to increase and mortality to decrease.

Deoxyribonuclease (DNase), one of a group of enzymes that can cut a DNA molecule into fragments by breaking the sugar-phosphate bonds. It may work by cutting off one base at a time from one end (when it would be an >exonuclease) or by cutting at a specific point along the sequence (>endonuclease).

Deoxyribose, the sugar that forms part of the structure of >DNA.

Deoxyribonucleic acid. See DNA.

Dependent character, a character (i.e. trait) that occurs only if another character is also present. For example, there could be a trait for 'curly horn' in sheep that is expressed only if the individual sheep also has the gene for horns.

Sheep having the hornless gene are unable to show the 'curly' type whether or not they carry the gene for it. See also sex-limited trait.

Derepression, activation of a gene when the >repressor protein is removed. See gene regulation.

Developmental genetics, the study of the genetic control of embryogenesis, growth, regeneration and degeneration. A key concept is the idea that a developmental gene is not a 'description' of, say, a tall horse; rather, it is an 'instruction' for how to make a tall horse. Many developmental genes appear to work by altering the timing or duration of production of some critical substance.

Diabetes mellitus, a metabolic disorder in which sugar, i.e. glucose, is not properly metabolized owing to the lack of the hormone >insulin. Early onset (or insulin-dependent) diabetes is an >auto-immune disease in which the insulin-producing cells in the pancreas are destroyed; its pattern of inheritance is not clear, but it is closely associated with HLA types DR3 AND DR4. The maturity-onset form also has a genetic component but cannot be attributed to a single gene. The gene for insulin in humans has been found (on chromosome 11), but it does not seem that a mutation in this gene is the cause of diabetes. Diabetes affects up to 1 per cent of adults in white populations.

Diakinesis, one of the stages in >meiosis (one of the two processes of cell division). See diagram under meiosis.

Dicentric, a chromosome with two >centromeres. A dicentric is an abnormality that tends to break up because its two ends try to go in opposite directions during cell division. See acentric.

Dihybrid, an organism that is heterozygous at two different loci. The word is also used for a cross between two such individuals. In such a cross with independently assorting characters (with dominance of one allele over the other at each locus), there are four phenotypes among the resulting offspring, in the ratio 9:3:3:1, as shown in the diagram (known as a >Punnett square).

Assume that the cross is between two rabbits: A is the allele for black fur, and is dominant to a for white; Z is the allele for long ears and is dominant to z for short ears. Both parents (with genotypes Aa Zz) are black with long ears.

Every possible genotype combining the four alleles occurs, and there are 16 of them. The most frequent (at 4 out of 16) is the dihybrid - double heterozygote (Aa Az) - like the two parents, and the least frequent (1 out of 16) is double homozygote (aa zz). This is exactly comparable to the position in the basic or >monohybrid cross, only with the numbers squared. See also hybrid; trihybrid.

Diminution, the reduction in the number of chromosomes in somatic cells. This is a feature of some nematode worms and some species of flies.

Dimorphism, the existence in a population of two distinct types, genetically determined. One example is sheepdogs, which are either roughcoated (like Lassie) or smooth. A much more common situation is the existence of more than two types, which is called >polymorphism. The most important form of dimorphism is sex (see sexual dimorphism), though it is a special case, being controlled by the sex chromosomes; most instances of dimorphism are controlled by single genes.

Dioecious, adjective used to describe a plant species that has male and female flowers borne on separate plants. See also monoecious; flowering plants.

Diploid, adjective (from the Greek diploos, 'double') used to describe a cell or organism that has two copies of each chromosome. The single set of chromosomes is present only in the cells that are gametes (ovum or sperm), whereas every cell formed from the moment of fertilization has a double set, i.e. it is diploid. All species of animals are diploid; some plants have multiple sets of chromosomes (i.e. they are >polyploid) and some yeasts and fungi have only the single set (>haploid) for the greater part of their life cycle. Sometimes the diploid number of chromosomes of a species is expressed as 2n, n being the haploid number as found in the gametes.

Diplotene, one of the stages in >meiosis (one of the two processes of cell division).

Directional selection, natural selection that is acting to change the prevailing phenotype in a population in one particular direction. There is then an increase in the frequency of the gene(s) responsible. An example would be directional selection for longer fur on mammoths as the climate grew colder. See also disruptive selection; stabilizing selection.

Discontinuous variation, variation between one individual and another such that each type falls into a separate class. This was the type of variation studied by Mendel, who looked at such discontinuous traits as flower or seed colour in peas. See also continuous variation.

Disjunction, the separation of homologous chromosomes during either of the two processes of cell division, >meiosis or >mitosis. See also non disjunction.

Disposable soma theory, a theory that attempts to explain ageing. Since, by definition, natural selection favours reproduction rather than an individual's survival per se, therefore an organism cannot afford(in bioenergetic terms) to spend more effort in repairing its body than it does in reproduction. This results in the selection of organisms that cease to carry out repairs (i.e. show senescence, or ageing) once their reproductive capacity has diminished. Soma is the Greek for 'body'; the name of the theory implies that the organism's body is expended in favour of reproductive performance.

Disruptive selection, >natural selection that is acting to maintain two or more phenotypically different types in a population. The sets of genes responsible for the selected phenotypes become more common, with any intermediate genetic combination selected against. An example of disruptive selection is >mimicry: in a (palatable) species of butterfly the normal phenotype is favoured, and so are phenotypes that are exact mimics of the other (unpalatable) species, but phenotypes that are only halfway to mimicking the other species are useless and are selected against. See also directional selection; polymorphism; stabilizing selection.

Distribution, principle of, the idea that related species that are geographically near to one another are more alike than those that are far apart. This not-surprising observation was one of the factors that led Wallace and Darwin to their theory of evolution. The creationist explanation was that God had placed each creature in the habitat that suited it best. The principle of distribution is the geographical equivalent of the geological principle of >succession.

Dizygotic twins, non-identical ('fraternal') twins. As the word implies, these twins develop from two separate >zygotes: there were two ova and two sperm and the twins are therefore genetically different individuals, no more alike than any other siblings. They can even be of different individuals, no more alike than any other siblings. They can even be of different sexes. It is curious that dizygotic twins tend to 'mirror' each other in things like which side the hair parts on or which is the preferred hand: a right-handed and left-handed pair of twins is quite usual. Like-sex dizygotic twins have been used in >twin studies as a basis for comparison with identical twins (see monozygotic twins). See also twinning rates.

DNA (deoxyribonucleic acid), the substance of which genes are made. It has three crucial properties: (1) it stores and faithfully transmits information; (2) it copies itself accurately; and (3) it can occasionally mutate, and when it does, the >mutation is inherited. DNA was identified as the genetic substance by Oswald >Avery in 1944, and though his discovery was at first ignored, it was confirmed by Alfred >Hershey in 1952.

The three-dimensional structure of DNA, defined the following year by Francis >Crick and James >Watson, is a double helix. It has been likened to a spiral staircase: the 'banisters' are composed of alternate phosphate and sugar molecules (the sugar being deoxyribose), but the key to DNA is in the 'step' made of pairs of nitrogenous bases locked together crosswise. Base A (adenosine) on one side must be joined to T (thymine) on the other or vice versa, and C (cytosine) must be joined to G (guanine) or vice versa. This means that DNA is able to self-replicate: the double-stranded molecule splits lengthwise, and each single-stranded half is able to reconstitute its opposite strand by following the base-pairing rules. This is what happens every time a cell divides, when the complete set of chromosomes has to be copied. The information-carrying capacity of DNA also comes from the bases: they are 'read' as if they were letters making up words, each three bases long. These words give the information needed for building proteins and for organizing the activity of the cell. See also central dogma; chromosome; genetic code; protein synthesis; replication.

DNA amplification. See polymerase chain reaction.

DNA cloning, the technique of isolating, maintaining and faithfully reproducing many identical copies (i.e. clones) of a piece of DNA such as a gene. This can be done by the transfer of DNA from an external source into a host bacterial cell via a vector. The extraneous DNA may be prepared by (a) physical breakage of complete chromosomes or large DNA molecules by stirring at 0°c (32°F); (b) digestion by a restriction endonuclease; (c) transcription from an RNA template by reverse transcriptase (cDNA cloning); (d) direct synthesis. The artificially produced DNA resulting from this is called >recombinant DNA.

DNA library, a collection of fragments of cloned DNA (see DNA cloning) that together represent all of the >genome of the organism that the DNA came from. A cDNA library can be constructed by transcribing the mRNA to DNA using reverse transcriptase, and then cloning. This provides a library of the genes that are active in the cells from which the mRNA was isolated.

DNA repair, any of the processes by which the correct sequence is restored after a mutational change in DNA. One is photoreactivation, which repairs the typical erroneous T-T base-pairs induced by exposure to ultraviolet radiation, and is carried out by an enzyme that utilizes visible light. So-called dark repair is similar but carried out by enzymes that work best in the absense of light. In humans the DNA repair system probably consists of the products of seven separate genes. For example, the disease *xeroderma pigmentosum* results from a failure to repair DNA damage in skin cells, due to homozygosity for a recessive mutant allele at any one of these seven loci. If DNA repair were 100 per cent efficient, there would be no transmission of mutations, and therefore no evolutionary progress. See also mismatch repair; proofreading; repair synthesis.

Dobzhansky, Theodosius (1900-75), Russian-born American geneticist; professor of genetics at

Columbia University, New York for many years. Dobzhansky researched the genetics of *Drosophila* in wild populations, and his findings provided the best illustrations of the Darwinian hypothesis. He was an excellent writer, responsible both for standard textbooks on genetics and evolution and for such popular works as Mankind Evolving (1962). Dobzhansky was highly sceptical of eugenics, and especially of the idea of >sperm banks.

Domestication, the transformation of wild plants or animals into forms that are useful to humans; the first applied genetics. Dogs may have been domesticated in China by *Homo erectus* half a million years ago. The first plants to be domesticated were primitive wheats (emmer and einkorn), found on sites dated to about 13,000 BC in Israel, with true wheat first found in Syria and dating from about 9000 BC. The domestication of sheep and goats dates from about 11,000 BC and took place in the Middle East; cattle were domesticated rather later, and in North America, the wild turkey was domesticted from about 5000 years ago. Yeast and the silk worm have both been domesticated since ancient times.

Domestication always involves selective breeding, with consequent reduction of the gene pool of the animal or plant involved. None the less, there are impressive numbers of different varieties of each domesticated organism: consider the domestic dog, all varieties of which are members of the same species, *Canis familiaris*. The sort of change that occurs under domestication (e.g. increased size) was one piece of evidence that suggested the theory of evolution to Lamarck, Wallace and Darwin.

Dominant, an allele that expresses its effect even when it is present only in a single dose. A familiar example is brown versus blue eyes: if the allele for brown eyes is B and that for blue eyes is b, then a person who has the genotype Bp will have brown eyes, because B is dominant to b. The opposite (b in this example) is called >recessive; alleles can also be >codominant, >semi-dominant or >overdominant. The notation using capital letters for dominat alleles was devised by Mendel, who was the first to explain the dominance effect (though it had been observed for literally ages).

Dosage compensation, a mechanism to equalize the genetic effects of having two X chromosomes in females compared with one in males. In mammals this is accounted for by the total inactivation of one X chromosome in each cell (see Lyon hypothesis). h *Drosophila* compensation does occur but both the Xs retain some of their activity; the mechanism is not understood.

Down's syndrome, a condition in which a baby is born with a number of defects including mental retardation and heart abnormalities; the typical flattened face gave rise to the syndrome's former name 'mongolism'. Down's syndrome children are notably sociable and affectionate, and the less severely affected can go to ordinary schools and later take sheltered work. About 50 per cent of those with Down's syndrome die before the age of five, and the rest mostly in their teens or twenties, from heart or lung complications. Boys with Down's syndrome do not grow up to be fertile, but girls do, and their babies are either normal or have Down's syndrome. The disorder is due to the presence of three copies (see trisomy) of chromosome 21 (hence the alternate name for the syndrome: trisomy 21). There is a strong association with maternal age, the risk of a baby being born with Down's syndrome increasing from about 1 in 2200 when the mother is 20 to about 1 in 40 when she is 45; none the less maternal age cannot account for all the cases of Down's syndrome, and is not a factor in about 40 per cent of them. It has been estimated that of all the foetuses conceived that have trisomy 21, twice as many are lost through spontaneous abortion (miscarriage) as are carried to term. The condition can be diagnosed prenatally, and it is now routine to offer prenatal diagnosis to pregnant mothers aged 35 or more.

Downstream, further ahead on a DNA molecule. in respect of the direction in which the sequence is being read. See replication.

Drosophila, a genus of fruit fly, used extensively in genetic research. There are 1467 species, found all over the world in warm countries. The species used in laboratories is ***D. melanogaster***, and the one whose population genetics has been most studied (see Dobzhansky, Theodosius) is ***D. pseudoobscura***. The typical ***Drosophila*** is 4-5 mm long, dark brwn, with two wings. In the wild, its diet is usually rotting fruit. Generation time (i.e. from fertilization of egg to mature adult) is 2 weeks, and the normal lifespan is 4-5 weeks. In the laboratory flies are kept in bottles, which can contain up to 4000 individuals, and are fed on a solution of yeast, sugar and other nutrients. For experiments on gene frequencies, whole 'populations' of flies are needed, and these are kept in population cages, 30 cm cubes made of transparent plastic and containing up to 10,000 flies. There are four reasons why ***Drosophila*** became one of the main organisms for genetic research: (1) it is easy to keep; (2) its generation time is short; (3) it has only four pairs of chromosomes; and (4) it has 'giant' chromosomes in its salivary gland cells which are easily studied microscopically. There is an immense literature on the genetics of ***Drosophila***, in which over 3000 mutants have been identified and mapped.

Ducheme muscular dystrophy, a human disease in which there is progressive weakening of the muscles, leading to death at about the age of 20. The cause is an X-linked recessive gene, so only boys are affected, although some of the female carriers do show some signs of progressive muscular weakening. The disease occurs at rate of about 2 per 10,000 male babies, and in about two-thirds of these cases the gene has been inherited from a carrier mother, while in the rest it has arisen as a new mutation. A woman who knows she is a carrier, because she has already had an affected son, can have prenatal diagnosis to ascertain the sex of her subsequent babies; she may then choose to abort any males, but half of these would have been normal. Boys with Duchenne muscular dystrophy are never well enough to be fertile, so the gene dies out each time it is manifested in a boy, although it can travel indefinitely down generations in females. The gene itself was located by reverse genetics, and has been found to be exceptionally large: at 2 million base-pairs, it accounts for over 1 per cent of the X chromosome and is one of the longest genes in the human genome. The protein encoded by this gene is called dystrophin, and it is believed to be involved in muscle contraction; boys with the mutant gene do not have this protein.

Duffy blood group system, a system of genetically determined antigens on human red blood cells (discovered in 1950). There are three alleles: Fy, Fya and Fyb. Fy is at very high frequency in black populations in Africa (and in people of African descent everywhere); some African populations have close to 100 per cent frequency of this allele. The reason for this very high incidence seems to be that the allele gives protection against one form of malaria (but without the deleterious effects that come with sicklecell anaemia, also associated with protection against malaria).

Duplication, a type of chromosomal mutation in which a segment of a chromosome is repeated. If the duplicated segment is next to the original one, it is a tandem duplication, but the duplication may end up anywhere, even on a different chromosome or not attached to a chromosome at all.

Dwarfism, the condition of having very short stature, usually due to excessively short limb bones. The most common form is >achondroplasia. Some other forms are due to non-production of growth hormone, and children born with these can be treated with this hormone. Formerly the growth hormone had to be obtained from the pituitary glands of human corpses, but now the gene has been cloned and so human growth hrmone can be produced in bulk by bacteria.

E

Early gene, a gene in the genome of a bacteriophage or an animal virus that is transcribed early, during the time when the virus is reproducing itself within a bacterial cell. The products of early genes are necessary for regulating or transcribing the later genes necessary for the virus's reproductive cycle.

Ecad, an organism whose phenotype has been adapted to suit a particular environment. An ecad's adaptation is environmentally produced and therefore not heritable (though the idea that it was heritable was the mainstay of the theory of the inheritance of acquired characteristics).

Ecological niche, an available 'space' for a species where it does not compete with others for such resources as food and shelter. There can be many ecological niches within the same geographical location, the species finding their own spaces within it by having different diets or nesting requirements, or perhaps by occupying it at different times (day vs night, winter vs summer). See competitive-exclusion principle.

Ecology, the study of the interrelationships of species and their environment, including both inter-and intra-species relationships. Genetics is an important factor in ecological studies, as every ecological situation is fluid and represents mutual or competitive genetical adjustments being made by the species involved. This is why a geneticist would say that the study of ecology is really the study of evolution.

Ectogenesis, developing an embryo outside the body of the female. This is what happens in frogs, newts and others that have spawn developing into tadpoles, etc. If artificial ectogenesis were ever to be achieved in humans, the result would have more claim to be called a 'test-tube baby' than the babies produced by in vitro fertilization who are now so called. Partial ectogenesis for humans already exists at both ends of the process, in that eggs can be fertilized outside the body and the embryo grown for several weeks, and that premature babies as small as 470 grams (just over 1 lb) can survive.

Effective population size, the number of individuals in a population who contribute offspring to the next generation. It is considerably less than the absolute population size, in that some individuals are not fertile, and in some populations dominance hierarchies restrict the chance to breed to a few individuals. Very few real-life effective population sizes in animal species have been measured: some examples are about 10 for wild groups of the mouse ***Mus musculus*** (with male dominance), and form 19 to 90 in species of lizard (no male dominance).

Electrophoresis, a technique for identifying the different molecules present in a mixture, by placing them in a gel in an electric field; molecules such as proteins and DNA are negatively charged and therefore move towards the positive pole. Smaller molecules move through the gel faster than large ones, so that after a certain time they spread out and can be identified by how far they have moved, which can be seen when the gel is treated with an appropriate stain.

Electrophoresis can also be combined with other techniques. If the sample contains a radioactively labelled molecule, its position on the gel after electrophoresis can be found by autoradiography or scintillation counting. In immunoelectrophoresis, the samples are first separated and then identified by their reaction to specific antibodies.

Embryo, the organism in its early stages after fertilization. See embryogenesis.

Embryo transfer, the technique of artificially implanting early embryos into a surrogate mother. In humans, it is the key technique in >surrogacy. Embryo transfer is often used in cattle-breeding: a cow is given a fertility drug to make her go into multiple ovulation (i.e. produce more than one egg at the same time); she is artificially inseminated; and the resulting embryos are removed (and stored frozen if necessary) before being implanted, preferably two at a time, into surrogate mothers. In this way, the best cow in the herd can have seven or eight calves in a season instead of just one.

Embryo transfer is also being introduced for attempts to get rare or endangered species to raise more offspring, by getting a female of the rare species to go into superovulation and implanting the resulting embryos into females of a related but commoner species with a similar gestation time. There are practical problems. First, the stress caused to all the females may reduce their reproductive ability. Second, an infant brought up by a mother of the wrong species may not be able to breed with its own species.

Embryogenesis, the process of development of the organism from fertilization to birth. The details vary from one type of animal to the next; the following is the outline of the stages that the human embryo goes through.

After >fertilization, the egg immediately divides into two cells, four cells and so on. After about four days, it is called the morula and consists of a ball of 16 of more undifferentiated cells called blastomeres. At about five days the cells making up the ball differentiate and become the blastocyst, which comprises an outside layer (the trophectoderm) around a hollow centre containing the inner cell mass (totalling 36 cells at the beginning of this stage). After implantation into the wall of the uterus (which begins at about 7 days and is completed about 14 days after fertilization), the embryo resembles a flattened disc and consists of three layers of cells; the mesoderm (also known as the >primitive streak, from which, among other things, the skeleton, musculature, heart and blood system and kidneys will develop); the endoderm (digestive system, lungs, liver and other things, the skeleton, musculature, heart and blood system and kidneys will develop); the endoderm (digestive system, lungs, liver and other glands); and the ectoderm (brain and nervous system, skin, hair, etc.).

At 20 days, the embryo, now with a definite head end, is about 1.5 mm long and begins to acquire a segmented appearance; the segments are called somites. The brain and spinal cord begin to be formed (and are in place by 30 days); at 25 days, the lungs and the gut canal are formed and differentiated; and at 28 days, the heart is functioning and the same organs are formed. The embryo is now about 5 mm long. The limb buds (arms first) appear at about 30 days, when the embryo is about 7 mm long. The internal sex organs develop from about 25 days, and by 45 days, the external genitalia are visible.

At the end of the second month (i.e. at 60 days), the embryo is about 35 mm ($1\frac{2}{5}$ in.) long and weighs about 1 gram. From this point, the individual is called a >foetus. All its essential features are present, and it grows very fast; about 1.5 mm per day in length, and increasing its weight so that, at birth (about 266 days from fertilization), it weighs about 3300 times what it did at 60 days.

Emergent evolution, sudden evolutionary change through rearrangement of existing genetic elements. See also saltation.

Endogamy, a cross, or a pattern of mating within a population, in which the parents are closely related. The opposite is >exogamy. See also inbreeding.

Endomitosis, process in which chromosomes replicate but remain within the same cell nucleus. One example of this is in the giant >polytene chromosomes in *Drosophila*: the chromosomes replicate dozens of times but the new copies are kept lying exactly alongside the old so that a sort of chromosome ribbon results.

Endonuclease, an enzyme that cuts a molecule of DNA or RNA somewhere in the middle. Some DNA endonucleases cut both strands; others make a nick, i.e. cut one strand only. RNA is usually single-stranded.
> Restriction endonucleaes cut only at specific sequences. See also exonuclease.

Endoplasmic reticulum, a net-like structure in the cytoplasm within a cell which has ribosomes embedded in it. It is thought to be the site of protein synthesis and modification. Newly synthesized proteins that are to be exported from the cell are pinched off the endoplasmic reticulum and move to the cell membrane from where they are released.

End-product inhibition. See feedback inhibition.

Enhancer element, a DNA sequence in eukaryotes which increases the transcription rate of genes near it. The element does not have to be exactly adjacent to the genes whose activity it is enhancing, and the sequence can be facing either way. Enhancer elements can be specific both to species and to tissue. For example, the enhancer elements associated with the transcription of immunoglobulin genes will only function in Lymphoid tissue.

Environment, in genetics, anything that influences the phenotype of the individual, apart from its genotype (see nature versus nurture). It is also of course the setting in which the phenotype must function, and in which natural selection operates.

Environmentalism, the belief that the environment is more important than the genotype in determining human traits of social importance. The opposite is hereditarianism. See nature versus nurture.

Enzyme, any protein whose function is to catalyse a particular chemical reaction while remaining unaltered itself. Enzymes work by means of their three-dimensional shape, with sites that bind the molecules (referred to as >substrates) that will be involved in the reaction. Their names end in -ase, and they are called after their substrate and/or the type of reaction that they initiate, e.g. adenosine deaminase. It is a tenet of genetics that each enzyme is coded for by a single gene. See Beadle, G.W.; Garrod, Sir Archibald; Tatum, E.L.

Epigenesis, the idea that the development of the organism from the fertilized egg consists of ever greater elaboration of structures and specialization of tissues. The opposite is (or was, for epigenesis is now totally accepted and has ousted its rival) >preformation.

Epigenetic landscape, an idea put forward in the 1950s by the Scottish geneticist C.H. Waddington. He proposed that the fate of a cell during embryogenesis is like the journey of a pebble down a hill: for most of the time it just rolls down the valley in front of it, but at times the valley forks and the pebble must choose one way or the other - a small choice at the time but one that will perhaps have a great influence on where the pebble ends up. See also von Baer's law.

Epilepsy, a condition of variable severity, in which the individual suffers from intermittent attacks of convulsions. Most cases of epilepsy develop during childhood (70 per cent). Epilepsy has a genetic basis, although the mode of inheritance is not simple and there are also environmental factors involved. In European populations about 5 people out of every 1000 suffer from epilepsy.

Epistasis, the situation in which alleles at one locus have an effect on the expression of alleles at another locus. One classic example is the shape of the comb in hens, in which there are at least two alleles at each of two loci. The shape given by genotypes pp. Pp and PP at one locus depends on whether or not the dominant allele R is present at the other locus.

Epstein-Barr virus. See herpes-viruses.

Equilibrium, the situation in which gene frequencies do not change from one generation to the next.

Error catastrophe, a theory that attempts to explain ageing, proposing that there is positive feedback in the mechanism of the DNA that codes for RNA molecules concerned with DNA transcription, such that beyond a certain point transcription breaks down completely and the cell dies. This process, repeated in numerous cells in the body, could give the effects of ageing.

Escherichia coli (abbreviation; ***E. coli***), a bacterium found in the large intestine of humans and many other mammals. It is normally harmless, but can cause trouble if it gets into the bloodstream, the kidneys or the bladder (cystitis). An immense amount of basic research in genetics and biochemistry is done with ***E. coli***, partly because it is a >prokaryote and therefore its genetic machinery is simple. It is also a very convenient organism, as it reproduces quickly, is easy to keep and, if handled properly, is non-pathogenic (i.e. does not cause disease).

There are tens of thousands of different strains, and the ones most often used in the laboratory have become so'domesticated' (by living on prepared nutrient media) that they are unable to survive in the human gut. They are also deliberately mutated so that they cannot survive in the 'wild'. This minimizes risks arising from escapes of bacteria that can carry recombinant DNA molecules.

Some of the major discoveries that have been made in *E.coli* experiments include conjugation in bacteria (Joshua >Lederberg and E.L. >Tatum, 1946), semi-conservative replication of DNA (Matthew >Meselson and F.W.Stahl, 1957) and the operon (Francois >Jacob and Jacques >Monod, 1961). *E. coli* is one of the workhorses of genetic engineering.

Ethology, the study of the behaviour of animals in the wild. The main concepts are instinct and learned responses, both of which have genetic and environmental input. The subject was pioneered by Konrad >Lorenz and Nikolaas Tinbergen.

Euchromatin, chromatin (the substance that chromosomes are made of) that is loosely wound because the DNA is being actively transcribed, and is accessible to >mRNA in that state. Euchromatin shows up more lightly under the microscope than the condensed >heterochromatin. See also banding.

Eugenics, the belief in or policy of attempting to improve the genetic quality of a human population. 'Positive eugenics' is the encouragement of those believed to have good genotypes to have more children. 'Negative eugenics', far more widespread, arms at the eradication of 'bad' genes, either by discouraging persons with presumed inferior genotypes from breeding or by eliminating such persons altogether. Both are surprisingly ineffective from the point of view of removing undesirable genes from the human gene pool. In the case of dominant mutations, these are usually so severe that the affected people do not reproduce at all or only infrequently; far the greatest number of persons with such conditions represent new mutations, and eugenics can do nothing to alter the rate at which these will occur. In the case of recessive mutations, the frequency of heterozygous carries is far higher than the number of persons with the homozygous condition, and removing the two genes together each time they occur in one individual has virtually no effect on the gene's frequency in the whole gene pool. >Genetic counselling is not, strictly speaking, a manifestation of eugenics, in that it deals with the breeding policy of couples, not of the population as a whole.

Eukaryote, any organism that has within its cell or cells a membrane bound nucleus containing chromosomes and all the rest of the genetic apparatus. All creatures from Protoctista upwards (see classification) are eukaryotes; the earlier, simpler type is the >prokaryote.

Euphenics, the policy of altering the environment so that the deleterious effect of a particular gene is not expressed. The word was coined by the American geneticist Joshua >Lederberg. There are several examples among the inborn errors of metabolism. For example, babies born with phenylketonuria grow up perfectly normally if their environment is controlled, i.e. if their diet is kept low in phenylalanine. This is the converse of the approach of >eugenics, the policy of which is to remove undesirable genes from the human gene pool. Indeed eugenicists and others object to euphenics on the grounds that it deliberately maintains unfavourable genes. This need not always be the case, one counter-example being sicklecell anaemia: if the environment is altered by eradicating malaria, the gene for sickling loses its selective advantage and becomes rarer.

Euploidy, the condition in which the number of chromosomes in the cells of a >polyploid species is a precise multiple of the diploid number of the original non-polyploid species.

Evolution, the process by which populations of organisms change their genetic constitution through time. During this process, different populations become unlike enough to each other or to the form that preceded them to be thought of as new species. (Note that evolution is something that happens to a population, which is a collection of individuals, not to the individuals themselves.)

Within any species, new variant forms are constantly being produced by mutation, but in very low frequencies; as they are caused by genetic mutation, the changes shown by these variant forms are inherited. If the environment stays constant, natural selection will operate to eliminate these variants and to keep the species to its present type. But if the environment stays constant, natural selection will operate to eliminate these variants and to keep the species to its present type. But if the environment changes or if a new ecological niche suddenly presents itself, variants may be better able to reproduce offspring than the original type, and thus their numbers will increase proportionately in the next generation. Over many generations, the variant form entirely replaces the original type, so that a new species has evolved. What selection 'chooses' is not the ability to survive or not, but the ability to produce viable offspring; this is known as >fitness.

There are a number of pieces of evidence for evolution:

- The principle of >succession shows that, during geological time, one form has given way to another, and that forms have become more complex over time.
- The principle of >distribution shows that more closely related forms tend to be geographically near each other.
- Mutations are constantly seen to appear in natural populations.
- New species have arisen in the few centuries of observation carried out by biologists.

The theory of evolution by natural selection is accepted throughout the scientific world, but its old rival >creationism still survives in a few places.

See also acquired characteristics, inheritance of; natural selection. For the history of the idea to its formulation in 1859 by >Wallace and >Darwin,

Evolutionary rate, the rate at which the genetic characteristics of a species change. In times of rapid environmental change, it can be very fast, or in a constant environment, it can be zero. A species that is perfectly adapted to its environment does not evolve at all, though this is a very rare case, as evolutionary change in one species has an impact on the other species within its range. Human evolution has shown a very fast evolutionary rate, believed to be partly due to the interaction between genetic and >exogenetic inheritance. Evolutionary rate is measured in a unit called a >darwin.

Evolutionary stable strategy (abbreviation: ESS), an idea, derived from >game theory, that suggests that a population may contain a number of different types, but only in ratios such that selection does not favour one over the other. The idea was developed by the British mathematical biologist John Maynard Smith, and one of his examples best illustrates it. Let there be two types, 'hawks' and 'doves'. When 'hawks' meet, they fight to the death; when 'doves' meet, they perform a lengthy but ritualized fight; and when 'hawks' meets 'doves', the latter retreat. It is not an ESS for the population to consist entirely of 'doves', the latter retreat. It is not an ESS for the population to consist entirely of 'doves' mutants would spread (they have the advantage of not killing each other). Smith made calculations for which he assigned the following values: +50 for a win, 0 for losing, -100 for getting killed, -10 for wasting time in a ritualized fight. On these figures, the stable proportion (the ESS) worked out at 5 'doves' to 7 'hawks'. This happens entirely by natural selection acting upon the individuals; group selection is not involved.

Exobiology, the idea that there are living things on other planets than Earth. Philosophical speculation on the subject goes back to ancient times (Christian opinion settled down to the idea that the Almighty could have made other living creations but probably did not make human beings elsewhere). Present scientific opinion is that, although the findings of space probes so far rule out the existence of life (other than simple microorganisms) on any other planets in our solar system, there is a high probability that the right conditions (of temperature, atmospheric pressure, water, oxygen, salinity and acidity) are met by many thousands of planets in our galaxy. There is no reason to assume that advanced life cannot have evolved on one or many of these; but there is no direct evidence that it has. See also origin of life.

Exogamy, a cross, or a pattern of mating within a population, in which the parents are unrelated. The opposite is >endogamy.

Exogenetic inheritance, inheritance by means other than via genes. Although cultural evolution in a general sense had been discussed for some time (Herbert Spencer wrote at length about it), the idea of exogenetic inheritance as being precisely parallel to genetic inheritance was first proposed in 1932 by T.H.Morgan, the pioneer experimental geneticist:

There are, then, in man two processes of inheritance: one through the physical continuity of the germ-cells; and the other through the transmission of the experiences of one generation to the next by means of example and by spoken and written language. It is his ability to communicate with his fellows and train his offspring that has probably been the chief agency in the rapid social evolution of man.

[Quoted in Jean and Peter Medawar, Aristotle to Zoos: A Philosophical Dictionary of Biology, London, 1984.]

The same idea has been more recently developed by Richard Dawkins, who has proposed the term >meme for a unit of exogenetic inheritance parallel to a gene. Dawkins goes further, saying that 'We biologists have assimilated the idea of genetic evolution so deeply that we tend to forget that it is only one of many possible kinds of evolution.' The theory is far from universally accepted.

Exon, a sequence of DNA that codes for part of a >polypeptide chain (part of a protein), in contrast to an >intron, which is a non-coding sequence intervening between exons. The exon-intron arrangement is a characteristic of eukaryotes. See also cistron.

Exonuclease, an enzyme that removes bases one at a time from the end of a molecule of DNA or RNA. See endonuclease.

Expressivity, the degree to which the trait determined by a gene is expressed in one individual or another. The term usually applies to genes that cause >continuous variation. An example is the recessive allele 'eyeless' in >Drosophila, which in some individuals causes only a slight reduction in the size of the compound eye and in others results in complete absence of eyes. There may also be a marked difference between the extent to which the two sides of the body are affected. Variations in expressivity are caused either by environmental factors or by other genes in the genotype. See also penetrance.

Extinction, the disappearance of a species. This can happen (a) if all the individuals are killed in a catastrophe; (b) if the habitat or some element in it is destroyed; (c) if reproduction fails for some reason; or (d) if population density drops so low that the few survivors have an insufficient supply of potential mates. The term is properly only applied to species, but is sometimes used to refer to the loss of a particular local population. Estimates of the rate of extinction of species are hard to make, as the total number of species is not known (see species, for the numbers that have so far been discovered). One estimate, based on a total of 5-10 million species, was that about one per day becomes extinct. Recently E.O. >Wilson has estimated that, if there are about 30 million species, the annual losses may be as high as 17,500 per year - which is nearly

50 per day. It is apparently considered morally acceptable to bring about the extinction of organisms that are harmful to humans.

Eye colour is usually taken as the classic example of Mendelian inheritance in humans. It is true that the allele for brown eyes is dominant over the allele for blue: two brown-eyed parents, genotypes Bb and Bb, could have a child with blue eyes, genotype bb, whereas two blue-eyed parents, genotypes bb and bb, could not have a brown-eyed child, genotype Bb. There is, however, more to eye colour than mere browness or blueness, and the precise inheritance of the various shades of grey, hazel and so on is not so clear-cut; there seems to be at least one other locus involved, with alleles interacting with the main blue/brown one. Although in many ethnic groups black hair and brown eyes go together, the genes are actually independent, as can be seen in the 'Irish' colouring of blue eyes with black hair.

F

F_1, the first filial generation. This is the hybrid generation resulting from the cross between two pure strains. If one strain is homozygous ZZ and the other is homozygous zz, the F_1 genotype is the heterozygous Zz. The term F_1 can often be seen on seed packets, indicating that the seeds are hybrids; this is a good sign in that the resulting plants will have >hybrid vigour, but it also means that the gardener is not advised to keep seeds from these plants as they will not breed true (see F_2). See also hybrid.

F_2, the second filial generation. This is the generation obtained from crossing (or self-fertilizing) individuals of the >F_1 generation. If the F_1 individuals have genotype Zz, the individuals in the F2 generation will have genotypes ZZ, Zz and zz in the ratio 1:2:1. The table shows how quickly the number of possible genotypes in the F2 goes up as more gene pairs are added. See also hybrid; dihybrid; trihybrid.

Family tree, the type of diagram that genealogists draw. It shows 'ego' (Latin for 'I', the person whose family is being described) at the bottom, with parents in the row above, grandparents in the next row and so on, up through the generations. The same sort of thing that is used in medicine for diagnosis and genetic counselling is called a >pedigree. An evolutionary 'family tree' is a >dendrogram.

Fate map, a map of the cells or tissues in the fertilized egg or early embryo that shows where they will end up in the adult organism. The data can be collected by staining specific cells in the embryo with a non-diffusible dye; artificial >chimaeras can also be used.

Favism. See glucose-6-phosphate dehydrogenase deficiency.

Fecundity, total reproductive capacity, as measured by the number of gametes produced by the individual (or population). It is usually greater than >fertility, which is the measure of actual reproductive performance.

Feedback inhibition, or end-product inhibition, a control system in which the product (e.g. an enzyme) inhibits further production of itself. Generally, the final product in a sequential line of gene products inhibits further synthesis of the first product.

Table showing relationship between the number of gene pairs and the F_2 progeny's genotypes and phenotypes (assuming dominance)

Number of gene pairs	Number of different types of gametes produced by F_1	Number of possible combinations, i.e genotypes of the F_2 progeny	Number of phenotypes in F_2
1	2	3	2
2	4	9	4
3	8	27	8
4	16	81	16
5	32	243	32
6	64	729	64
7	128	2187	128
8	256	6561	256
9	512	19683	512
10	1024	59043	1024
n	2^n	3^n	2^n

Fertility, actual reproductive capacity, as measured by the number of viable offspring produced by an individual (or population). It is usually less than >fecundity, which is the measure of potential reproductive capacity.

Fertility, actual reproductive capacity, as measured by the number of viable offspring produced by an individual (or population). It is usually less than >fecundity, which is the measure of potential reproductive capacity.

Fertility drug, a drug given to a female to stimulate the ovary to produce ripe eggs. A compound that stimulates the body's own production of the hormone follicle-stimulating hormone (FSH) is used for this, both in cattle to produce multiple embryos for >embryo transfer, and in humans in the course of treatment for infertility, when it is often responsible for the woman becoming pregnant with more than one baby. See multiple births; quadruplets; triplets.

Fertilization, the creation of a new individual resulting from the fusion of two gametes. In animals, the gametes are the ovum and the sperm. When one sperm has penetrated the ovum's membrane, there is an immediate change in the membrane's chemistry so that other sperms are repelled (however, see polyspermy). Then the nucleus (strictly, the pronucleus, as it is haploid) of the one sperm arrives at and joins with the pronucleus of the ovum; the result is one diploid cell, which is the zygote. It is also the very beginning of the new and genetically unique individual. Protein synthesis begins immediately, proceeding to cell division and the growth of the embryo.

Fibrocystic disease. See cystic fibrosis.

Finger prints. Both these and the prints on the palms of the hands and the soles of the feet are almost entirely genetically determined, and no two people, not even identical twins, have exactly the same finger prints. The first person to study finger prints systematically was Galton, who saw that the key to the patterns was the triangular 'partings' that he termed triradii. He classified the patterns on the finger tips as (a) arches (no triradius), (b) loop (one triradius), and (c) whorl (two triradii).

An individual's finger prints (and also their palm patterns) can be described not only in terms of these patterns but also the number of ridges between the centre of the pattern and its triradii. The frequency of patterns differs between populations. In Britain, loops are the commonest (70%), then whorls (25%), then arches (5%), but Bushmen have more arches and North American Indians more whorls than Europeans.

There is no theory as to what selective advantage one pattern might have had over another in prehistoric times. Although there are strong correlations between the patterns within families, neither a simple mode of inheritance nor a multifactorial one gives a good explanation.

Fisher, Sir Ronald Aylmer (1890-1962), British mathematician and geneticist. Fisher's first work in genetics showed that quantitative variation (e.g. height) in humans must be due to a small number of Mendelian genes acting together; this was the first proof that quantitative traits are controlled by genes and not (as was thought at the time) by some 'blending' process. This work led to Fisher's apponitment in 1919 as the first-ever statistician at the Ministry of Agriculture's Rothamsted Experimental Station, where he found over 60 years of results of agricultural field trials awaiting analysis. He laid the foundations for all subsequent applications of statistics to experiments in the life sciences. His most important work was The Genetical Theory of Natural Selection (1930), which proved that Darwinian evolution was not only compatible with Mendelian inheritance but depended on it as the underlying mechanism; this had been much debated, but Fisher's mathematical proofs settled the matter. A convinced believer in eugenics, Fisher had eight children as a matter of principle.

Fitness, the relative ability of one >genotype to reproduce itself, compared to the reproductive ability of other genotypes. Fitness is a relative quantity: the fittest genotype in the population has a fitness of 1; a genotype with no fitness at all (because it is lethal or infertile) has O; others are somewhere is favoured by >natural selection. Fitness can also be expressed as 1 - s, where s is the >selection coefficient, a measure of how much the genotype is selected against. As has often been pointed out, the word 'fitness' is troublesome, as are many other words that are taken into scientific usage from general vocabulary. See also inclusive fitness.

5' ('five prime'), one end of a molecule of RNA or DNA, the name referring to the number of the carbon atom in the sugar that comes at the end (joined to a phosphate). The other end is >3', the links between the sugars throughout the molecule being between the 5' and the 3' carbons. A code sequence of DNA written out starts at the 5' end.

Fixation, the situation where for a particular locus there is only one allele present in the gene pool, i.e. the locus is monomorphic. Until a new mutation arises, all individuals are homozygous for that allele.

Flowering plants, the higher forms of plants which reproduce by means of the sexual parts contained in the flowers. A flower's pollen, carried by the anthers which are at the end of the stamens, is equivalent to sperm in an animal. The female part is the pistil, which consists of the stigma, which receives the pollen grain, and the ovule (equivalent to the ovum in an animal) contained inside the ovary (which becomes the fruit). In many plants, anthers and ovules are present in the same flower, and in others (e.g. cucumbrs), there are separate male flowers and female flowers on the same shoot; some plants (e.g. holly) have the tow sexes on different plants. Insects, especially bees, are important in pollen transfer, but some species rely on the wind to carry it. Many species have evolved devices to prevent >self-fertilization. The flowering plants include virtually all the economically significant plants, notably all the grain crops and the nonconiferous trees and shrubs. See also gymnosperms; pteridophytes.

Fluctuation test, a test to determine whether mutations in bacteria occur at random or whether they are evoked in response to environmental factors. Using statistical analysis of the fluctuations in numbers of favourable mutations, the American geneticists Luria and Delbruck proved (in 1943) that mutations are random with respect to the environment.

Foetus (pl. foetuses), the unborn mammalian embryo from the time that it has the recognizable features of its final form. In humans this is from the end of the second month after conception. The word can also be applied to animals, though it is much more often used in connection with humans. See embryogenesis.

Footprinting, a technique for finding DNA sequences that have a protein bound to them which protects them from being cut by endonucleases. Two samples of the whole DNA molecule are prepared, one with the protein and one without; both are digested with a restriction endonuclease and the resulting fragments are runon an >electrophoretic gel. When the two gels are compared, the bands that have resisted cutting by the enzyme (because of being protected by the protein) will show up because they do not match the bands on the other gel. In this way, DNA sequences that bind proteins, such as regulatory sequences, promoters, and so on, can be identified.

Forward mutation, a change from the normal, or wild-type, version of an allele to a mutant form. The vast majority of mutations are of this type, the opposite being >back mutation.

Founder effect, the principle in population genetics that a small group of individuals colonizing a new habitat are unlikely to carry a representative sample of the gene pool of the population that they come from. The new population therefore has different gene frequencies from the start. Founder effect is believed to be an important factor in the formation of new species. For instance, the extraordinary diversity of the forms in the Galapagos Islands is probably due to a considerable extent to it. See quantum evolution.

Fragile X syndrome, a genetic disorder causing mental retardation in males. It is the result of a mutation in the region at the very end of the X chromosome, which is, for some unknown reason, particularly liable to lose pieces - hence the name 'fragile X'. Some female carriers are also slightly affected.

Frameshift mutation, a >mutation in which one or more base-pairs are inserted into or deleted from a gene, so that the >reading frame of every codon after that point is shifted along. For example, if the original code was UCU-CAA-AGG-UUA, and the mutation put an extra U at the beginning, the message would be read as UUC-UCA-AAG-GGU and so on.

Franklin, Rosalind (1920-58), British X-ray crystallographer. Franklin graduated in chemistry at Cambridge, and did war work on coal utilization. In 1951 she joined the Medical Research Council's Biophysics Unit at King's College, London, where she worked on the study of >DNA via its X-ray diffraction pattern. She discovered that DNA occurs in two forms, and suggested a helical structure for the form that occurs in the natural state. Her calculation of how much water this form contains was one of the clues that Francis >Crick and James >Watson used to find DNA's structure, and her X-ray photographs were even more important to them. Franklin's early death from cancer prevented her from sharing the 1962 Nobel prize with Crick, Watson and her colleague Maurice Wilkins. The portrait of her in Watson's The Double Helix is grossly unfair, as Watson later admitted.

Freemartin, the female of a pair of twins in cattle, sheep, goats and pigs, when the other twin is male. The placentas for the two twins become joined and allow sex hormones of the male twin to interfere with the development of the female twin's ovaries and other sexual parts. Thus a freemartin is hermaphrodite in appearance and is almost always sterile. The male twin is unaffected.

Fungi, one of the five kingdoms of living things (see classification), containing a number of genera that are economically important. The most important of all is yeast (see *Saccharomyces*), Much genetic research has been done on other fungi, including the moulds >*Aspergillus* and >*Neurospora*. Although they are plants, fungi are unable to photosynthesize (i.e. utilize energy from sunlight to synthesize organic compounds), so they have to obtain their food by living off other plants or animals, alive or dead. Fungi can reproduce either by cell division (equal or unequal, the latter including either budding or the production of spores) or by sexual reproduction, in which case they have >alternation of generations, with haploid and diploid forms.

Fusion gene, a gene resulting from the accidental mixing of parts of two different genes. This may happen as the result of a small deletion occurring between two genes, leaving their ends joined up. Fusion genes can also be made in vitro using recombinant DNA technology.

G

G_1,G_2, See cell cycle.

Galactosaemia, a genetically determined disease in which the liver degenerates from early infancy owing to the inability to metabolize the sugar galactose; there is also serious mental deterioraton. If the condition is diagnosed at birth, the baby can be put on a galactose-free (i.e, non-milk) diet, in which case there are no ill effects. This diet can be relaxed after the age of five. The genetic cause is an autosomal recessive, and about 1 in 50,000 babies is affected. Heterozygotes (i.e. >carriers) among the siblings of an affected child can be identified by biochemical tests.

Galton, Sir Francis (1822-1911), British scientist. Galton, a cousin of Charles Darwin, was a scientist of wide-ranging interest. He stressed the importance of quantitative evidence, and was first a geographer and meteorologist (he discovered and named anticyclones) before turning to biology. Taking up Darwin's theme of evolutionary progress, he applied it to humans, saying that it stood to reason that the human species could be improved by selective breeding. His general idea, for which he coined the term >eugenics, was that 'better' people should breed more, and he cannot be held responsible for the later type of eugenics whose thrust was that 'worse' people should breed more, and he cannot be held responsible for the later type of eugenics whose thrust was that 'worse' people should breed less (or not at all). He carried out many practical studies, the results of which included Hereditary Genius (1869) which attempted to show, in pre-Mendelian terms, that intellectual achievement is inherited, and proved that an impecunious peer who marries an heiress is highly likely to have got himself a barren wife (being the only child of subfertile parents, she may well be entirely infertile herself).

Galton was the first person to realize that the study of identical twins could give evidence on the >nature versus nurture question, and he collected much data in this field (see also twin studies). His system of classifying finger prints is still in use throughout the world today, but perhaps his greatest achievement was the calculation of the >correlation coefficient which is one of the basic tools of biometrical genetics.

Galton's law, principle stated by >Galton concerning the degree of genetic similarity between relatives. The same idea has been restated in post Mendelian terms as the coefficient of >relationship.

Game theory, a mathematical way of assessing strategies in real-life situations, originally devised by the Hungarian-American mathematician John von Neumann in 1944. The real-life situation, whatever it may be (e.g. a military confrontation, an economic crisis), is likened to a game in which the players make choices between possible strategies, which are limited by the rules of the game. Each player is deemed to want to win, or at least to prevent an opponent winning; player A does not know what player B is going to do, and vice versa. By analysing the possible outcomes algebraically in terms of points won or lost, one can work out what strategy is statistically the most likely to be the winning one.

The idea of using game theory to analyse what happens in the evolution of species was proposed by the British geneticist C.H. Waddington in his book The Strategy of the Genes (1957). When used for this purpose, there is one substantial disadvantage: game theory is based on choice - i.e. upon purposive behaviour - but clearly there is no sense in saying that an animal or species 'chooses' one evolutionary strategy rather than another. In the simplest applications of game theory to evolution, one of the 'players' is the environment (and this also cannot be thought of as having any intentions). In other examples, the environment is considered to be a static background against which the 'players' compete with each other; sometimes the players are different genotypes within a species (see evolutionary stable strategy), sometimes species competing for a resource (see K selection; r selection; red queen hypothesis). See also mimicry.

Gamete, one of the reproductive cells; also called a germ cell. In animals they are the ovum and the sperm; in plants, the comparable cells (though not quite the same) are the ovule and the pollen grain. A gamete is haploid, so that when fertilization takes place, the newly formed individual (the zygote) is diploid.

Gametophyte, the >haploid part of a plant's life cycle. In higher plants it consists of only a few cells, which make up the pollen grain or embryo sac. See alternation of generations.

Gamma-globulin, a class of proteins, comprising the >antibody molecules in the immune system.

Garrod, Sir Archibald (1857-1936), British physician, pioneer of medical genetics. In 1903 he identified four inborn errors of metabolism - albinism, alkaptonuria, cystinuria (not in fact an enzyme deficiency) and pentosuria - and realized that they were examples of Mendelian inheritance in humans. He also saw them as evidence that one gene codes for one enzyme, a hypothesis ignored at the time and only taken up (by >Beadle and >Tatum) some 40 years afterwards, when it became the cornerstone of biochemical genetics.

Gaussian distribution. See normal distribution.

Gene, the unit of heredity. A gene is a sequence of >DNA, occupying its own place (see locus) on a chromosome. Most genes are structural, i.e. they code for a particular protein; these genes are divided into >housekeeping and >luxury genes. Other genes code for the RNA molecules that are necessary for protein synthesis, and others again provide recognition markers for the polymerase enzymes involved in >gene regulation. A gene may exist in two or more alternative forms, called >alleles. Genes do not change except in the very rare event of >mutation; there are many genes in the human genome that have been passed unchanged down the generations for millions of years, since before we became humans (see molecular evolution). The word 'gene' was coined (in German, as Gen) by >Johannsen in 1909, quite a time after Mendel's republished work had identified the existence of such a thing. Before that, there had been an appreciable linguistic difficulty; for example, in the course of a single article, T.H. >Morgan used five different words for what is now called a gene.

Gene amplification, the process by which extra copies of a gene that is needed temporarily are made. This can occur naturally, as in >*Drosophila* females who multiply the genes that code for the proteins that compose the membranes of their eggs: all the eggs are matured within about five hours, and to achieve this, the relevant genes are multiplied about ten times. Gene amplification can also be artificially induced, and has been observed in mice, who respond to a sub-lethal dose of cadmium by multiplying copies of the genes coding for the protein metallothionein that neutralizes the poison. See also silk.

Gene cloning, the technique of taking a complete gene from an organism's genome and inserting it, via a plasmid, into a host bacterial cell. To make sure that the gene will be expressed by the bacterial cell, regulatory regions as well as the coding sequence of the gene must be included. This technique, perhaps the most basic one in genetic engineering, can be used to produce large amounts of the product of any gene that has been isolated. Human genes have been cloned in >*Escherichia coli* to produce once scarce proteins for treatment of people whose own gene is defective; these proteins include growth hormone and insulin.

Gene divergence, the process by which a sequence of related genes arises from an original single gene that has been duplicated a number of times and then undergone divergent changes in the different copies.

Gene dosage, the number of times that a gene is represented in an organism's cells, or total genome. It can range from as low as one, if there is a single-copy gene on the X chromosome in animals, to many thousands for a highly duplicated gene, as exemplified by the genes coding for rRNA in amphibians.

Gene flow, the movement of genes from one population into a neighouring one. It may happen by interbreeding or migration or both.

Gene frequency, the frequency with which a particular allele occurs at a given locus. It is expressed as a decimal: if there are only two alleles, the gene frequencies could be, for example, 0.1 and 0.9 (the total must add up to 1). (See Hardy-Weinberg equilibrium.) Evolution is really a matter of changing gene frequencies: the old allele gets rarer as the new one comes in, and finally the new one may become fixed (see fixation), so that there is a complete change in the genetic constitution of the population in respect of that locus. Taken over a sufficient number of loci (which could perhaps be rather few), this gives rise to a new species. Changes in gene frequency are brought about by >natural selection, >gene flow, >founder effect and >genetic drift.

Gene library, a collection of the genes from a species, maintained as gene clones. It is different from a >DNA library in that the items stored are whole genes, not DNA fragments.

Gene pool, the total genetic material present in a sexually reproducing population. To put it the other way round, a population is a group of individuals who share the same gene pool.

Gene product, the molecule that is made from the information in a single gene. In the majority of cases it is a protein (or polypeptide, which is just a small protein), but some genes code for a sequence of RNA. See protein synthesis.

Gene regulation, the various systems by which genes that are not constitutive (i.e. in use all the time) are switched on and off. Gene regulation is the means by which cells specialize into the very many different cell types found in even quite simple organisms, and by which organisms respond moment by moment to changes in their external and internal environment. Most of what is known about regulation has been discovered in prokaryotes, and eukaryote gene regulation is still hardly understood. The key feature in prokaryotes is the >operon, a sequence of DNA in which the regulatory genes and the several structural genes (e.g. for a metabolic process) are in a line, one next to the other. Some operons are inducible, that is they are not transcribed until they are switched on by the presence of some molecule within the cell.

Gene switch, the moment when a developing organism changes over from one gene (or set of genes) to another related one. A classic example is the change in humans from foetal haemoglobin to the adult equivalent. The series of 'choices' made by gene switching are irreversible. See epigenetic landscape.

Gene therapy, two possible techniques for curing genetic disease. In germ-line gene therapy, a normal gene to replace the defective one would be implanted at a very early stage of embryogenesis, so that the new gene was incorporated into the individual's germ line (i.e. ova or sperm); this is generally considered unethical for humans, and it is not being developed for clinical use. However, some of the research on human embryos now being considered in Britain and elsewhere will be into techniques that could be used for germ-line gene therapy.

Somatic-cell gene therapy is different in that the new gene is inserted into developed tissue and so cannot be incorporated into the germ line (and is therefore not heritable). Suppose that the patient is homozygous for a recessive gene that causes an inability to produce a vital enzyme, and that the normal form of the gene has been found and isolated. The normal gene would be cloned so that many copies of the DNA sequence were available. Next, a sample of the cells that should make the enzyme is taken from easily accessible tissue in the patient (e.g. bone marrow), and cultured. The cloned DNA is added to the cell culture in conditions likely to result in the cells taking up the DNA and are making the enzyme. Then these cells can be put back into the patient who should be able to integrate them into their own tissue and sustain them as a functional cell line. Genetic disorders that could possibly be suitable for this approach include Lesch-Nyhan syndrome, haemophilia, sickle-cell anaemia, thalassaemia and severe combined immune deficiency. An attempt in 1980 to use gene therapy to cure two patients with thalassaemia was a failure, but much technical progress has been made since then and it is realistic to expect gene therapy to be in clinical use by early next millenium.

Generation time, the average time in a species between the birth of an individual and its production of offspring.

Genet, an organism raised clonally (see clone) from a single zygote. See also ortet; ramet.

Genetic code, the meaning of the set of triplets (or codons) of nitrogenous bases within a DNA or RNA molecule. Each codon codes for one of the 20 >amino acids from which proteins are made. However, since there are four bases arranged in threes, there are 64 (i.e. 43) possible combinations; therefore most of the amino acids have alternative codons, which nearly always involve an alternative base in the last place in the codon (see wobble). There is no 'punctuation' within the genetic code, but three codons act as 'stop' codes (see protein synthesis for how the 'start' signal works). The genetic code is non overlapping, comma-less, degenerate and universal.

The genetic code is universal, the meaning of the codons being the same in every organism from amoebas to zebras. The code was deciphered in the late 1950s and 1960s by, among others, >Brenner, >Crick, >Khorana and >Nirenberg.

Genetic counselling, the advice given to people about the risk of their having a baby with a genetic disorder. Except in a very few cases where whole populations are screened for a harmful gene and carriiers identified, genetic counselling usually takes place when a couple have reason to believe that they are at risk, because they have already had one baby with a defect or because a relative is affected.

The genetic counsellor begins by studying the >pedigree of the family and identifying the nature of the genetic problem. In some cases, e.g. when there is a dominant >mutation in a relative of one partner but that parter is him/herself normal, the couple can be assured that they are at no risk of having an affected child. Equally, if both partners are themselves normal but have had one child with a dominant defect, they can be reassured that they will not have another such child, as this must have been a case of a fresh mutation. If, however, one of the partners has a dominant defect, they will be told that there is a 1 in 2 chance of them passing it on to any child.

Recessive defects are more complicated, as the gene must be inherited from both parents; if a relative of one partner has such a condition, the question is whether that partner is a carrier. Diagnostic techniques are now available to ascertain for many such traits whether or not a person is a carrier; if there is no such test for the trait, then a probability can be calculated from the pedigree. If a couple has had one child with a recessive defect, then it is clear that they are both carriers, and that the risk of their having another is 1 in 4. The options that such couples can take are: (a) take the chance; (b) start another pregnancy but have prenatal diagnosis; (c) artificial insemination by donor; (d) egg donation; (e) embryo donation; (f) adoption; (g) have no more children.

Genetic death, the death of a gene, in the sense that the individual who carried it failed to reproduce. This may have been because the individual itself died as a result of carrying the gene, but not necessarily. From the genetical point of view, it is the fact that fertility was reduced to zero that counts.

Genetic distance, a measure of the amount of relatedness of two individuals or populations, in terms of the probability of their possessing the same alleles at a locus or loci. See consanguinity.

Genetic drift, random fluctuation in the frequency of a gene. According to some theorists (known as'neutralists'), genetic drift is very widespread and accounts for much of the polymorphism in natural populations. It becomes an important effect in small populations when drift can cause the loss of an allele from the gene pool even though it was a useful one. Genetic drift was first described by the American geneticist Sewall >Wright.

Genetic engineering, the group of techniques that involve altering the natural state of an organism's genome. The core activity is transferring DNA (either a whole gene or part of one) from one organism to another. (The DNA so produced is called >recombinant DNA, and the term 'recombinant DNA technology' is usually used by geneticists rather than 'genetic engineering'.) There are various ways of isolating DNA required for transfer. For example, it can be found by digesting with >restriction endonucleases and cloning, or by >polymerase chain reaction. In the case of eukaryote genes, the messenger RNA produced during the primary transcription process can be isolated, and reverse transcriptase used to make a corresponding DNA sequence.

The foreign DNA cannot simply be placed just as it is in the host cell, since it would not be expressed or replicated (gene expression being controlled by a position-specific mechanism); it must be carried in via a particular DNA molecule that already has the property of being able to get itself expressed in the host cell. Such a carrier is called the vector, and it is usually a >plasmid, a bacteriophage or a virus. The DNA to be inserted is joined to the DNA of the vector, either by a sticky-end join (when a restriction endonuclease has left complementary tails on each), by bluntend ligation or by adding a poly-A tail to one and a poly-T tail to the other so that they anneal. In addition to carrying the required DNA sequence, the vector must confer some property on the host cell that can be detected, e.g. antibiotic resistance or a marker gene, so that the experimenter can identify which host cells have assimilated the vector.

A gene transferred from one prokaryote (e.g. a bacterium) to another will generally be synthesized without difficulty, but owing to the different gene regulation systems of eukaryote cells there are problems with eukaryote genes that are to be expressed in a bacterium. These can be overcome by using a >shuttle vector or by attaching the DNA to a plasmid containing the lac operon.

The above techniques are all used when the end result is (1) production of particular RNA or protein molecules in large quantities, (2) creation of novel bacteria with commercially useful properties, or (3) synthesis of vaccines. An example of (1) is the production of human insulin by synthesis in *Escherichia coli*, and of (2) the modification of a marine bacterium so that it metabolizes petroleum and can be used to break up oil slicks. Progress with (3) is very encouraging, with vaccines against hepatitis B, herpes and Epstein-Barr virus having been successfully made, and an anti-AIDS vaccine is a real possibility.

At a different order of magnitude is the problem of incorporating permanent genetic change in eukaryotes - that is, getting the foreign DNA incorporated into the germ line at an early stage of >embryogenesis. With plants, the gametes are fairly accessible, and genes can be transferred for instance, by means of a virus. There is also the advantage of being able to regenerate whole plants from single cells that have had recombinant DNA inserted in vitro. Gene transfer has been done with animals: microinjection of plasmids carrying recombinant DNA into the pronucleus of recently fertilized mouse eggs works well in terms of getting the DNA into every cell of the mouse embryo's body, but this is far from guaranteeing that the gene will be expressed. Genes can be transferred into animal cells in vitro using viruses as vectors, but again the problem is getting the gene expressed. See also gene therapy.

Genetic fingerprinting, a technique which distinguishes the individuality of a person's DNA. Devised by the British geneticist Alec Jeffreys, genetic fingerprinting is the analysis of the >Alu sequence that is present in many repeats in the human genome.The number of repeats, their position in the genome and their exact sequence differs from one person to another. The repeats are digested with restriction endonucleases and identified by Southern blotting. The technique has been used in forensic science in two ways. First it can identify with certainty whether a sample of blood, saliva or semem came from a particular individual. Second, it can establish whether a child is or is not the offspring of an alleged mother or father, given that the genetic fingerprint of the other parent is available; this is used by the Home Office in Britain to establish the relationship of dependants of immigrants. Fingerprinting has also been carred out on dogs and horses, to establish the genuineness of their pedigrees.

Genetic load, the hidden burden on a population of deleterious recessive mutations. While these alleles are in the heterozygous state, they do no harm, but once they meet in the homozygous state, they damage the individual and therefore the population.

Genetic map, a diagram that shows the position on a chromosome of the various genes that are known to be on it. A.H. >Sturtevant devised the method of using the frequency of >crossing-over between two genes as a measure of how far apart they are on a chromosome: two genes exactly next to each other will very seldom have a cross-over between them, but genes at opposite ends of a chromosome will be separated by any cross-over). The basic technique of mapping is therefore to make large numbers of crosses involving two or more loci at which there are distinguishable variant alleles, and to record the number of times that recombination of the original types has taken place. The frequency of the cross-overs is expressed as a percentage, and one percentage point is known as a >map unit. When three or more loci are mapped in this way, the map units add up as they should. For example, if locus A and locus B have a 7 per cent cross-over frequency between them, and locus B and locus C have 10 per cent, loci A and C have cross-overs in 17 per cent of instances. Another is only passing on the wherewithal for the next generation to run the body's functions. Of all the DNA molecules that a human being makes in a lifetime, the vast majority (many millions to one) will be involved in making the proteins in the body's cells; only a few will go into the gametes. To make an imperfect analogy, DNA is like a small businessman, who does all the jobs around the business himself, some things on some days, some on others, but of all the working days of his life, he spends only one or two telling his son what to do when he takes over.

Essential Genetics or Genetics Today

Transcription and translation and Genetic code.

When DNA makes a protein, it does so via an intermediate molecule called ribonucleic acid (>RNA) which is very similar to DNA. Its sugar backbone consists of ribose, not deoxyribose, and the pairs of bases are adenine (A) and uracil (U) (instead of thymine (T), and cytosine (C) and guanine (G). RNA is also usually in the form of a single strand, not a double-stranded helix like DNA. A molecule of RNA can be made to an exact and repeatable pattern by reading off the base sequence of a stretch of DNA; this process is called >transcription. The strand of RNA produced in this way is known as >messenger RNA, or mRNA, since it carries the message telling what protein is to be made from the particular stretch of DNA that was copied (see protein synthesis).

All proteins consist of a chain of sub-units called >amino acids. There are some 20 amino acids commonly occurring in protein; every protein has its own sequence of amino acids, from as few as half a dozen (which is not really a protein, but a >polypeptide) up to as many as a thousand or more. The sequence determines not only the chemical activity of the protein but also its physical shape; proteins are wound into their own characteristic balls or spirals. The sequence of bases in the DNA, as was immediately obvious when the Crick-Watson structure was proposed.

Much of the research on working out the details was done, as had been Crick and Watson's, at the Medical Research Council's Laboratory of Molecular Biology in Cambridge, England, which in the 1950s was a leading centre of biological research. A South African, Sydney >Brenner, worked with Crick on deciphering the DNA code, and they discovered that it is written in 'words' of three letters. For example, a sequence of DNA in which the bases are GCATAGATC is to be read as GCA-TAG-ATC (these groups of three are known as >codons, or triplets). It seemed likely that one codon coded for one amino acid, and the truth of this was first demonstrated in 1961 by the American scientist Marshall >Nirenberg who, having made a strand of synthetic mRNA consisting of one amino acid (phenylalanine) similarly repeated. There followed a worldwide race to decipher the other codons, with many of the results coming from the Indian-American geneticist Har Gobind >Khorana. As there are four possible bases occurring in groups of three there are 64 (i.e. 4 x 4 x 4) possible combinations, but only 20 amino acids to be coded for. It turned out that most amino acids are coded for by more than one codon, and that there are three codons which do not represent any amino acid but are 'stop' signals where the protein-coding message ends.

Further research demonstrated that this genetic code is the same in all living things. This is the strongest possible evidence that all life on earth is descended from a single origin.

Apart from some minor differences between >prokaryotes (which include very simple organisms such as bacteria) and >eukaryotes (the more advanced creatures, including humans, that have a proper nucleus in their cells), the way in which proteins are made is the same in all living things. The code sequence of the DNA gene is used to build a molecule of messenger RNA; this is assembled by an enzyme called >RNA polymerase, and this part of the process is called transcription.

There are also many molecules of a different kind of RNA known as >transfer RNA, or tRNA molecule consists of only three bases, that is, it is a single triplet. These three bases form an >anticodon, and each of these matches on to a codon in the mRNA. Thus if the mRNA reads GCA-AAG-AAA, it will attract to itself three tRNA molecules having the anticodons CGU-UUC-UUU, and each of these has the appropriate amino acid attached to its other end. This part of the process, known as >translation, takes place on one of the many lumps of protein-and RNA in the cell called >ribosomes, which act as a sort of assembly frame for building proteins. The tRNA molecules form a line, and the amino acids join up in the specified order to form a line, and the amino acids join up in the specified order to form the protein chain.

Once this machinery was understood, both mutation and dominance became explicable in molecular terms. If just one base in the DNA changes (in a mutation), the codon also changes and is very likely to code for a different amino acid. The difference that this new amino acid, being only one out of so many, makes to the resulting protein is very small but it will probably have a bad effect on its function, even to the point where the protein does not work at all. (Very occasionally, such a change will make the protein more efficient than it was before.) Sometimes the mutation alters the intended codon to a 'stop' signal, in which case the formation of the protein ends before it is complete. There are also >frameshift mutations, in which the process of transcription starts in the middle of a codon rather than at the beginning, so that the entire message is garbled and a useless protein results.

Equally, it became clear why most mutations are recessive. When a mutation occurs on one chromosome of a pair, which then starts coding for a defective protein, it does not matter so long as the gene on the other chromosome can produce adequate quantities of the normal protein. This is normally the case if the protein in question is an enzyme, which is only needed in small amounts, but for structural proteins, such as the ones that form bone tissue, the output of a single gene will probably not be enough. Thus the enzyme-defect mutations resulting in diseases such as >phenylketonuria are recessive, while structural-protein mutations such as those causing >achondroplasia are dominant.

Gene regulation

Even after the statement of the central dogma of molecular biology, there was still a major difficulty in understanding how genes work. Every cell in the body has the same set of genes, located on the same paired set of chromosomes that have been copied exactly in every cell division since fertilization. Yet cells are specialized: some are skin cells, some line the intestines, some are nerve cells and so on. It must be that genes can be switched off when they are not needed (any probably most genes, about 90 per cent, are switched off most of the time). How does the switch work?

This was first investigated in the common bacterium >*Escherichia coli*, which has a simple metabolism but needs different enzymes at different times according to what nutrients it has to digest. In 1961 two French molecular biologists, Jacques >Monod and Francois >Jacob, discovered that the switch mechanism consists of what they called an >operon: a group of genes with related functions have an 'operator' sequence next to them; this is switched off by a 'repressor' molecule which physically combines with the operator, thereby preventing mRNA being formed to transcribe the genes. Repressors (like genes themselves 60 years before) remained only hypothetical entities until in 1968 the American molecular biologists Walter >Gilbert and H.Muller-Hill successfully isolated a repressor and showed it to be a protein.

Operons and repressors are now well known in prokaryotes, but they have not been definitely shown to work in the same way, or indeed to exist, in eukaryotes. Prokaryotes have another way of switching genes on and off, called >attenuation, but this also has not been proved to exist in eukaryotes.

In some insects (including *Drosophila*), all of which are eukaryotes, the chromosomes in the cells of the salivary glands are so large that they can easily be observed, and in the areas where DNA is being transcribed they can be seen to swell up into a >puff. Although these puffs are known to be site of gene activity, that does not explain how these are switched on and off.

One possibility is that gene regulation in eukaryotes is connected with the phenomenon of >jumping genes. Discovered in maize plants in the 1940s by the American geneticist Barbara >McClintock, jumping genes are segments of DNA within a chromosome which have flipped over so as to alter their position in relation to neighbouring genes on the same chromosome. McClintock proposed that this might enable the genes that 'jumped' to act as switches for the neighbouring 'working' genes. Her idea was ignored at the time, but later research in the 1970s showed that jumping genes, now known as >transposons, are common in both prokaryotes and eukaryotes, and it is possible that they are indeed involved in the regulation of gene activity.

Movable genes and cancer

It may also be that movement of genetic elements within the chromosomes is a factor in cancer. The clue came from the fact that some cancers are caused by viruses; this was first discovered in chickens, who develop tumours when infected with the Rous sarcoma virus. The virus introduces a gene - known as an >oncogene - into the chicken cell, which transforms it into a cancer cell. It was then discovered that all normal cells contain many genes that are exactly comparable to viral oncogenes; these are known as protooncogenes, and they seem to play a part in the control of cell division and differentiation. There is evidence to suggest that different types of cancer are caused by the activation of different proto-oncogenes; this activation might in turn be caused by a viral oncogene or by a mutation of the cell's own proto-oncogene on contact with a harmful substance.

For example, Burkitt's lymphoma, a cancer causing tumours of the jaw, which is common in Africa, occurs in children who have been infected with the Epstein-Barr virus; the tumour cells have a chromosome abnormality, with a translocation broken off from chromosome abnormality, with a translocation broken off from chromosome 8 at exactly the point where the proto-oncogene in such a way as to cause chromosome breakage and translocation; this moves the proto-oncogene into a new position where it is not under its normal control mechanism, so the cell in which this occurs proliferates and gives rise to a mass of cancer cells. A similar situation occurs with chronic myeloid >leukaemia, in which all the cancer cells have an abnormal chromosome, the Philadelphia chromosome, which again is a translocation that moves a protooncogene to a new position where it is de-regulated. Some experiments suggest that there must be two oncogenes involved, one to enable the cells to proliferate indefinitely (the 'immortalizing' gene) and the other to cause the cells to lose their own special shape and behaviour (the 'transforming' gene).

Genetic engineering

The ever-increasing amount that is known about gene regulation in prokaryotes (and in simple eukaryotes such as yeast) has become the basis for the fast-developing science of >genetic engineering. The organism most often used for genetic engineering is the laboratory workhorse *E. coli* and the enzymes that occur naturally in it. The latter are employed for 'snippin' DNA (when the enzymes are called >restriction endonucleases) and for sticking lengths of DNA together (when the enzymes are called ligases; see >ligation). Single genes or whole sequences can be put into the bacteria; this is called gene >splicing. The genes or gene fragments being inserted might be from another strain of *E. coli*, or they might be artificially produced DNA, or they might be genes from a different species (even a human being). The new DNA is called >recombinant DNA. But the recombined 'thing' in such a case - let us say that it is recombinant *E. coli*/human DNA - is only a piece of DNA; it is in no way a 'cross' between a bacterium and a human being.

Once a gene has been inserted into *E. coli's* own DNA machinery (actually on a >plasmid rather than in the chromosome), it functions as an ordinary working gene, so that the bacterium finds itself making an alien protein. Thus human proteins can be produced by bacteria by means of a technique called >gene cloning, which is now in use on a commercial scale to make several useful proteins, such as insulin to treat diabetics.

The techniques of genetic engineering are also useful for improving strains of food plants, such as rice, wheat and maize. It is a feature of many plants that they can have doubled or tripled sets of chromosomes (polyploidy). If a useful gene, such as one that helps make a plant resistant to drought, is found in a wild species, it can be crossed into a good commercial strain by 'crossing in' a whole set of chromosomes. This technique has been used since the early days of genetics (Tschermak was a pioneer), but it is now much more sophisticated. As well as crossing-in useful genes found in the wild, genes from other species of plants can be spliced in by recombinant DNA techniques.

Genetic engineering as such has not yet been used to improve farm animals, though selective breeding has always been used and still is. Selective breeding is speeded up by techniques (biotechnological rather than genetic) such as >artificial insemination and >embryo transfer, the former used very widely particularly in the dairy industry, and the latter used sometimes when establishing herds of cattle or flocks of sheep of breeds new to a particular country. One technique that would be of overwhelming importance to cattle breeders has not yet been devised, even though the matter has been intensively pursued for over 50 years: this is the separation of X and Y sperm. If the sperm. If the sperm collected from a bull for artificial insemination could be separated, then the sex of the calf could be chosen, and all dairy calves would be female, and all beef calves male.

It is technically possible to >clone animals, and it has been done with sheep and mice, but the technique is not yet used for producing commercial farm stock, and is likely to remain too expensive in relation to its benefits, compared with the simple and effective technique of artificial insemination. In theory it might be possible to produce better farm animals by inserting single desirable genes into the genome, as is done with plants. However, gene transfer with animals has so far worked only occasionally, it being a hit-and-miss affair with more miss than hit. And so it will remain until the mechanisms of gene control in higher organisms are better understood.

Recombinant DNA techniques, though very useful, are also potentially extremely hazardous because of the risk that they could give rise to new forms of viruses or bacteria so virulent that there would be no cure for the diseases they caused and no possibility of wiping them out. (It is certain that such novel forms of bacteria and viruses are being developed for biological warfare, though no government will admit to it.) For this reason recombinant DNA research is subject to controls in all countries, though the controls and restraints are in most countries voluntary, not legally enforceable. Laboratory practice for such research is carefully controlled.

Genes cannot (at present) be created to order; the genes involved in genetic engineering as well as in conventional selective breeding must be ones that exist in real organisms. For this reason, it is extremely important to preserve all the different genes that exist in the wild and in cultivation. To take an extreme example, if all the apples in the world were Granny Smiths, and these were suddenly threatened by an infection that they were genetically susceptible to, there would be no chance of breeding resistant apples unless there was a source of non-Granny Smith genetic stock to select from. Modern specialized agriculture has already done too much in the way of narrowing the genetic range of commercial plants (and animals), but now 'gene banks' are being built up to prevent and more loss of genetic variety.

The Human Genome Project

By far the most ambitious research programme that the biological sciences have ever staged is the Human Genome Project (>genome means the total of the DNA, including both genes and >junk DNA). The plan is no less than to map the entire human DNA sequence base by base. There are four stages of investigation (though they will not necessarily be carried out consecutively).

1. By using linkage to already-known markers (many of them >restriction-fragment-length polymorphism loci), a linkage map with a resolution of about 1 centimorgan will be built up for all the chromosomes. At the outset of the project, such a map already existed but only with a resolution of 8-10 centimorgans.

2. A corresponding physical map of the chromosomes will be made, using observable morphological variants, so that the genes whose linkage relationships are known can be physically located, and their actual distance from one another assessed.

3. Each region of each chromosome will be studied by cutting it into short stretches with restriction enzymes and reproducing these as clones. Producing these clones is not difficult, though the number of separate clones is daunting; the problem will come in identifying the overlaps among many thousands of clones for each region and finding their correct order.

4. Once this has been done, the sequence of the overlapping clones will be analysed. This is the most labour-intensive and expensive stage of all because present techniques mean that it takes one person one year to sequence 100,000 base-pairs: at that rate the sequencing of the approximately 3 billion base-pairs in the human genome will take 30,000 person years.

The challenge to present biological research capabilities is immense, and the challenge to information technology is almost as great, with existing database technique having to be greatly expanded to cope with the analysis of vast amounts of overlapping data and with disseminating up-to-date information to research teams.

Appropriately, it is a global project. The idea started in 1985, when the molecular biologist Walter Sinsheimer wanted to found an Institute to Sequence the Human Genome at the Santa Cruz campus of the University of California. Many biologists were attracted by the idea of their science at last having the sort of money spent on it that is taken for granted in the physical sciences, and the project became the subject of top-level political sciences, and the project became the subject of top-level political manoeuvring in American research circles. It was fortunate that at the end of day leadership of the project was in the hands of James Watson, and equally unfortunate when internal wrangling led to his sudden resignation in 1992. International co-ordination is being managed by the Human Genome Organization (HUGO), headed by Sir Walter Bodmer. The work is parcelled out to research teams all round the world, each of which is assigned to work on one particular chromosome. The databases collating the results are in the United States, Britain, Germany and Japan. The total cost was at first estimated at $3 billion, but later this had increased to $10 billion.

The project is not without its critics. There have been three types of objection. The earliest was that it would absorb much of the money available for biological research, thus endangering work on more fundamental issues. After all, it was said, the Human Genome Project breaks no new ground technologically; it is just a huge data-crunching exercise. This criticism was made strongly by the American molecular biologist David Baltimore, a Nobel Prize winner. Next, there were worries about the sensitivity of the project to commercial exploitation: could an international co-operative venture of pure research really work if at the end of the day there were potentially huge profits to be made somewhere? This came to a head in 1989 when the Japanese were slow to pay their contribution to HUGO: Watson wrote in the strongest terms (perhaps unwisely, referring to events in World War II) to say that the United States would not hesitate to withhold data from the Japanese unless they paid up. Finally, objections have been made on ethical grounds. Full knowledge of the human genome will mean that in principle all humans, born and unborn, can be screened for all genes, and the abuses that this could lead to are enormous. Research into ethical implications has been allocated 3 per cent of the project's budget.

At the same time that the Human Genome Project is absorbing so much money and effort in the geneticists' community, other similar projects of great importance are up and running. There is an independent British project, the object of which is to sequence all the genes in the human genome. This is a lesser task than tackling the entire genome, as it ignores all the sequences that are noncoding (>junk DNA), which actually comprise the greater part of the DNA in humans as in other higher eukaryotes. Much of the work is being carried out at the Medical Research Council's Molecular Genetics Unit, whose director Sydney Brenner was one of the prime movers of the research project. Co-ordination of work done throughout the country is carried out at the Human Genome Mapping Project Resource Centre at Northwick Park, near London.

Another project, devoted to the yeast genome, is being carried out in various laboratories in the countries of the European Community. The scale of the task is modest compared with the human equivalent, as the yeast genome is only 1/200 the size of the human one, but the genes are packed more closely, there being some 6,500 genes in yeast. The project is budgeted at ECU 20 million, and is due completim. A project to map the pig's genome bas also begun, with 16 countries contributing and the work being co-ordinated at Edinburgh University. Although the pig's genome is comparable in size to the human one, some aspects of research can proceed faster owing to the facts that pigs have a shorter generation time, they have more offspring at one time, and there are no significant vetoes on experimental procedures.

Geneticism: Genes and Politics

From the time of the Darwin-Wallace discovery of evolution in the mid-nineteenth century, and even more since the discoveries of Mendel and Crick and Watson, there has been an almost supernatural belief in the power of the gene. Genes are invisible, but if science says that they exist and that they exert enormous influence, then they will be brought into practical human life. Belief in the power of genes has been termed 'geneticism', and it has been going strong for well over a hundred years, although it has changed the focus of its attention from time to time to keep pace with intellectual fashion. The coinage of the term is a recent one, and is somewhat unfortunate because of the possible implication that 'geneticism' is what geneticists do or think. As will become clear, this is far from the case. Geneticism is purely political, whether in the public or private sphere.

The curse on Ham

Geneticism is an idea as old Testament. In Genesis, Noah curses his son Ham, saying that Ham's descendants, i.e. the Africans, will be subservient to those of his other two sons. This curse was taken literally by Europeans as an excuse for the practice classical method of locating genes on chromosomes is

>deletion mapping, while a newer technique is >restriction mapping. See also Morgan, Thomas Hunt.

Genetic marker. See marker gene.

Genetic screening. There are two different types of genetic screening. The first is the search within a well-defined population or community for >carriers of a deleterious recessive gene, in order to advise them of the risk of having an affected child (see genetic counselling). This can only be done if (a) the community can be identified as having a significantly high frequency of the mutant gene; (b) there is a simple and cheap test to identify carriers; (c) there is a prenatal test to be used for pregnancies at risk; and (d) the procedure is acceptable to the community. Given these conditions, genetic screening can be successful in reducing the number of babies born with a particular defect; one notable example was the screening of the Jewish communities in Washington D.C. and Baltimore, Maryland for Tay-Sachs disease (for the >thalassaemia screenings. An example of a screening programme that went wrong happened when, in the 1970s, black people in some American states were screened for sickle-cell anaemia. This failed because the screening was not acceptable to the community (point d above), largely because it was related to repressive legislation, including restrictions on carriers getting married.

The second type of genetic screening is the genetic inspection of persons for employment, insurance assessment and other commercial transactions. This is more a problem for the future than for the present, as genetic factors affecting employment (e.g. susceptibilities to particular workplace hazards) are not, at present, understood. The ethical issue will have to be confronted, as to whether it is justifiable to exclude a person from a job or benefit because of his or her genotype. On the face of it, there does not seem to be much difference between genetic screening and ordinary medical screening, where the civil rights of the person not to be discriminated against has to be weighed against the civil rights of other people who may be put at risk by that person's medical condition.

Genetic variance, the proportion of the total variance of a trait that is due to genes. It may have several components, due to the additive effects of separate genes, or dominance, or interaction. The rest of the variance is environmentally produced. The relative contribution of genetic factors to variance is not a fixed feature of any trait (see heritability).

Geneticism, the belief that human characteristics are shaped wholly or largely by genes. See heritability; nature versus nurture; IQ;

Genetics, the study of heredity and variation, which as a unified branch of biology dates back only to the rediscovery of Mendel's writings in

1900. Since the ability to reproduce is the criterion of what is or is not a living thing, genetics as the study of the machinery of reproduction is the central life science, and from it all the others are derived. The range of subjects that have a direct relationship to genetics is enormous, from the more obvious ones such as medicine or microbiology to the (equally directly related) ecology and politics include 'Essential Genetics'

Genocopy, a genetically determined character that mimics the appearance of another genetically determined character, though caused by an allele at a different locus. See also phenocopy.

Genome, the total amount of genetic information that an organism possesses; the sum of its genes. In eukaryotes, this means the entire set of genes in a haploid set of chromosomes. The genome of a virus may be made up of their >DNA (single or double stranded) or RNA; it can be as small as 3500 base-pairs and may contain as few as three genes.

The genome of a prokaryote (bacteria, etc.) is contained in a double-stranded DNA molecule (not strictly a chromosome), and contains from about 750,000 to 4,500,000 base-pairs, comprising up to about 1500 genes. This molecule is extremely compact: in *Escherichia coli*, the circular DNA is 13 pm long and is packed into a cell 1 pm x 2 pm.

The eukaryotes have even larger genomes, and in general, the higher up the evolutionary scale, the larger the genome (but with exceptions: some species of amphibians and fish have genomes nearly 40 times larger than the average for mammals). The size of the genomes bears no simple may have either a large number of small chromosomes or vice versa, for roughly the same-size genome. Humans have a genome of about 2900 million basepairs of DNA, which, if it were all in one DNA molecule, would be 1 metre long. (This is the haploid total; each of our body cells has twice this amount.) It would be physically possible for there to be as many as one million genes encoded in that much DNA, but in common with other eukaryotes, humans have many instances of repetitive sequences of DNA that do not seem to have any meaning, and the actual number of working genes is probably much lower. There are also many genes that are present in multiple copies.

Genotype, the genetic characteristics of an organism. This may refer either to the total of all the genes or, more usually, to the alleles present at one or two >loci of interest. The genotype is contrasted to the >phenotype (the appearance and physical characteristics of the organism) which may or may not be a direct reflection of the genotype. For example, where one allele is dominant, the phenotypes of both genotypes AA and Aa are identical.

Germ cell. See gamete.

Germ plasm theory, an early idea of inheritance, foreshadowing chromosomal inheritance. Put forward in 1875 by the German biologist August >Weismann, the theory states that the material of which the germ cells (sperm and ova) are made is self-perpetuating and is entirely separate from that of all the other cells in the body (which Weismann termed 'somatoplasm'), which are mortal. This is the idea so neatly summed up by Samuel Butler's aphorism 'A hen is only an egg's way of making another egg.' Weismann's experiments to support his theory included cutting the tails off mouse pups for many successive generations; their offspring, of course, still had tails. Although crude, these experiments established Weismann's theory among scientists, simultaneously disposing of the theories of the inheritance of acquired characteristics and preformation.

German measles. See rubella.

Gigantism. See growth hormone.

Gilbert, Walter (1932-), American molecular biologist; professor of molecular biology at Harvard University since 1968, having been professor of biophysics 1964-8; Gilbert is one of several important molecular biologists who began as physicists (Crick is another). In 1961 Jacob and Monod had proposed that gene activity is controlled by a hypothetical 'repressor' that switches the gene off when it is not needed. Gilbert was successful in isolating and identifying a repressor, which was difficult because each repressor is only present in a cell in minute quantities. Later he devised a method for analysing the sequence of single bases on a strand of DNA. He shared the Nobel prize for chemistry in 1980.

Globins, a group of proteins involved in the transport of oxygen in blood, muscles, etc. They are genetically well understood. Changes in their constitution during evolution have been studied (see molecular evolution), as has their role in the hereditary anaemias (see sickle-cell anaemia; thalassaemia). See haemoglobin.

Glucose-6-phosphate dehydrogenase deficiency (G6PD deficiency), a hereditary lack of an enzyme involved in the metabolism of carbohydrates. Inability to produce this enzyme is inherited as an X-linked recessive. Under normal circumstances, people with this gene show no ill effects, but when they eat broad beans (also known as fava beans), they suffer a severe bout of haemolytic anaemia (anaemia caused by the destruction of red blood cells).

The gene causing G6PD deficiency is found at frequencies of up to 30 per cent in countries around the Mediterranean (Sardinia, Greece, Israel), in parts of Africa (especially the Congo Basin) and in India, and it seems likely that its distribution is connected with that of malaria, because the trait gives some resistance to malarial infection (as do the genes for >sickle-cell anaemia and >thalassaemia).

Gonad (from the Greek gone, 'generation' or 'seed'), an organ in an animal that produces gametes. The female gonads are the ovaries, and the male the testes.

Gradualism, or phyletic gradualism, the idea that the formation of a new species in geological time is a gradual process, proceeding at a more or less steady speed. See also catastrophism; punctuated evolution.

Graft, the transfer by surgery of any tissue or organ from one position on or in the body to another (an autograft), or from one organism into another. (The word >transplant is now more usual where a whole organ is involved.) A homograft is one carried out between two different species. Any graft except an autograft or a graft between monozygotic twins is likely to be rejected by the recipient's immune system, unless treatment with >immunosuppressive drugs is given. See also graft-versus-host reaction.

Graft-versus-host reaction, the situation in which, after a graft has been made, the immune tissue within the graft produces antibody against the cells of the recipient. The only time when this is likely to occur is after a patient has been given a bone-narrow transplant; in other circumstances, there are not likely to be any immunologically competent cells within the graft.

Group selection, >natural selection that acts upon a group of organisms, rather than upon the individual, so as to produce an evolutionary change that benefits the group (or its successors). >Darwin's theory of evolution was based on individual selection, and he never accepted group selection, though Wallace, co-originator of the theory, did so. The idea of group selection was revived in 1962 by the British zoologist V.C. Wynne Edwards, who suggested that it could explain various aspects of population density in animals (including >altruism). After a brisk controversy, genetical opinion settled to the viewpoint that group selection cannot happen because the mathematics do not work out. The idea has since been replaced with >kin selection, also a contentious matter.

Growth, increase in size of an organism, either by an increase in cell size or (much more usually) by an increase in cell number. Most organisms have a species-typical growth rate and final size, both of which are, to some extent, genetically controlled (as is shown by the fact that they can be changed by artificial selection) but are also much influenced by environmental factors. Some fishes and reptiles grow all their lives.

Growth hormone, in animals, a protein hormone produced by the pituitary gland in the base of the skull. Either deficiency of it or excess production can be genetically determined, the former causing a form of dwarfism and the latter gigantism (excess growth in childhood) and acromegaly (enlargement of the jaw, hands and feet in adult life). The dwarfism can be treated with human growth hormone, which formerly had to be obtained from human cadavers but can now be made by gene cloning.

Guanine (abbreviation: G), one of the four nitrogenous bases which comprise the core of DNA and RNA. It forms pairs with cytosine (C). See also replication.

Gymnosperm, a plant that has its seeds exposed rather than encased in an ovary (as >angiosperm plants do). Gymnosperms, which include coniferous trees, are wind-pollinated. See also pteridophytes (ferns, etc.).

H

Haeckel's law 'Ontogeny recapitulates phylogeny.' Haeckel's theory was that each organism, during its embryonic life (>ontogeny) literally goes through 'being' each one of its primitive ancestors (>phylogeny). If this were so it would imply that evolutionary change can only take the form of adding extra chapters to the life histories already written. It is however true that the early stages of embryogenesis are more similar between groups of animals than their later stages are (see von Baer's law). Ernst Haeckel (1834-1919) was professor of zoology at Jena; his law was stated in his *Naturliche Schopfungsgeschichte* (1867, English translation *History of Creation*, 1902).

Haemoglobin, the molecule that carries oxygen on the red blood cells, consisting of a complex of the protein globin and the haem group which actually binds the oxygen; it gives the red blood cells their colour. Genetically determined defects in the haemoglobin genes are the cause of some anaemias. See sickle-cell anaemia; thalassaemia.

Haemolytic disease of the newborn, a kind of jaundice that happens when a second or subsequent >Rhesus-positive (Rh+) baby is born to a Rhesus-negative (Rh-) mother. There is no problem during the first pregnancy because the foetal and maternal blood cells are kept apart by the placenta. But at the time of delivery the placental barrier breaks down, and some of the baby's Rh+ red blood cells get into the mother's bloodstream. The mother, whose body identification these Rh+ (this does not happen instantaneously, so the first baby is not damaged while being born). If the mother has a second Rh+ baby, all goes well during pregnancy because of the placenta separating the baby's blood from the mother's, until the very last minute when the placental barrier breaks down; at that moment, the mother's cells get into the baby's bloodstream and attack its red blood cells, causing severe haemolytic anaemia (a lack of red blood cells because of their destruction). The newborn baby will probably die unless given an immediate total-exchange blood transfusion. The same thing can happen with ABO incompatibility between baby and mother, but it is very rare.

Haemophilia, a disease in which the blood does not clot normally. The gene responsible is the best known of the > X-linked recessive genes, partly because it occurred in the pedigree of the British royal family. Haemophilia affects about 1 in 10,000 babies; as with all X-linked inheritance, boys who have the gene show the condition while girls who have one mutant gene and one normal are unaffected and are known as carriers (it would be possible for a girl to have haemophilia if her father was a haemophiliac and her mother was a carrier). Queen Victoria was a carrier, the mutant having probably arisen either in her mother or not much further back down the maternal line, as there were no earlier haemophiliac relatives. Victoria passed the gene on to at least three of her nine children: one was Leopold, duke of Albany, who died of the disease at the age of 29: another was the Princess Alice of Hesse, whose daughter Alexandra, Czarina of Russia, was the mother of the haemophiliac Czarevitch Alexis. In the past, mothers known to be carriers have had the option of aborting all male foetuse, even though this meant aborting a normal individual in 50 per cent of cases, but now genetic markers on the X chromosome can be used in certain families to predict whether a male foetus of a carrier mother will be haemophiliac. Treatment is also now possible with the injection of the clotting factor (Factor VII) prepared from donated blood; unfortunately, unless the blood is rigorously tested and treated, it carries a risk of infection with blood-borne diseases such as hepatitis and AIDS.

Hairpin loop, a loop formed within a molecule of RNA when matching adjacent sequences become bonded to each other. This can also happen within a DNA molecule, with sequences on a single strand becoming joined to each other.

Haldane, John Burden Sanderson (1892-1964), British biologist. Haldane is one of the major figures in twentieth-century British biology, partly because of his wide-ranging interests and partly because of his determination to bring politics into science (a Marxist, Haldane contributed hundreds of articles on science to the *Daily Worker*). His father was the leading physiologist of the time, specializing in respiration, and that was Haldane's own first field; he used himself as an experimental animal (as he described in the 1927 paper 'On being one's own rabbit') in research that was later applied to saving lives in submarine disasters. As a biochemist, he gave the first proof that enzyme reactions in the body obey the laws of thermodynamics. Then, in the 1930s, he turned to genetics and studied the inheritance of genetic disorders in humans, publishing (in 1935), the first human linkage map, of the X chromosome.

Half-siblings (half-sibs), brothers, sisters or a brother and sister having one parent in common but not both. Human half-siblings are of some usefulness in nature versus nurture studies, in that they give a comparison with full siblings who are twice as closely related, while the environment they share is broadly similar (though not exactly similar, it being unlikely that half-siblings are treated as much alike as full siblings).

Handedness, the innate tendency to use one hand in preference to the other. About 6 per cent of people in European population are left-handed. There is evidence that left-handedness is inherited through the mother (i.e. by a >maternal effect), though it is not known how this works; it is also true that older women are more likely to have left-handed children. Left-handedness is more frequent in twins, owing to their general tendency to mirror-imaging, and curiously this effect is seen in dizygotic (fraternal) twins as often as monozygotic (identical) twins. Other species often show a favouring of one hand (or foot), but in all of them, this is randomly the right or left, with humans alone having a species-wide overall preference. Also inherited in complex fashion are two other behavioural asymmetries: the preferences as to which thumb goes on top in clasping the hands together and which arm goes on top when folding the arms. These are not related to each other or to handedness.

Haploid, adjective (from the Greek *haploos*, 'single') used to describe any cell or organism that has only one copy of each chromosome. The gametes (ovum and sperm) are haploid, in contrast to all the other cells of the organism, which have two copies of each chromosome and are called >diploid. Some algae, yeasts and fungi are haploid for the greater part of their life cycles. The haploid number of a species' chromosomes is referred to as n.

Haplotype, a set of alleles at very closely linked loci, coding for related products. The term is used especially for the HLA alleles of the human major histocompatibility complex.

Hardy-Weinberg equilibrium, a theorem which forms the basis of population genetics. It states that, for a two-allele locus, if the gene frequency of one alleles is p, the frequency of the other is 1-p, which is called q. The frequency of the three possible genotypes is $p^2 + 2pq + q^2$. If the population breeds with random mating, these proportions will be maintained in succeeding generations, other things being equal (i.e. with none of the factors than can cause changes in >gene frequency in a population). The proportions are not affected by >recombination. The Hardy - Weinberg formula can be used to work out gene frequencies from observed phenotypes: if the recessive homozygotes (q^2) are 1 in 100 of the population, i.e. 0.01, then q, the gene frequency of the recessive allele, is the square root of 0.01, i.e. 0.1 or 1 in 10.

Hayflick phenomenon. See tissue culture.

Heart disease, an imprecise term covering a number of disorders of the heart, some of which are genetically determined. Heart disease, in the sense of deformities of the heart, is found as one of the effects of quite a few genetic syndromes, including Down's syndrome. Heart disease in the more general sense of the various degenerative diseases of the heart and circulation that start in middle age also seems to have a genetic component, which interacts with such obvious external factors as diet, exercise and smoking. In particular, a genetic pattern has been discovered for heart disease resulting from too much cholesterol in the blood (see hypercholesterolaemia).

Hela cells, a line of human cancer cells that has been maintained in culture since 1951 and has been much used in cancer research (see cell line). It is named after the patient from whom the cells were taken: Henrietta Lacks, who had cancer of the cervix.

Helix, a spiral shape. It occurs in nature as the three-dimensional structure of >DNA, which is a double helix. The alpha helix, described by Linus Pauling, is the secondary structure of many proteins, i.e. they are twisted threads that are subsequently folded into a particular three-dimensional shape.

Hemizygote (adj. 'hemizygous'; noun 'hemizygosity'; from the Greek hemi, 'half , and zugon, 'yoke'), an organism that has a gene present in only one dose (see gene dosage), usually in the case where the gene is on the sex chromosome and the individual is of the heterogametic sex (the one that has two different sex chromosomes, e.g. mammalian males). See also heterozygote; homozygote; X linkage.

Hereditarianism, the belief that the genotype has more effect than the environment in determining human traits of social importance. The opposite is environmentalism. See nature versus nurture.

Heredity, the process by which characteristics are transmitted from one generation to the next. The term is only useful if it is confined to genetic transmission; an alternative word that covers all modes of transmission of traits is 'inheritance' (e.g. cytoplasmic inheritance, cultural inheritance).

Heritability, the proportion of the variability of the >phenotype of a trait that is attributable to genetic factors. Heritability is not a fixed property of the genes concerned but depends in each situation on how much variation there is both in genotypes and in the environment. For example, in one population everybody is >homozygous for the gene for black hair, but nevertheless some children have reddish-brown hair as the result of malnutrition; in another population there is little malnutrition, but the population consists of people with every combination of four different alleles for hair colour, so there are many shades of hair. In the first case, heritability of hair colour is effectively 0 per cent, and in the second case, 100 per cent.

Most of the research on heritability has been done on plants and animals used in agriculture, because heritability gives a measure of how much improvement in a trait can be achieved by selective breeding. The agriculturalist starts with a heterogeneous population of pigs, say, in which growth rate is, at the outset, highly heritable; then as the genetically better pigs are selected and become a higher proportion of the pig population, environmental factors become relatively more important in the difference between individual pigs, and heritability goes down (see selection limit).

Hermaphrodite, an animal that has the reproductive organs of both sexes (the equivalent in plants is >monoecious). Animals that are normally hermaphroditic include flatworms, leeches, land snails, some fish and some flies. They are capable of self-fertilization, unless they are sequential hermaphrodites (i.e. they have first one set of sex organs, then the other). Hermaphrodites do not have sex chromosomes (see sex determination). In humans, hermaphroditism is a rare abnormality. The person has both ovarian and testicular tissue, which may occur as (a) one ovary plus one testis, (b) two ovotestes or (c) one ovotestis plus either an ovary or a testis; the external genitalia are ambiguous. The condition is the result of a single ovum being fertilized simultaneously by an X and a Y sperm, so that the person is a >chimaera of XX and XY cells. See also intersex.

Herpes viruses, a group of viruses that infect animals, including humans. Their genome is double-stranded DNA. Herpes simplex causes a number of conditions in humans, ranging from cold sores to blisters on the genitalia to encephalitis, an inflammation of the brain that can be fatal; Herpes zoster causes shingles and chicken pox. Epstein-Barr virus was the first to be proved to be associated with cancer in humans, being one element in the causation of >Burkitt's lymphoma.

Hershey, Alfred Day (1908-), American microbiologist. Following Avery's suggestion that DNA is the carrier of genetic information, Hershey and his colleague Martha Chase proved that this is so, in experiments on the DNA in bacteriophages (viruses that infect bacteria). Hershey shared (with Max Delbruck and S.E. Luria) the 1969 Nobel prize for physiology or medicine for his work on how viruses reproduce.

Hertwig, Oskar (1849-1922), German embryologist; professor of zoology at Munich for 40 years. In 1879 Hertwig made the first observation of fertilization, when he saw an egg being fertilized by a single sperm (the species in question was a sea urchin). The fact that only one sperm was involved was crucial to the understanding that the sexes make equal contributions to heredity. Hertwig was also the first to understand the significance of meiosis. His brother Richard was also a distinguished zoologist, and they wrote a number of joint publications.

Heterochromatin, >chromatin (the substance of which chromosomes are made) that is in a highly condensed state because the DNA, not being actively transcribed, can be packed densely around the >histones. Examples are DNA in the >centromeres and in >Barr bodies. Because of its more condensed state, heterochromatin in more intensely coloured when stained than ordinary chromatin (>euchromatin).

Heteroduplex, a >nucleic acid molecule that consists of two strands of different genetic origins. It can be either a double-stranded DNA molecule or a DNA-RNA combination. Where the sequences are homologous, the two strands will link; non-homologous sequences can then be seen under the electron microscope looking like bubbles. Examination of a heteroduplex made from DNA strands from wild-type and mutant-type DNA will show where there are deletions or insertions; this is called 'heteroduplex mapping'.

Heterogametic, adjective used to describe the sex that has the non-matching pair of sex chromosomes, X and Y. This sex therefore produces two different types of gametes, one with the X and one with the Y, and is responsible for the sex of the offspring. In mammals and most insects, the heterogametic sex is the male, but in birds, reptiles and some amphibia and insects, it is the female. Fish are variable, and some species even have different strains, some with heterogametic males, others with heterogametic females. The opposite is >homogametic.

Heterogeneous nuclear RNA (abbreviation: hnRNA), the >RNA molecule that reads the DNA code in eukaryotes. See messenger RNA.

Heterograft. See graft.

Heterokaryon, a cell or individual that contains two nuclei, each of different genetic origin. It can happen naturally when cells of two different fungi fuse, and it can be artificially induced by mixing the cells of higher organisms in vitro.

Heterosis, the technical term for the phenomenon more generally known as >hybrid vigour.

Heterozygote (adj. 'heterozygous'; noun 'heterozygosity'; from the Greek hetero, 'other', and zugon, 'yoke'), an organism that, at a given locus, has two unlike alleles. The term 'double heterozygote' refers to an individual who, at two given loci, has unlike alleles. Heterozygotes do not breed true with respect of the traits(s) in question. See also hemizygote; homozygote; F_1.

Heterozygous advantage, the situation where an organism that is heterozygous at a given locus has greater genetic fitness than either of the two homozygotes. It is one way in which two different, alleles can be maintained within a population at stable frequencies: neither is better on its own, but the two in combination are the most successful genotype, and therefore neither allele is lost from the population (see polymorphism). The classic example of this is >sickle-cell anaemia in humans, in which homozygotes for the mutant allele have severe anaemia, homozygotes for the normal allele are liable to get malaria. In a more general sense, organisms that are heterozygous at many loci are at an advantage over more homozygous ones; this is known as >hybrid vigour.

Himalayan rabbit, a rabbit that has a white body and dark-coloured ears, nose and feet. This coloration is an example of an environmentally caused gene switch: all the hair cells in the rabbit's skin have the same genes, but the pigment-forming enzyme that they code for is heat-sensitive and will only work in cooler tissues. Thus the gene is not expressed in warmer parts of the body, but only in the extremities. This first happens during the first week after the rabbit's birth, when the fur is growing, but the gene can be switched on or off in later life: if part of the skin is shaved, and the rabbit is kept in a cold place, the new fur growing in will be dark even though that patch was white before.

Hinny, the offspring of a male horse or pony and a female ass or donkey (the reverse is a >mule). The diploid chromosome number in the horse or pony (*Equus caballus*) is 64, and in the ass or donkey (*Equus asinus*), it is 62, so the hinny has 63 chromosomes, and is infertile.

Hippocrates, c. 460-c. 357 BC, Greek physician, 'the father of medicine'. Hippocrates taught at Kos, and was both famous in his lifetime and influential thereafter. His medical ideas were based on observation (his are the first real case-notes) and reason, a huge advance on the superstitions of the time. His ideas on genetics were also ahead of their time. Hippocrates knew that the male contribution to heredity was in the semen and he believed that there must be similar fluid in the female. Thus the contributions of the parents were equal, and what kind of baby was born was the result of a blanding of the two fluids. Though not correct, this theory was much closer to the truth than >Aristotle's. Not everything that is in the writings known as the Hippocratic Collection can be Hippocrates' own work, as they cover a span of over 100 years, nor can the Hippocratic Oath be credited to him, though it does reflect his principles.

Histocompatibility complex, the system whereby the body recognizes as foreign any tissue other than its own. It is medically important in that difference between donor and recipient lead to the rejection of >transplants.

Each of the individual's own cells has on its surface a number of specific proteins, the histocompatibility antigens (also known as >HLA, human leucocyte antigens), there of which are six different types, each with dozens of allelic variants. The number of possible combinations is many thousands of millions, and the probability of any two people (apart from identical twins) having exactly the same combination is correspondingly remote, except that, as the genes involved are very closely linked, there is good correspondence between near relatives. Mismatch in some of them is apparently acceptable to the recipient's immune system, otherwise transplantation would be impossible. The genes controlling the antigens are grouped together into what is termed the major histocompatibility complex, which in humans is located on chromosome 6.

Histones, the DNA-binding proteins that are present in chromosomes and form the core of the >nucleosomes. There are five types that occur in eukaryotes: H1, H2A, H2B, H3 and H4. The last two are the most evolutionarily conservative proteins known. See molecular evolution.

HIV, or human immunodeficiency virus, the virus responsible for >AIDS. It was formerly known as HTLV or LAV.

HLA system (human leucocyte antigen system), the group of antigens that are involved in tissue recognition, enabling the body to recognize its own tissue and reject foreign tissue (as in transplants and grafts). There are six loci (known as A, B, C, DR, DQ and DP) and at each there are large numbers of alleles, designated by numbers (e.g. DR4). The loci are found in a group on chromosome 6, and are closely linked. It has been found that some HLA type have an association with auto-immune diseases. For example, ankylosing spondylitis is 120 times more likely to occur in people with allele B27, and multiple sclerosis is associated with there different HLA types: A3, with a risk factor of x24; and B7 and B18, both with a risk factor of x 12.

Holandric, occurring only in males, i.e. a gene that is on the Y chromosome.

Homeostasis (from the Greek homoio, 'equal', stasis, 'state'), the tendency of a physiological system to maintain itself in balance, whatever the changes in the internal or external environment. Homeostasis is an evolved mechanism, operated by the switching off and on of many genes. An example is internal body temperature, which has homeostatic control in mammals but not in reptiles, whose body temperature varies according to the environmental temperature.

Homeotic mutation, a mutation that replaces a normal structure with another structure that is also normal but is in the wrong place. For example, there is a mutation in *Drosophila* that causes legs to grow where the antennae should be.

Homo erectus. See human evolution.

Homo sapiens, the species to which all human beings alive belong. In the aftermath of the publication of Darwin's On the Origin of Species in 1859, there were attempts to show that not all humans were of the same species, but the matter is settled by the fact of the ability of all human beings to interbreed (as stated in the eighteenth century by the philosopher Immanuel Kant, among others). We are classified as a member of the order Primates within the class Mammalia of the phylum Chordata in the kingdom Animalia.

Homo sapiens has 22 pairs of autosomal chromosomes plus two sex chromosomes (XX or XY); the amount of DNA in each human somatic cell is about 6 picograms, and if fully extended would be about 1.9 metres long. The number of genes is estimated at somewhere in the region of 100,000. Humans also have an extensive (but not unique) system of >exogenetic inheritance. See also human evolution.

Homogametic, adjective used to describe the sex that has the matching pair of sex chromosomes, i.e. two Xs. This sex can only produce one type of gamete, with a single X chromosome, and therefore does not determine the sex of the offspring. In humans (and all other mammals) the female is the homogametic sex. See heterogametic; Barr body.

Homograft. See graft.

Homologous chromosomes, the members of a pair of chromosomes that carry the equivalent set of genes. All the >autosomes come in homologous pairs. In the case of the sex chromosomes, the X and the Y are not homologous as they do not carry the same gene sequence (but two X chromosomes are homologous with each other). See meiosis.

Homology, the correspondence of structures between one species and another, even if they are no longer performing the same functions. Homologous structures are derived from an ancestral form. The human wrist and the cow's knee are homologous, being one among many examples that could be drawn form the fact that the vertebrate skeleton has one basic pattern.

Homosexuality. A gift, a normality, a condition, a disease, a vice, a crime, a sin? Homosexuality has been all of there things at one time or another. For >sociobiologists, it has taken yet another form: it is an evolutionary advantage. Starting, as sociobiology always does, from the standpoint that all social behaviour can be understood as being controlled by genes which have evolved because they are advantageous, three quite different theories of a genetic basis for homosexuality have been put forward. First, there is the idea that there is a single gene for homosexuality. It has a curious dominance pattern, in that persons who have one such gene coupled with one gene for heterosexuality are even more heterosexually active than people with two heteosexuality genes (persons with two homosexuality genes are more or less exclusively homosexual). This means that the people with one of each gene have the most babies, and that is how the homosexuality gene continues to exist in the gene pool and to be beneficial to it (this is heterozgous advantage).

The second and third theories both relate to kin selection. A single gene for homosexuality might be of advantage in a kinship because the person with the gene is able to be a significantly useful helper of their non-homosexual siblings' reproductive performance. This can be imagined in a scenario such as a small proto-human troop, with the dominant male mating with the females and a number of homosexual males assisting the troop by keeping watch for predators, foraging and so on. The third theory is a variant of this, in that the gene is supposed to be one that induce parents to bring up some of their offspring homosexual so that they will help the non-homosexual children with child raising as above.

Any of these ideas could be true, but they are mutually exclusive, and there is absolutely no evidence for any of them. This kind of reductionist analysis of something as complex and manifestly socially conditioned as human sexual preference causes anthropologists, psychologists and others to despair.

Homosexuality, as a sexual preference for one's own sex, does not seem to occur in animals, though it is a commonplace that mock homosexual activity of various sorts does take place, e.g. heifers mounting one another when they are on heat, or male dogs mounting one another when they have nothing else to do.

Homozygote (adj. 'homozygous'; noun 'homozygosity'; form the Greek homo, 'same', and zugon, 'yoke'), an organism that, at a given locus, has two alleles the same. The term double homozygote refers to an individual who, at two given loci, has matched alleles. Homozygotes breed true with respect to the traits in question. Obviously any organism is a mixture, being a homozygote at many loci and a heterozygote at others, so that it does not strictly make sense to speak of a homozygous genotype if the total genotype is meant. None the less one can refer to increases or decreases in the average amount of homozygosity within populations, which is influenced by the amount of >inbreeding or >outbreeding. See also hemizygote.

Homunculus, the complete miniature human being that used to be thought to be contained in a sperm cell. The existence of homunculi was part of the theory of >preformation. The word homunculus comes from the Latin for 'tiny man'. Since presumably other species were believed to have their preformed miniatures too, these must have been referred to with similar names, e.g. caniculus, 'tiny dog'.

Horizontal transmission, the transmission of DNA by viral infection into cells (see transduction). Normal genetic transmission is >vertical transmission.

Hormones, any of the chemical messengers in the body. They are produced by the endocrine glands and carried round in the bloodstream to alter the behaviour of other specific target cells or tissues. Steroid hormones are small lipid molecules that enter the target cells; they include the >sex hormones. Other hormones are polypeptides (chains of amino acids shorter than proteins) which bind on to the outside of the target cell; they include >growth hormone and >insulin. Many hormones can be synthesized to provide treatment for patients with a deficiency; some can now also be produced by gene cloning. Genetic defects in a hormone system can produce either a fault in the production of the hormone or a fault in the receptor cell; both are likely to be single recessive alleles.

In insects, there are hormones that trigger the phases of metamorphosis, as well as the physiological sort described above. Plants have three types of hormones: gibberellins (stimulating stem growth), auxins (stimulating regeneration) and cytokinins (promoting cell division). See also allomones; pheromones.

Host. There are two uses of this word in genetics: (a) the recipient of a graft or transplant; and (b) the cell or organism that is infected by some pathogen or parasite, particularly a bacterium invaded by a virus.

Hot spot, a site in a DNA molecule where mutations or recombinations occur with much higher frequency than elsewhere.

Housekeeping genes, a gene that has a product essential for the normal metabolic requirements of any cell. These genes are active in most or all cells, in contrast to the cell-specific >luxury genes. The parts of the chromosome where they are situated show up as light-staining interbands (see banding).

Human evolution, or, as >Darwin had it, the 'descent of man'. The human (or hominid, a term that means the group including >***Homo sapiens*** and its extinct human-like precursors) stock diverged from the ancestral stock of the great apes in the Miocene age, about 8 to 14 million years ago. For some reason, the environment at that time was extremely favourable to ape-like creatures, and many of them evolved and were successful all over the Asian land-mass as well as in Africa and in Europe. In about 10 million BC there was a split in the lineage, which separated the Asian apes (the orang-outang and the closely related gibbon and siamang, with their Miocene ancestor ***Ramapithecus***) from the African apes, including the ancestors of humans and the African great apes. Somewhere around 8 million years ago, the African lineage also split, along ecological lines, with the forest-dwelling apes (ancestors to chimpanzees and gorillas) diverging from savannah-dwelling hominid ancestors.

The fist fossils that can definitely be called hominid come from East Africa: ***Australopithecus afarensis*** is dated to 3-4 million BC (the best-known specimen is 'Lucy', comprising about half of the skeleton of a creature about 3ft/1m high). The genus *Australopithecus* (from the Greek for 'southern ape') stood upright, walked on its feet with no support from its hands, and was probably omnivorous (though not much of a hunter). A successful genus found widely throughout the savannah lands of Africa, it occurred in severa, different types (including ***A. africanus***, the one ancestral to humans) whose ranges overlapped; they must have occupied different ecological niches.

However, the heavier species (***A. robustus***) eventually lost out to the lighter type represented by the lineage that led from *Homo habilis* (the first fossil hominid to show substantial enlargement of the brain, and to use tools) to *Homo erectus*. This last was a very successful species: starting from East Africa in about 1.6 million BC, within 600,000 years it had reached South-east Asia, within another 500,000 years, it had travelled to China (where the remains of the famous Peking Man were found) and perhaps to Europe. *Homo erectus* had a brain at least two-thirds the size of that of modern humans, and had a skeleton almost entirely the same as the modern version (evolutionary improvements having been made particularly in the legs and pelvis); it was able to use fire and a variety of tools, and was probably an efficient hunter.

The era between 400,000 and 200,000 years ago, during which >Homo sapiens emerged, was characterized by the development of large regional differences. One of the predominant types was ***Homo sapiens neanderthalensis***, Neanderthal Man, of which the type specimen came from Germany but which was found throughout Europe, the Middle East and North Africa; similar early ***Homo sapiens*** subspecies (e.g ***H.s.rhodesiensis***) were found elsewhere. (It is not true that Neanderthal Man walked with bent knees; this misunderstanding arose because one of the specimens had severe arthritis.)

Truly modern *Homo sapiens* first appeared about 120,000 years ago, with the earliest specimens coming from Ethiopia and South Africa. Within 80,000-90,000 years, *Homo sapiens* was established in Australia; the colonization of the Americas by people from Asia crossing the landbridge at the Bering Straits probably occurred at about the same time, and the whole of the two American continents were populated by 10,000 BC. The earliest fully modern fossil found in Europe is 'Cro-Magnon Man', about 25,000 years old.

Human Genome Project Refer Genome.

Hunter's syndrome, a human disease caused by an X-linked recessive allele. The biochemical problem (as with >Hurler's syndrome) is that the patient is unable to break down mucopolysaccharides, large compounds of sugars that combine with proteins and are important in nerve cell function. A build-up of these molecules leads to mental deterioraton and deafness, as well as various disfiguring symptoms. Women in families where the disease has occurred can be tested to see whether they are carriers; and prenatal diagnosis is reliable. Huntington's chorea, a human disease caused by a single dominant allele. The underlying mechanism is not known, but the disease us a distressing one consisting of involuntary movements and mental deterioration; it does not appear until the suffer is 40 years old or more, and is at present incurable, leading to death within five to ten years. Before diagnosis, the symptoms may be mistaken for drunkenness, and this can lead to much distress for sufferers and their families, especially if they lose their jobs because of it. Diagnosis is no comfort, however, for not only is it the death sentence on the person who has the disease, it may mean the same for the children that they had before they knew about it.

It is now possible to use a genetic marker on chromosome 4 to detect with about 95 per cent accuracy whether a person whose parent had Huntington's chorea is going to develop the disease. However, this can only be done in families with enough members in preceding generations available to provide DNA samples, and only if the relative with the disease has a marker gene that is different when compared to those of the ones without Huntington's chorea. Similarly, a prenatal test can detect whether a foetus is risk-free or not. The decision whether or not to abort a high-risk foetus is complicated by two things; (1) even if it has the gene, it will have at least 40 years of normal healthy life; and (2) a cure for the disease could be found during its lifetime. In white populations, about 1 in 10,000 has the gene for Huntington's chorea.

Hurler's syndrome, a human disease caused by a recessive allele. The biochemical problem is the same as in >Hunter's syndrome, but the effects are worse, with dwarfism and heart problems leading to death at the age of 8-10. Carrier females can be identified biochemically, as can affected foetuses. About 1 in 100,000 babies has Hurler's syndrome, with little difference in frequency in different ethnic populations.

Huxley, Thomas Henry (1825-95), British biologist. Huxley trained as a doctor and served as a naval surgeon. He went on the voyage of HMS Rattlesnake round Australia (1846-50), and did research on plankton, which established his scientific reputation (he won the Royal Society's Gold Medal for it). When first he read On The Origin of Species, he exclaimed, 'How silly of me not to have thought of that!' and he became a passionate supporter of >Darwin's theory of evolution, earning himself the nickname of 'Darwin's bulldog'. In Man's Place in Nature (1863), he made the first explicit statement of the non-unique biological position of humans. Huxley refused to believe anything not susceptible of logical analysis, and therefore rejected both atheism and belief in God. He became a member of the London School Board and was highly influential in getting rid of rote learning and substituting basic skills of numeracy. His son Leonard and grandsons Julian, Aldous and Andrew have all been intellectually distinguished.

Hybrid, broadly, the offspring of any cross between two (fairly pure) strains. In a narrower sense, the term is used to mean the offspring of a cross between parental strais that are homozygous at the given locus or loci, such that the hybrid individual is heterozygous for these (see F1; monohybrid; dihybrid; trihybrid). A hybrid can also be a cross between different species or genera, e.g. a mule or hinny.

Hybrid DNA that is artificially made in vitro by mixing two strands of DNA (or one of DNA and one of RNA) of different genetic origins (see heteroduplex). It is not the same as >recombinant DNA.

Hybrid vigour, the phenomenon (technically called 'heterosis') in which the hybrid is more vigorous (larger, more prolific, more disease-resistant, etc.) than either of its parental strains. The genetic explanation is that each of the parental strains has become inbred to the point where, at a significant number of loci, recessive alleles with a bad effect on the genotype have become fixed (i.e. all the individuals are homozygous at these loci). The hybrid between two such strains has a good chance of having dominant alleles from the other strain to cover the poor effects of the recessive in at least some of the loci concerned. See F_1.

Hybridoma, an artificially produced hybrid cell, combining a normal Lymphocyte and a myeloma (cancerous) Lymph cell, and used in the preparation of monoclonal antibodies.

Hypercholesterolaemia, familial, an inherited condition in which too much cholesterol is deposited in the blood vessels, leading to a high risk of coronary artery disease. The physiological problem is that the affected person has too much of a type of protein called LDL which is involved in cholesterol deposition;

the genetic factor is a defective gene for the receptor molecule that normally removes excess LDL and therefore also excess cholesterol. This defective gene is inherited as a dominant (there may in fact be a number of different mutations of the receptor gene, all of which have the same effect). Hypercholesterolaemia occurs in white populations at a frequency of about 1 per 500, and is responsible for about 5 per cent of early deaths from heart attacks. People who have the gene will not inevitably get heart attacks. as a low-fat diet has a strong preventive effect, particularly if started early in life.

Hypersensitivity, over-reaction by the >immune system. It involves the mounting of a large-scale immunological defence against antigens that are in low concentration and are possibly not harmful to the organism in that low concentration though they would be at a higher one. See allergy.

Hypha (pl. hyphae), a filament that forms the substance, or mycelium, of some fungi. Hyphae cells are haploid and can produce asexual spores called >conidia. See also ***Neurospora crassa***.

I

I_1, I_2, I_3, **etc.**, the first, second, third, etc, generations obtained from an outcross, i.e. a cross between strains that are themselves already heterozygous. Compare this to >F_1 where the initial cross is between inbred strains.

Identical twins. See monozygotic twins.

Idiogram, a formal representation of all the chromosomes in an individual's >karyotype, usually consisting of photographed chromosomes form a single cell-division, which have been cut out and arranged in order. Preparation of an idiogram is an essential step in diagnosing >chromosome abnormalities.

IgA, IgD, IgE, IgG, IgM, the various different types of >immunoglobulin found in the human immune system.

Immigration. See migration.

Immune system, the system by which the body evercomes infectious and other invasions by foreign bodies. Natural, or non-specific, immunity works by phagocytosis (literally, 'cell-eating', in which cells surround and dissolve particles or bacteria and by the action of the protein interferon (the mode of which is still not clear). Specific immunity, i.e. a response to one particular infective agent (all of which, whether viruses, bacteria, parasites, or non-living particles are known as antigens in this context), is most highly developed in vertebrates, including humans. It is of two kinds: humoral immunity, in which antibody molecules circulate in the lymph and blood; and cell-mediated immunity, in which lymph cells directly bind to the antigens. In humoral immunity, the arrival of the antigen stimulates the appropriate line of >B lymphocytes in the bone marrow (see clonal selection theory); these cells, and their descendants in the lymph nodes and spleen, produce the specific antibody molecules which bind on to the antigens, either making them incapable of infective action or holding them in lumps so that they can be easily engulfed and dissolved by the large macrophage cells (this is facilitated by a group of proteins collectively known as >complement, activated by the antibody-antigen complex. Cell-mediated immunity is similar, with the antigen binding being done by >T lymphocytes which originate in the thymus gland. The humoral immunity produced by circulating antibody is most effective against infections by bacteria and viruses while they are outside the cells; cell-mediated immunity is better against viruses inside cells, parasites, cancer cells and foreign tissue. See auto-immune diseases; immunization.

Immunization, the technique of artificially creating >active immunity by injecting the patient with non-virulent antigens that resemble the disease-causing pathogens (see vaccination). Some diseases have a more lasting immunity than others: the acquired immunity to measles and mumps is generally lifelong, but in some cases (e.g. flu) it is very short-lived. This is probably because of changes in the virus responsible, which affect its antigenic properties.

The term is also used for giving >passive immunity by the injection of antiserum. Immunization can also mean any experimental process in which animals are injected with some substance to which they acquire immunity.

Immunoglobulins (abbreviation Ig), the proteins that are produced as antibodies in the immune system. There are several types, each consisting of two pairs of chains, two heavy (H) and two light (L). Each Ig chain, whether heavy or light, consists of a length that is constant (the Cregion, occupying about half of an L and three-quarters of an H chain), a joining region (J region), and at the other end a variable part (V regions). Specificity to the antigen is due to the sequence in the variable regions, and to the combinations of the variable-region sequences in the variable regions, and to the combinations of the variable-region sequences in the heavy and light chains. Although each immunoglobulin-producing cell has the same DNA sequence for immunoglobulin, each will have a different phenotype, in that it suppresses all but a few sequences to produce its own unique combination of V and J regions. The five types of immunoglobulins found in humans are:

- IgA, found in various fluids that protect mucous surfaces (in tears, gut and urinogenital fluids and sweat).
- IgD, found on the surface of B lymphocytes, especially in the foetus.
- IgE, effective against helminth (worm) infections and involved in the >hypersensitivity reaction (see allergy).
- IgG, the main active antibody, which can also be passed, through the placenta so that the foetus can receive some immunity from its mother.
- IgM, the largest (consisting of five of the normal immunoglobulin units) and the first to be made after an antigenic challenge.

Immunosuppression, artificial suppression of the immune system, used to help to prevent rejection of a transplant. Drugs such as cortisone decrease the activity of the T-lymphocytes. There is always the risk that a patient who is receiving immunosuppressive drugs to protect a transplant will die from is receiving immuno-suppressive drugs to protect a transplant will die from an infection; in the early days of heart transplants there was a high rate of deaths from pneumonia because of this.

In vitro, from the Latin, 'in glass'. A term for processes that happen in a controlled environment outside of a living organism, usually in tissue culture. It is strictly speaking an adverb but is often used adjectivally. The opposite is >in vivo.

In vitro fertilization, the technique of fertilizing one or several ova with sperm outside the body of the female, in a glass culture dish (hence the term in vitro, 'in glass'). It is used as a treatment of infertility, where the problem is blocked or absent Fallopian tubes (which lead from the ovaries to the uterus) which make it impossible for an egg to pass from the ovary to the uterus) which make it impossible for an egg to pass from the ovary to the tubes). The first baby to be conceived by in vitro fertilization was Louise Brown, born in 1978. She was the first of thousands of normal babies who prove that there is not, as was at first feared, a risk that in vitro babies could be abnormal.

To increase the chances of pregnancy, it is usual to give the mother a >fertility drug to increase ovulation so that several eggs can be collected, fertilized and implanted at once; this has resulted in many cases of mothers becoming pregnant with more babies than were intended (see multiple births). The eggs must be removed surgically from the mother and put into a fluid in a culture dish. Shortly before the eggs are collected, the father has to provide a fresh sample of sperm by masturbating. Fertilization takes place in the fluid, and progress of the embryo is carefully watched under a microscope, until three or four cell divisions have taken place (within 48-72 hours); then the embryo is transferred into the mother's uterus. The most uncertain part of the procedure is whether the embryo will successfully implant itself in the uterus, which only happens in one out of five or six attempts. In vitro fertilization is also used when the mother is unable to ovulate, and an egg from another woman is used. The technique has had the support of public opinion since it was introduced, except that there is much misgiving about what should be done with embryos that are created in vitro but are surplus to the number wanted for implantation; they can be frozen and used if a second attempt at implantation has to be made, but if this does not happen, is it acceptable either to throw them away or to use them for experi-mentation? In Britain the Human Fertilization of Embryo Research Act 1990 has legalized experi-mentation on embryos up to 14 days after fertilization.

in vivo, from the Latin, 'in a living thing', A term used for processes that happen in a living organism. It does not imply that the process described is as it occurs in natural conditions, because the word is applied to what happens in experiments; it is used solely in contrast to what happens in tissue culture (see in vitro). It is strictly speaking an adverb, but is often used adjectivally.

Inborn errors of metabolism, a group of disorders in which the basic cause is the lack of an enzyme involved in normal metabolism. These disorders are almost always inherited as recessives. For example, if the normal gene coding for the enzyme is T, and the mutant gene is t, a heterozygous person Tt will be healthy because the T gene can produce sufficient quantities of the enzyme; whereas the person with genotype it will be without the genetic means to produce the enzyme and will therefore have a block in a metabolic pathway. The term 'inborn error of metabolism' was coined by Sir Archibald >Garrod who identified four of them in 1903; over 100 such conditions have now been described. Most patients with these disorders can be treated with diets that avoid taking in the substance that cannot be metabolized. See alkaptonuria; cystinuria; phenylketonuria.

Inbreeding, mating between related individuals. It is a system that increases the amount of homozygosity within the genotypes of the succeeding generations. Under the most extreme form of inbreeding - self-fertilization - the proportion of heterozygotes is halved at every generation. In the case of the next-closest form of inbreeding - parent-child or brother-sister- the reduction is only half as fast. Inbreeding has a bad reputation, partly arising from its association with >incest (which is a social concept, not a biological one). In fact, whether or not inbreeding produces bad results is entirely a matter of how many deleterious recessives there are in the stock to begin with: a stock with a large number will decline as these come into the homozygous state in more and more individuals, but if there are few or none, there will be no such effect. Indeed, if the recessives in the original stock produce beneficial effects, an inbred strain will actually improve, a phenomenon that has been seen again and again with laboratory animals (e.g. the standard strains of laboratory mice are all much bigger and more prolific than wild mice). It is also a principle much used in agricultural breeding where it is known by the more positive name of >line breeding).

Inbreeding, coefficient of; a measure of the increase of >homozygosity as a result of inbreeding, as compared to what would be expected with random mating. The coefficient of inbreeding, F, devised by Sewall

Wright, can be taken to be either a property of inbred populations, being the average proportion of homozygosity in that population, or a property of inbred individuals, being the proportion of loci at which the individual carries two alleles identical by descent.

Incest, sexual relations between closely related individuals. Almost all human societies forbid incest, although the relationships prohibited vary: parent-child and brother-sister incest is practically always forbidden, but sexual relations between uncle and nice, aunt and nephew or cousin and cousin are viewed very differently, being sometimes banned, sometimes encouraged, sometimes obligatory. The genetic consequence of incest is to make it more likely that any deleterious recessive in the family will come into the homozygous state (see inbreeding). The sociobiological view of incest is that since the genetic consequences of close inbreeding are unfavourable, any gene that caused aversion to mating with one's close relatives would be favoured by natural selection to mating with one's close relatives would be favoured by natural selection. The existence of such a gene is, of course, unproven, notwithstanding the fact that our nearest evolutionary relatives, the great apes, sometimes (but not always) avoid incest. The sociobiological theory does not explain why there should be so many societies in which first-cousin marriage, which is quite close inbreeding and known to cause an excess of genetic disorders in the off-spring, should be encouraged. The anthropological view of incest, given the diverse forms in which it occurs, is that it is a cultural not a biological phenomenon. (And that it is absurd to worry that children born by artificial insemination from the same donor may unwittingly marry and so commit incest.) It is certainly the case that a great deal of illicit incestuous mating goes on in developed societies, mostly father-daughter or father-stepdaughter. The fact that the latter comes under the incest taboo fits with either the sociobiological or the anthropological viewpoint, but the fact that there is so much of this non-admitted activity is strongly supportive of the idea that the incest taboo is under the control of social sanctions rather than genes.

Inclusive fitness, the total >fitness (i.e. comparative reproductive performance) of an individual, including its own reproduction and that of its near relatives (their contribution diminishing with the distance of the relationship). It is an important idea in the arguments surrounding >kin selection.

Independent assortment, the principle that genes are inherited independently of one another; >Mendel's Second Law. When Mendel was doing his experiments on peas, he found that the inheritance of one trait was independent of another: a pea plant might have purple or white flowers, but the colour of the flowers had no effect on whether the plant produced round or wrinkled peas. The underlying mechanism is that genes will behave in this way they are located on separate chromosomes. F1 hybrid expresses the dorminant gene and in F2 generation a 3:1 or 9:3:3:1 ratio will be obtained with one or two genes respectively.

Inducer, a molecule that binds to a repressor to turn on the transcription of an inducible >operon. The inducer is usually a small molecule, smaller than the repressor (a protein), but able to alter the latter's configuration so that it dissociates from the operator region and allows transcription to proceed. The inducer is often the substrate of the enzyme whose production is being induced. See gene regulation.

Inducible enzyme, an enzyme which is produced only when its substrate is present to act as inducer. The opposite is a repressible enzyme. See operon; gene regulation.

Infertility, inability to produce viable offspring. It has been estimated that about 10 per cent of human sexual partnerships are infertile. The problem may be in either partner or (more rarely) may be a genetic incompatibility between them. The commonest causes are (in men) low sperm count or non-viable sperm, and (in women) blockage of the Fallopian tubes. Treatments are available for all forms of infertility. Sperm problems can be treated by artificial insemination; Fallopian tube blockage by in vitro fertilization; failure to ovulate by egg donation; failure to sustain pregnancy by surrogacy. See also sterility.

Insertion, the addition of one or more base-pairs into a DNA sequence. The effect of a small insertion may be a >frameshift mutation, but a larger insertion can upset gene regulation around it, either by activating genes that should be inactive or vice versa. See insertional inactivation.

Insertional inactivation, a technique used in genetic engineering to prevent expression of a gene by inserting a foreign DNA sequence adjacent to it or in its coding region.

Instinct, a behaviour pattern that is inherited. The term is not much employed by ethologists and psychologists, as it has been used in so many ways that its meaning has become elusive. But there is still a useful distinction to be made between instinctive behaviour, where the stereotyped pattern of actions is genetically programmed and is prompted by a specific stimulus and learned behaviour which, however typical of the species, is the property of the individual who has learned it. This is not to deny that some instinctive behaviour patterns require a certain learning process; in many singing birds, the urge to sing is innate but the bird needs to hear the song of its particular species.

The mode of inheritance of instinctive behaviour patterns is still unknown. The pioneers of the study of instinct were Konrad >Lorenz and the Dutch ethologist Nikolaas Tinbergen.

Insulin, the hormone involved in sugar (more specifically glucose) metabolism. Insulin deficiency causes >diabetes, which is treated by diet or with injections of insulin. Formerly, insulin had to be extracted from pigs or cattle, but human insulin can now be produced by gene cloning.

Intercross, a cross between two individuals or strains that are themselves hybrid. See F2.

Interferons, a group of proteins that are produced by cells in response to a viral attack. They seem to enable other cells to resist viruses, but the mechanism is not understood. It is believed that interferons may be anticancer agents, if they are able to inhibit oncogenes in the same way as they do viruses, but clinical trials have not confirmed this. Genes for human interferons have been cloned so the proteins can now be produced in bulk.

Interphase, the stage on the >cell cycle between divisions. This is the phase during which the chromosomes are active in synthesizing proteins.

Intersex, a broad term covering various conditions in which an individual does not have normal reproductive organs but is not a hermaphrodite. Male pseudohermaphroditism is the condition in which there are testes (probably internal) but the external genitalia appear more or less female; the cause can be chromosome abnormality, a single gene (as in testicular feminization) or hormone imbalance in utero (see also Klinefelter's syndrome). Female pseudol. hermaphroitism, in which there is a combination of normal ovaries and more or less male external genitalia, is not caused by chromosome disorders, but by a series of recessive alleles, with a combined frequency of about one in 7000 births. Hormone imbalance in utero can also be responsible.

Interspecific competition, competition between different species for a resource. It may lead to the extinction of one of them, or at least to its exclusion from a particular area.

Intrachromosomal recombination, exchange of sequences of DNA between sister chromatids (i.e. two halves of the same chromosome) during meiosis. Because the sister chromatids are identical, there is no genetic effect, unlike in the normal process of recombination, when non-sister chromatids from homologous chromosomes exchange genetic material.

Intraspecific competition, competition for a resource between members of the same species. This is the 'struggle for existence' that Darwin and Wallace identified as the driving force of evolution.

Intron, a sequence of DNA that does not code for a gene product (in contrast to an >exon). The sequence is initially transcribed into RNA, but all the introns are cut out before the messenger RNA is finalized. This is called 'intron splicing'. Introns do not occur in prokaryotes, but they are very common in eukaryotes, though their function is obscure.

Inversion, a form of mutation in which a sequence of a chromosome is removed and replaced the other way round. Because the genes remain the same, there is usually no genetic consequence, but sometimes a gene's activity may be altered due to its being separated from its normal neighbours, which are be part of its control mechanism: see position effect. Inversions can be either pericentric, i.e. involving the >centromere, or paracentric, involving only one of the arms.

Invertebrates, all animals without backbones. These include sponges, jellyfish, sea anemones, worms, molluscs, starfish, spiders, crustaceans and insects. See classification.

IQ (intelligence quotient), a measure of mental capacity, defined as a person's mental age x 100 divided by its chronological age. Thus a child of 10 with a mental age of 9xxx has an IQ of 95. The idea of intelligence as being something that could be measured and the result represented by a single figure - even one that, like IQ, is derived from a whole series of tests - has been criticized from the outset. The contents of the tests used by pioneers of IQ testing are to modern minds quite ludicrously sociocultrally biased. None the less the fact that IQ was a 'measurement' gave it a spurious reality and respectability, and battle-lines were drawn up as to whether the observed difference between children. (and groups of children) were genetically determined or not.

The important thing is not to try to discuss whether 'nature' is more important than 'nurture' - the geneticist's position is that this is a non-question (see nature versus nurture; heritability) - but to ask why the matter has been so hotly and emotionally debated. Broadly, hereditarians (i.e. those who subscribe to the 'nature' side of the argument) have been politically of the right, and have been anxious to prove that their monopoly of educational and employment opportunities has been justified by their superior genetic endowment. Environmentalists (adherents of 'nurture') have been of the left, viewing the human infant as a tabula rasa, and keen to show that inequalities of intelligence are the result rather than the cause of inequalities of opportunity. Legitimate methods of research have included the study of adopted children (who should show more resemblance to their natural or adoptive parents, according to hereditarians and environmental respectively), and the study of identical twins, measured both against non-identical twins, and, if reared apart, against the different environmental conditions (see twin studies). The many studies along these lines have, not surprisingly, come up with different conclusions. Sir Cyril >Burt was so anxious to prove his hereditarian position that he falsified his research, making up pairs of twins that did not exist. In the late 1960s there was an attempt by the American psychologist Arthur Jensen and others to show that there were differences in average IQ between races, specifically between white and black children in the United States. It was bad science, and bad politics, but what was perhaps saddest of all was the response by people who should have known better: rational scientists and others apparently felt that all that was needed to meet this idea was a torrent of abuse, denouncing the perpetrators as fascists.

Isochromosome, an abnormal chromosome in which two identical arms have become joined at the >centromere. Thus half of the genes that ought to be on the chromosome are present in double dose, while half are missing altogether.

Isoenzymes (also isozymes), a set of enzymes all carrying out the same chemical reaction. Several isoenzymes may be present within the same organism, as they are coded for by genes at different loci. Compare >allozyme.

Isogamy, the production of two equal-sized types of gametes by a species, i.e. the male and female gametes are identical. This is a primitive feature, not found in the higher eukaryotes whose sperm and ova are specialized in shape and function (see anisogamy).

Isolation, in genetic terms, the situation in which a given population is unable to exchange genes with any others, owing to barriers (geographical or other). See isolating mechanism.

Isolating mechanism, any barrier that prevents interbreeding between one population and another. The simplest form is a geographical barrier, but a behavioural, physiological or anatomical one is just as effective in separating populations, which is the first step towards the formation of new species.

J

J region, or joining region, the part of an >immunoglobulin molecule that lies between the constant and the variable regions.

Jacob, Francois (1920-), French microbiologist; director of the Pasteur Institute since 1982; co-proposer of the >operon mechanism of gene control. In 1961, Jacob and E. Wallach announced the discovery of extranuclear genetic elements that they called episomes (now known as plasmids), and in the same year, Jacob and Jacques >Monod proposed the theory that gene action is controlled in a unit called an operon, and also predicted the existence of the type of RNA molecule now known as messenger RNA. For this work Jacob shared the 1965 Nobel prize for physiology or medicine with Monod and Andre Lwoff.

Johannsen, Wilhelm Ludwig (1857-1927), Danish geneticist; inventor of the word 'gene'. Too poor to go to university, Johannsn became a pharmacist, but through his private researches, he became the leading theoretical geneticist of his time. He began to study heredity in the 1890s, and was influenced by Galton; he proposed a particulate theory of inheritance which he published before the rediscovery of Mendel's very similar theory. He was the first to make the distinction between the genotype and the phenotype; in the same paper, published in 1909, he suggested the word 'gene'. Johannsen later became rector of the University of Copenhagen.

Jumping genes, genes that move within chromosomes. They influence the regulation of gene activity, probably by physically removing a gene from its promoter (see operon). Jumping genes were discovered by Barbara >McClintock. See also transposon.

Junk DNA (also known as >selfish DNA), sequences of DNA in the genome that have no apparent genetic function.

K

K selection, the type of natural selection that operates when the environment is relatively constant and the population is at or near the carrying capacity. In these circumstances, high reproductive rates will not pay off, but good adaptation, particularly efficency in food and other resources, will be favoured. Longer parental care and small numbers of offspring, a high proportion of which are successfully raised are also favoured. See also r selection; MacArthur, Robert Helmer.

Karyotype, the chromosomal constitution of an individual as seen in the nucleus of a somatic cell.The chromosomes are classified by their size and centromere position, and for convenience are usually shown in a specific order in an >idiogram (this is also sometimes referred to as the karyotype). The karyotype for a normal human male would be notated'46, XY'. (Note the ambiguity: this means 46 chromosomes in total - i.e. 44 autosomes and two sex chromosomes - but it looks as if it could mean 46 autosomes plus two sex chromosomes.)

Khorana, Har Gobind (1920-), Indian-born American molecular biologist; professor of biology and chemistry at the Massachusetts Institute ofTechnology since 1970. By 1961 Francis >Crick and Sydney >Brenner had found that the >genetic code consists of codons, each of which were three bases long. Khorana applied his skill in synthesizing biological molecules to the problem of deciphering the code: by creating each codon artificially, he was able to show which one coded for which amino acid. He then achieved the first-ever synthesis of a gene, when he reproduced artificially the DNA sequence of an >Escherichia coli gene. Khorana shared the 1968 Nobel prize for physiology or medicine with Marshall >Nirenberg and Robert W.Holley.

Kilobase, literally, 1000 bases. A unit of length of DNA or RNA, used in describing genes or shorter sequences. When measuring double-stranded DNA, it is actually base-pairs that are counted, so that the unit is comparable with counting bases in a single-stranded molecule.

Kin selection, natural selection that acts upon a group of kin rather than upon the individual, and produces evolutionary changes of the sort that favour kinship groups rather than individuals. After the demise of the theory of >group selection, kin selection was proposed in 1964 by W.D.Hamilton as the best explanation of how altruism could have evolved. It is true that 'altruistic' genes will be preferentially represented in the succeeding generation as a result of altruism towards close kin, given that relatives share many genes. Full siblings have 50 per cent of their genes in common, so that if the benefit to sister A, it pays sister A to help sister B. In evolutionary terms, the help that counts is help towards reproducing successfully, i.e. towards getting one's own (or one's sister's) genes preferentially represented in the next generation. By the same calculation, it pays to help a first cousin (one-eighth of genes in common) if the benefit to her is more than eight times the benefit of the same effort to oneself. From this, it appears that kin selection is likely to be much more effective between very close kin, and sociobiologists point out that small groups of close kin are the actual breeding units of very many species.

Klinefelter's syndrome, the condition in which a male person has genotype XXY (i.e. two X chromosomes and one Y chromosome). Boys with Klinefelter's syndrome look normal until puberty, but in adulthood, they have some female features such as enlarged breasts and sparse body hair, and their testes are small and never produce sperm. These features are due to the conflicting genetic information in the sex chromosomes as to which sex the individual is. Less obviously, they are taller than average, and a significant percentage are mentally retarded, for which there is no physiological explanation. The extra chromosome usually comes from the mother (i.e. she produced an ovum with two X chromosomes), and there is an increasing risk of this with maternal age, but in a few cases, the extra X comes from the father (i.e. he produced a sperm with an X and a Y). Klinefelter's syndrome is the commonest of the sex chromosome disorders, occuring at a rate of about 1 per 1000 male babies. Hormone replacement therapy, consisting of injections of the male sex hormone testosterone, is effective, particularly if it can be started at puberty (though many cases are not noticed during childhood).

Kornberg, Arthur (1918-), American biochemist; professor of biochemistry at Stanford University, California since 1959. Kornberg specialized in enzyme biochemistry, and isolated the enzyme DNA polymerase 1 from >Escherichia coli. This is one of the two enzymes that normally build the second strand of DNA during replication, but Kornberg found that it could also assemble a DNA molecule >in vitro, using only unconnected nucleotide molecules, provided that there was some natural DNA in the mixture to act as a template. This became an important technique in later DNA research. Kornberg shared the 1959 Nobel prize for physiology or medicine with Severo Ochoa.

L

Labelling, a technique for tagging molecules by introducing a radioactive isotope (e.g. carbon-14, iodine-125) into their structure. The presence of the radioisotope does not affect the chemical activity of the molecule. The fate of the labelled molecule(s) can then be tracked, in either in >in vivo or >in vitro experiments, by autoradiography or scintillation counting.

Lac operon, the group of genes in ***Escherichia coli*** that controls the metabolism of lactose. It was the first >operon to bè discovered, and was found by Jacob and Monod in 1961. They identified the following parts. First is the gene lacl, which codes for the >repressor; then the control region, consisting of the >promoter region lacP and the operator region lacO; lastly come the three structural genes lacZ, lacY and lacA, which code for the three enzymes needed for metabolizing lactose.

In the absence of lactose (i.e. in normal conditions for **E. coli**), a molecule of the repressor protein is bound to the operator region, preventing RNA polymerase from proceeding downstream from the promoter where it attaches. When molecules of lactose attach to the repressor, the repressor is lifted off the operator and transcription of the three enzymes occurs. One of these (Y) transports lactose through the cell membrane; another (Z) hydrolyses it; the function of the third (A) is not clear. When all the lactose in the cell has been metabolized, the repressor binds once more to the operator and synthesis of the enzymes stops. The system of having the substrate of the enzymes act as the inducer of the operon is commonly found. Jacob and Monod put together this scheme of gene regulation by deduction from the effects of various mutations that interfered with its normal working.

Lamarck, Jean-Baptiste Pierre Antoine de Monet, Chevalier de(1744-1829), French biologist, one of the first people to propose a theory of evolution. After a medical training, Lamarck became interested in botany and wrote Flore francoise, proposing a system of classifying plants by a succession of either/or divisions. This impressed the influential biologist Comte >Buffon who, in 1781, helped Lamarck to get a job as botanist to the king. After the Revolution, in 1793, Lamarck became professor of zoology at Paris, where he did important work in the >classification of the animal kingdom. He was the first person to notice that some animals have backbones while others do not, and he invented the term 'Invertebrata' for the latter. He also identified and named the Crustacea (crabs, lobsters, woodlice, etc.), Arachnida (spiders, scorpions, mites, etc.) and Annelida (earthworms, lugworms, leeches, etc.). He is remembered for his ideas on evolution. He was quite correct in following Buffon and Erasmus Darwin in saying that species are not fixed but have evolved from earlier forms (he based this on the fact that domesticated plants and animals are changed from the wild varieties). But his most important misconception, and the one that bears his name as 'Lamarckism', was that acquired charateristics can be inherited

Lambda, one of the most intensively studied of the >bacteriophages. It attacks >*Escherichia coli*. The genome is double-stranded DNA, but with single-stranded, mutually complementary 'tails' (see sticky end) that join up to make a circular DNA after entry into the host cell.

Lampbrush chromosome, an unusual type of chromosome found in the oocytes (immature eggs) of newts and other amphibians. There are two chromatids which form many paired loops joined back to the chromosome core. The loops are segments of DNA that have become decondensed soas to be transcribed, and each one probably represents a single gene.

Landsteiner, Karl (1868-1943), Austrian-born American immunologist. Between 1900 and 1902 Landsteiner discovered the >ABO blood group system in humans. Later, in 1927, he discovered the similar >MN blood group system, which is not as important in practical terms because blood transfusions are not affected by it. For these discoveries, he was awarded the 1930 Nobel prize for physiology or medicine. In 1940, at the end of his long career, he and his colleague A.S. Wiener discovered the >Rhesus blood groups.

Lat genes, genes that are expressed late in the process of a virus infecting a bacterial cell. They usually code for the protein(s) that make up the virus's coat. See also early gene.

Laurence-Moon-Biedi syndrome. See obesity.

LD50, a term used to denote the dose of a substance that will kill 50 per cent of the organisms receiving it. LD50 tests are used on laboratory animals for testing the toxicity of drugs, toilet preparations, cosmetics, etc. designed for humans. When a group of bacteria or larger organisms are used, which are well matched for size (so that the relative dose they receive is comparable), the survivors may well be genetically different from the non-survivors, so that the result says as much about genetic resistance as it does about the substance being tested.

Leader sequence, the part of a molecule of messenger RNA between the 5' end and the start of the coding sequence. It contains sequences that are concerned with binding to the ribosome.

Lederberg, Joshua (1925-), American geneticist; professor of genetics at Stanford University, California, 1959-78, then president of Rockefeller University, New York. Working with E.L. >Tatum. Lederberg discovered that in some strains of >*Escherichia coli* sexual conjugation takes place (up till then it had been thought that bacteria could only reproduce by simple division). This made it possible to map bacterial chromosomes, and was the starting ponint for bacterial genetics. Then, with N.D. Zinder, he discovered transduction, the process by which a virus can transfer a piece of genetic material from one bacterium to another. He also found that a B lymphocyte produces only one antibody, which was the basis for the development of monoclonal antibodies. Lederberg shared the 1958 Nobel prize for physiology or medicine with Tatum and George >Beadle.

Leeuwenhoek, Antony van (1632-1723), Dutch naturalist; pioneer of the microscope. Van Leeuwenhoek had no formal training in science. Compound microscopes had already been invented by his time, but he used instead single lenses with a very short focus (i.e. magnifying glasses) with which he obtained magnifications as great as x 250. (It is believed that he had the advantage of unusually good close-range vision.) He gave the first descriptions of red blood corpuscles, >sperm (which he supposed contained homunculi, bacteria, protozoa and rotifers. All these, and many other observations, were described in his 375 letters to the Royal Society in London (of which he was elected a Fellow in 1680). Perhaps his most important contribution was his study of the life cycle of the flea, which he announced to be 'endowed with as great perfection in its kind as any large animal'. His work proved that fleas reproduce sexually like any other insect or animal, and were not spontaneously generated, as was believed at the time.

Left-handedness. See handedness.

Leptotene, one of the stages in >meiosis (one of the two processes of cell division).

Lesbianism. See homosexuality.

Lesch-Nyhan syndrome, a human disease inherited as an X-linked recessive. It therefore effects boys who have inherited it from carrier mothers. The actual mutation is a failure to produce the enzyme hypoxanthine phosphoribosyl transferase (HPRT), which is part of the process for metabolizing purines; build-up of only partly metabolized purines leads to mental retardation and cerebral palsy. there is also a bizarre and totally unexplained behavioural effects: boys who suffer from this syndrom have a compulsive urge to bite their own lips and fingers, and if they are not restrained, they will actually bite pieces off. The condition is very rare. Female relatives of a sufferer who may be carriers can have a foetus prenatally diagnosed, as the lack of HPRT can be detected from conception. Lesch-Nyhan syndrome is a candidate for gene therapy, as the normal gene for human HPRT has been isolated and cloned.

Lethal mutation, any mutation that has such a severe effect on the organism as to cause early death. From the genetic point of view, a mutation is considered 'lethal' if the death occurs at any time before the individual reaches reproductive age. Common usage in relation to humans, however, includes only those mutations that cause death before birth or in early infancy. Lethal mutations can be dominant, recessive or X-linked, and there are some known as >semi-lethals. Lethal mutations are quite common in humans: about 20 per cent of pregnancies end in spontaneous abortion (miscarriage), with either lethal mutations or chromosome abnormalities responsible for most cases.

Leukaemia, a group of blood diseases in which the white blood cells or their precursors become cancerous, and go into uncontrolled proliferation (the word 'leukaemia' comes from the Greek for 'white blood'). The white cells often show a chromosomal translocation, and it is liely that what happens is that a single white cell gets this chromosome damage which altrs the genetic control of cell division; all the cancerous white cells are descendants of this one damaged cell (so they are clonal and have inherited the same damaged chromosome).

Not all the leukaemias have the same cause. Acute myeloid leukaemia seems to arise spontaneusly and does not appear to run in families. Chronic lymphoid leukaemia, on the other hand, may have a genetic component, in that it has an association with >multiple sclerosis. Acute lymphoblastic leukaemia has a possible genetic link with the >HLA locus, but it has also been suggested that environmental factors are most important: this is the type of leukaemia that shows a high incidence near the Sellafield nuclear plant in the English country of Cumbria. See also Philadelphia chromosome.

Life cycle, in genetical terms, the cycle form fertilization through maturation to the fertilization of the next generation. It can also mean the stages through which any individual organisms progresses, from fertilization to death.

Ligation, the joining together of separate sequences of DNA by DNA ligase enzymes. It occurs during replication (see >Okazaki fragments), DNA repair, and is also widely used in genetic engineering to stick DNA molecules together in vitro.

Line breeding, the technique of breeding from animals as closely related as possible, much used in improving agricultural stocks. It is not unusual for cows, for example, to be mated to their own sire or to a brother. See inbreeding.

Lineage, a sequence of species through evolutionary time, from the ancestral species to its present-day descendants.

Linkage, the tendency of genes that are on the same chromosome to be inherited together. Linkage of two loci provides an exception to Mendel's law of >independent assortment, which proposes that all genes are independent from one another. Clearly this cannot be so, given that genes are physically located in strings along chromosomes; those that are on the same chromosome are linked. To prove whether two given loci are linked, an F1 cross is made between two strains, one breeding pure for the dominant alleles at each locus (AABB), the other for the recessives (aabb). The F1 (AaBb) is then >backcrossed to the double-recessive strain, and if there is no linkage (i.e. the loci are on different chromosomes), there are four types of offspring in equal quantities: AaBb, Aabb, aaBb, aabb. If the loci are linked, however, the F1 individuals will pass on only AB chromosomes or ab ones, so that in the offspring of the backcross, there are only AaBb and aabb. This is the result with very close linkage; but more usually there is a measurable proportion of cases in which linked alleles become separated from one another by the process of >crossing-over during meiosis.

Linkage group, a group of genes that can be observed to be linked together rather than to assort independently. In practice, a linkage group means a group of genes all on the same chromosome.

Linnaeus, Carl (also known as Carl von Linne) (1707-78), Swedish botanist. Linnaeus began a medical training but soon turned to botanical research. His first project was a system of >classification of flowering plants based on their sexual parts (Systema naturae, 1735). Economic necessity forced him to complete his medical training, and from 1741, he was professor of both medicine and botany at Uppsala university. He continued his work in classification, and perfected binomial nomenclature, the system under which all creatures have a genus name (noun) and a species name (adjective). His passion for classification blinded Linnaeus to the possibility that species could change, and he was sure that there had been no alteration in the living world since the Creation.

Locus (pl. 'loci'), the point on a >chromosome where a given gene is situated.

Lorenz, Konrad (1903-89), Austrian zoologist; pioneer in the study of animal behaviour. Lorenz trained as a doctor, but was prosperous enough to give up work and study bird behaviour. He believed that this could only validly be done in the wild (against the mainstream of psychological research at the time, which concentrated on reproducible experiments in the laboratory). Lorenz proposed that much behaviour is controlled by instinct, foreshadowing the line now taken by sociobiology. He discovered 'imprinting', the process whereby a bird learns what species it is from what creature it sees immediately after hatching (in many of Lorenz's experiments with geese, this creature proved to be Lorenz himself). Lorenz had >eugenicist ideas that fitted in well with those of the Nazis, whom he supported, although he came to regret his involvement with them. After the war, he continued his bird researches and also applied his ideas about instinct to human behaviour. His non-technical writings include King Solomon's Ring (1925) and the influential On Aggression (English edition 1966) (see aggression). Lorenz shared the 1973 Nobel prize for physiology or medicine with Nikolaas Tinbergen and Karl von Frisch, these three being the first people in the behavioural sciences to win the prize.

Lucy, the name given to the earliest known 'human' being, who lived in Ethiopia nearly 4 million years ago. See human evolution.

Luxury gene (a term rarely used nowadays), a gene that codes for a product specific to one particular type of cell, in contrast to >housekeeping genes. An example of a luxury gene would be the gene for insulin, which is active only in certain cells in the pancreas (though present in all cells). Luxury genes are switched off in all cells except the ones where their function is needed. While they are inactive, the DNA of luxury genes is highly condensed as >heterochromatin, and it shows up as a dark-staining band (see banding).

Lyell, Sir Charles (1797-1875), British geologist. Lyell took up the idea previously put forward by James Hutton (see p. 15) that the formation of the Earth had not been a single event but was a continuous process that was still in action and would continue for ever (see p. 12). He wrote the popular Principles of Geology (1830), which influenced Darwin's thought about evolution. Lyell was one of the friends who encouraged Darwin to publish On the Origin of Species, though he himself never accepted Darwin's theory.

Lyon hypothesis, the theory that, in female mammals (sex chromosomes XY), one X in each cell is inactivated so as to achieve >dosage compensation, the individual then having the same amount of X-chromosome activity as the male (XY). The X chromosome to be inactivated is chosen at random, and this happens in different tissues at different times during the development of the embryo (>embryogenesis). Once one X has been inactivated in one cell, all cells, derived from that cell will have the same X inactivated. A good example of this is the >tortoise shell cat (always female) which has patches of black and orang fur; the gene for fur colour is on the X chromosome, and the black patches derive from a cell in which the orange-gene X had been inactivated and vice versa. (In marsupials, X inactivation is not at random but always affects the X inherited from the father. There could never be a tortoiseshell wombat.) The female mammal is a naturally occurring >mosaic. The inactivated X chromosome can be seen microscopically as the >Barr body. The theory is called after the British geneticist Mary F. Lyon.

Lysenko, Trofim Denisovich (1898-1976), Soviet agricultural botanist. Lysenko was president of the Lenin Academy of the Agricultural Sciences from 1938, and also director of the Institute of Genetics at the Soviet Academy of Sciences from 1940 until 1945. He refused to accept the Mendelian genetics current in the rest of the scientific world, preferring the >Lamarckian system of the inheritance of acquired characteristics. His science was accepted as orthodoxy by Stalin (as it seemed to fit with a Marxist viewpoint), and Lysenko enjoyed great political patronage, though the great majority of Soviet scientists and notably >Vavilov knew he was wrong. After Stalin's death, criticism of Lysenkoism became possible, but he continued to occupy a highly influential position; the damage that he did to Soviet science and its international reputation has still not been repaired.

Lysis, the break-up of a cell after the rupture of its cell wall. This can be the result of attack by chemicals (e.g. detergents) or viral infection (as happens when bacteria are invaded by bacteriophages), or the cell may simply dissolve itself (autolysis). See also virus.

Lysogeny the colonization of a host cell by a >virus, which gets itself replicated by the host's genetic machinery.

M

MacArthur, Robert Helmer (1930-72), Canadian-born American ecological geneticist. Formerly a mathematician, MacArthur brought new theoretical insights to ecology. He proposed the theory of two different evolutionay strategies - opportunistic (r selection) and stable (K selection) - and gave the mathematical proof that both are compatible with Darwinian natural selection.

McClintock, Barbara (1902-90), American geneticist. McClintock's research into varieties of maize (sweetcorn) led, in the 1940s, to her discovery of jumping genes (now known as transposons), but the significance of these was not realized at the time. McClintock did further research on the role of jumping genes in the control of gene action, which was later developed by Monod. She belatedly won the Nobel prize for physiology or medicine in 1983.

Macroevolution, evolution viewed over a long time-span, especially concerning major trends. The development or disappearance of higher taxonomic groups is macroevolution. See also microevolution.

Major histocompatibilty complex, the group of genes that control the >HLA (human leucocyte antigen) system. See also immune system.

Malaria, an infectious disease in which the red blood cells are attacked by the protozoan parasite *Plasmodium*, which is transmitted from one infected perosn to another by *Anopheles* mosquitoes. Malaria is common in tropical areas, and used also to be common around the Mediterranean until mosquito numbers were reduced, partly through loss of their habitat as marshes were drained and partly by insecticidal campaigns. In those areas still affected by malaria, prevention is better than cure, and synthetic quinine-like drugs are effective in deterring the *Plasmodium* parasite.

Malaria, which causes high fever that often recurs for years after the original infection, is a dangerous disease, so it is not surprising that in areas where it is (or was) endemic genes that confer protection against it have been selected for and are common in the population. Unfortunately, in most cases these genes 'buy' protection against malaria at the cost of giving people who are >homozygous for the gene another disease: the two best examples are >sickle-cell anaemia and >thalassaemia, and another is >glucose-6-phosphate dehydrogenase deficiency (but one antigen in the >Duffy blood group series seems to give protection without deleterious effects).

Malthus, Thomas Robert (1766-1834), British clergyman and economist. Malthus graduated in mathematics from Cambridge University. His Essay on the Principle of Population (1798, 1803) stated the human populations increase in a geometric progression whereas the means for their subsistence (i.e. food) increase only in an arithmetic progression; and the factors that prevent an infinite increase of population are famine, pestilence and war. After the first edition brought a storm of controversy, Malthus produced a second five years later, containing a wealth of data from populations both historical and contemporary. In this edition Malthus suggested that 'moral restraint' (meaning late marriage) might also be a means of population control. His work had a seminal influence on >Darwin, and also on the economist David Ricardo. His work was also the first piece of serious research in demography, and is still a foundation of that science.

Manic-depression, a mental condition in which people are subject to extreme swings of mood not related to the events around them. In the depressive phase, they suffer wretched misery which may lead them to commit suicide. Manic-depression affects about 1 in 100 in European populations.

A breakthrough in the understanding of this disorder came through the study of the Amish, an American religious-fundamentalist community. In this small population there are quite frequent cases of manic-depression which run in families. An extraordinary finding was that all the 26 suicides among the Amish over the past 100 years had come from just four families. Analysis of the family pedigrees led to the conclusion that manic-depression is caused by a dominant gene, but that the gene has incomplete penetrance so that a person who inherits it has a chance (about a half to a third) of escaping the disease. By means of a genetic marker, the Amish manic-depression gene has been located on chromosome 11. Other researches into manic-depression have also found evidence of a dominant gene but without any connection with chromosome 11, so it is likely that there is more than one gene that can cause the same condition (as is suggested by its high frequency). There may also be an X-linked form. The mechanism of the gene's action is probably concerned with the neurotransmitters, molecules that are involved in communication between brain cells.

Map unit, a measure of the distance between two loci on the same chromosome. Loci are 1 map unit apart if cross-over occurs between them in 1 per cent of cases. A map unit has the formal name of >centimorgan.

Marfan's syndrome, an inherited condition with a number of effects, which are very variable: the person may be tall and thin, have extremely long limbs and digits, and have heart abnormalities. These various effects are the result of a single dominant gene, though the biochemical mode of action is not known (see pleiotropy). The symptoms do not cause any ill health until middle age, when the heart deformities often prove fatal; thus people with the syndrome have usually already had children by the time thay are diagnosed, and the gene has been passed had children by the time they are diagnosed, and the gene has been passed on to the next generation. About one person in 25,000 has Marfan's syndrome. It is said that the extraordinarily long fingers of the Russian composer and pianist Sergei Rachmaninov were caused by this syndrome.

Marriage. See monogamy.

Maternal effect, the situation in which the phenotype of the offspring is determined not by its own genotype but by its mother's. The classic example of this the coiling of the shell of the watersnail *Limnaea*. The alleles are D for right-handed coiling and d for left-handed, and are present in both sexes. To take a hypothetical pedigree, a DD snail (right-handed) has a Dd (right-handed) daughter who has a dd (right-handed) daughter who has a Dd (left-handed) daughter who has a Dd (right-handed) daughter. The system does obey Mendelian rules, but at a one-generation remove. There is some evidence to suggest that left handedness in humans is inherited in this way. See also maternal inheritance; Sturtevant, Alfred Henry.

Maternal inheritance, any effect on the offspring that is not genetically transmitted but is passed on through the cytoplasm of the egg, i.e. >cytoplasmic inheritance. Maternal inheritance is the explanation of the situation in which a cross between a Strain A female and a Strain B male gives a different result form a cross between a Strain B female and a Strain A male (assuming that >X-linkage has been ruled out). See also maternal effect.

Mating system, the system within a population by which mates are chosen. At one extreme are parthenogenesis and self-fertilization; at the other is random mating. There may be systematic inbreeding or outbreeding, or assortative mating in respect of one or more traits. Some mating systems involve equal numbers of both sexes; in others, members of one sex have multiple mates. All mating systems have an effect on the genetic constitution of the succeeding generations: under random matting, it stays the same; inbreeding and positive assortative mating increase homozygosity; outbreeding and negative assortative mating increase heterozygosity.

Mean, the arithmetical average, i.e. the sum of all the values divided by the number of individuals. See also median; mode; normal distribution.

Median, the value in a distribution that comes half-way between the highest and the lowest. See also mean; mode; normal distribution.

Meiosis, the central event in heredity. It is the process of cell division that takes place when gametes (ovum or sperm) are made. The key feature is that the number of chromosomes is reduced from a set consisting of pairs to a set consisting of one copy of each. It works as follows:

First meiotic division

PROPHASE I After being elongated during >interphase (the non-division part of the cell cycle), the chromosomes begin to condense themselves (leptotene stage). The >homologous chromosomes come together and lie alongside each other (zygotene stage). The chromosomes double themselves lengthwise into two so that each consists of two >chromatids held together at the >centromere. There are now four chromatids lying alongside each other (pachytene stage), and the non-identical chromatids continue to thicken and shorten (diakinesis).

METAPHASE I The membrane of the nucleus disappears and a fibrous structure known as the >spindle forms. The homologous pairs of chromosomes, now at their most condensed, orient themselves along the equator of the spindle.

ANAPHASE I The homologous chromosomes separate, the members of each pair moving towards opposite poles of the spindle (paternal and maternal chromosomes going either way at random)

TELOPHASE I The spindle disappears. Membrances for the two new nuclei may form.

Second meiotic division

PROPHASE II Each nucleus contains one member of each homologous pair, each consisting of two chromatids. The nuclear membranes disappear and spindles form.

METAPHASE II Chromosomes align themselves along the equators of the spindles.

ANAPHASE II The centromeres split, allowing the two chromatids of each chromosome to move to opposite poles of the spindle.

TELOPHASE II The spindles break down and membranes re-form round the four new nuclei; these each contain a set in which there is one representative of each chromosome (the erstwhile chromatids). If there has been any >crossing-over, which there usually has, each of the four versions of the chromosome is different.

The process of meiosis is the same in all eukaryotes. It is also the same in males and females, but whereas the end result in males is four spermatozoa, the female only brings one egg to maturity, the other three products of meiosis being lost as >polar bodies. The whole process takes several days.

The importance of meiosis in genetics is that it provides an endless source of variation over and above the variation contained in having a choice of alleles for most loci. Firstly, the maternal and paternal chromosomes are randomly divided at anaphase I. Even in a species with as few chromosomes as four pairs, this would mean only a 1 in 8 chance of an individual inheriting a complete set unchanged from those of a grandparent. However, for a human with 23 pairs, the chance is 1 in 4,194,304, and for a dog with 39 pairs, it is 1 in about 3 million (the formulais 1/2n-1 where n is the number of pairs of chromosomes).

Secondly, there is an immense opportunity for new genetic combinations to come about through the process of crossing-over between homologous chromosomes in the first meiotic division; see recombination. This is additional to the reshuffling due to whole chromosomes going one way or another, and means that each individual of a complex species is literally genetically unrepeatable. See also mitosis.

Meiotic drive, any mechanism that causes uneqal proportions of the possible gametes to be formed at >meiosis. One example is >B chromosomes, which tend to be included preferentially in the ovum during meiosis in the female.

Melanin, the black or dark brown pigment that gives colour to skin, hair and eyes in animals, including humans. Melanin production in skin cells is stimulated by ultraviolet radiation, either natural (i.e. from the sun) or artificial, and is catalysed by the enzyme tyrosinase, lack of which causes >albinism.

Melanism, Industrial, black or blackish coloration, used as camouflage by moths living in areas polluted with soot. The phenomenon was described by the British lepidopterist H.B.D.-Kettlewell in the 1950s. He observed that the peppered moth *Biston betularia*, which had been light-coloured when originally described, was commonly found in a black or dark variant, and he deduced that this was because dark forms survived better in areas with industrial pollution, being less conspicuous to predatory birds. He tested his idea by releasing equal numbers of normal and black moths into two woods, one unpolluted and the other heavily soot-contaminated. After a time, he recaptured as many moths as possible, and found that, as predicted, many more normal moths survive in the unpollute wood. Further study revealed that the melanism is caused by two mutant alleles at different loci, both dominant, one giving a darker colour than the other.

Melting profile, a graph that shows how much DNA in a sample melts over time in an increasing temperature. DNA melts in the range 60-80°C(140-175°F). DNA with a preponderance of A-T base-pairs melts at lower temperaturs than DNA with more G-Cs, so the melting profile gives a crude indication of the composition of the sample.

Meme, a unit of cultural transmission (see exogenetic inheritance). The idea of a cultural unit of inheritance comparable to the physical unit (i.e. the gene) was put forward by the British zoologist Richard Dawkins in The Selfish Gene (1976). Its name is derived from the Greek mimesis, 'imitation', because the way in which a meme is transmitted from one person or group to another is by imitation. Dawkins' own examples of memes are 'tunes, ideas, catch-phrases, clothrs, fashions, ways of making pots or of building arches'. He believes that memes are like genes in being able to replicate themselves, and in doing so for their own advantage rather than for the advantage of the individual carrying the meme. As an example of this, he cites the meme for belief in God, which propagates itself extremely readily, finding human culture a very comfortable environment. Humans are not the only species to have memes: two well-known examples from the animal kingdom are tits breaking milk-bottle tops and monkeys washing potatoes in the sea. The idea of memes is unacceptable to many geneticists and other scientists.

Mendel, Gregor (1822-84), Czechoslovakian (then Austrian) monk and botanist; discoverer of the laws of heredity. Mendel, a peasant's son, was interested in science from an early age and studied it at Olomouc before he was 21. His abbot sent him to the University of Vienna for two more years of scientific study (physics, chemistry, mathematics, zoology and botany), and on his return to the monastery in 1854, he taught mathematics and natural science at the local high school.

Mendel began his experiments in plant hybridization in 1856, and knew from the start that statistical results would be the key. He crossed over 20,000 plants of the edible pea and recorded the results. (Fisher noted that Mendel's data are just a little too good to be true; it seems possible that if in doubt he put a plant into the class that best fitted his theory.) Mendel revealed these results and propounded the simple two laws that bear his name _ the law of >segregation (Mendel's First Law) - and the law of >independent assortment (Mendel's Second Law) - and which are the foundation of all biology, in a paper which he read to the Natural History Society of Brno in 1865. It caused no stir either at the meeting or when it was published in the Society's journal the following year. Mendel was disappointed, but hoped that further research would bring more evidence so that his theory would be accepted. He experimented on other plants for the next two years until he was elected abbot of his monastery in 1868. After that his administrative duties took up all his time, and he died a respected abbot but an unrecognized scientist. Sixteen years later (1900), his name became known to the world when his theory was independently arrived at by three scientists - de >Vries, >Correns and >Tschermak - all of whom Mendel's paper of 1866 and publicized his findings.

Mendelize, a useful verb which according to the Oxford English Dictionary means 'to exhibit or transmit parental characters in conformity with Mendel's law of inheritance; to behave as Mendel's law predicts'. Apparently first used in 1906, it has now totally disappeared, which is a pity.

Menstruation, a physiological arrangement unique to human beings and the great apes. In all other animals there is a periodic breeding cycle controlled by hormones, so that there is a regulated time during which sexual activity occurs, the female body only being fertile and in the right physiological state to begin and maintain pregnancy at that time (>oestrus in mammals is an example). The system of menstruation is quite different, in that it enables females to be sexually receptive all the time.

The change from oestrus to menstruation is probably a very significant evolutionary change, enabling different social structures to emerge in the higher primates. The British zoologist Sir Gavin de Beer considered that it was a necessary precondition for the evolution of >monogamy. (De Beer is almost the only biologist to have paid any attention to menstruation. It appears to have been entirely ignored by sociobiologists, even though it is a clear case of a physiological system having behavioural consequences, just the kind of topic that sociobiology deals with.)

Mental illness is distinguished from mental handicap in that the latter involves retarded development or brain damage whereas the former involves inappropriate reactions on the part of a person with a normal mental development. Degenerative diseases, e.g. >Alzhernier's, are usually included under mental illness.

Certain forms of mental illness have a genetic component in their causation, though it is never as straightforward as a gene 'for' disease X - there are always environmental factors too. See manic-depression; schizophrenia.

Meristic variation, variation in the phenotype that goes by step, e.g. the number of fingers. Such variation may have the same sort of underlying genetic cause as >continuous variation, or there may be a single gene responsible (e.g. a gene in mice that reduces the number of toes on the front feet from four to three). Meristic variation includes such traits as the number of puppies in a litter.

Meselson, Matthew Stanley (1930-), American molecular biologist; professor of natural sciences at Harvard University since 1976. In 1957 Meselson and his colleague F.W.Stahl were the first to confirm Crick and Watson's suggestion that when DNA replicates it does so by unwinding down the middle so that each of the two new molecules contains one strand of the original and one new one (i.e. semi-conservative replication). Their technique involved the use of radioactively labelled samples of DNA in >*Escherichia coli.*

Messenger RNA (mRNA), the molecule that transcribes the genetic message from DNA at the start of >protein synthesis. In prokaryotes, mRNA reads directly off the DNA and had its front end already bound to a ribosome making the protein while, further down, the molecule transcription is still taking place. In eukaryotes, there is a precursor of mRNA which is called heterogeneous nuclear RNA, and it is this that actually reads the DNA. Then all the sequences on it that represent introns (non-informational DNA) are cut out, the result being an mRNA molecule containing only sequences that represent exons (coding sequences). Eukaryotic mRNA is synthesized in the cell nucleus, where it reads the DNA, but it must travel out of the nucleusfor the protein to be assembled on the ribosomes. After protein synthesis, the mRNA molecule may be reused for another translation or its constituent molecules may migrate back into the nucleus to be incorporated into another molecules may migrate back into the nucleus to be incorporated into another molecules of mRNA. The front (5') end of an mRNA molecule has a >cap, and the 3'end has a poly-A tail.

Metacentric, a chromosome having its >centromere at or near the middle.

Metafemale, a sterile female ***Drosophila***, which has two sets of autosomes and three X chromosomes. It provides evidence for the role of the balance between autosomes and sex chromosomes in >sex determination.

Metamale, a sterile male **Drosophila**, which has two sets of autosomes and three X chromosome. It provides evidence for the role of theY chromosome in >sex determination.

Metaphase, one of the phases of the two types of cell division, >meiosis and >mitosis. See diagram at meiosis.

Microevolution, evolution viewed over a relatively short time-span, and involving changes of gene frequency within populations. It can also concern the process of speciation, but not the development of higher taxonomic groups (see macroevolution). For an example of a microevolutionary study, see melanism, industrial.

Migration, movement of populations, groups or individuals. In genetic terms, migration inwards (immigration) is an important factor in maintaining genetic variability within a population's gene pool. Outwards migration (emigration) is a mechanism of population control, and can lead to the foundation of new populations.

Milstein, Cesar, (1927-), Argentinian-born British molecular biologist. Milstein taught chemistry in Buenos Aires before joining the Medical Research Council's Molecular Biology Unit at Cambridge, England, in 1963. His work on immunoglobulins (with Fred >Sanger) led to the discovery in 1975 of monoclonal antibodies, the single pure antibodies produced by a clone of >B lymphocytes (see immune system). Milstein shared the 1984 Nobel prize for physiology or medicine with his colleague George Koehler and the theoretician Niels Jerne who, following Milstein's results, developed the clonal selection theory of immunity.

Mimicry, the process which one species evolves so as to resemble another. It is found especially in butterflies that are preyed upon by birds. Many such species have evolved unpalatability to deter predators, but this system costs a certain loss of individuals as each new generation of predators has to learn the hard way (i.e. by attempting to eat one) that the members of a particular species are inedible. Mimicry can make unpalatability more efficient, and this can be done in one of two ways. In Mullerian mimicry, two or more unpalatable species come to look like each other, the advantage to each species being that a proportion of the sacrifice involved in educating, say, young birds is shared with the other species. The more alike the species look, and the more uniform the individuals within each species, the better it works. Conversely, Batesian mimicry is the system in which a species look, and the more uniform the individuals within each species, the better it works. Conversely, Batesian mimicry is the system in which a species that is edible comes to resemble one that is unpalatable. This only works if the mimetic (mimicking) species is substantially less common than the unpalatable one; otherwise the predators would not learn that butterflies with that pattern are inedible. Some species that practise Batesian mimicry have several different mimetic forms, each an imitation of a different unpalatable species, so as to keep the numbers of mimics suitably low. The genetic control of mimetic forms is often by means of a >supergene at a single locus, sometimes with modifier alleles at other loci.

Mismatch repair, enzymatic removal from a DNA molecule of a base that is an incorrect match with the one opposite, and therefore is not properly bonded. It is replaced with the right one. See also >DNA repair.

Missensemutation, a mutation in which a codon that codes for one particular amino acid is changed to a codon that codes for another, with a consequent change in the chemistry of the gene product. See genetic code.

Mitochondria (sing. mitochondrion), >organelles in the cytoplasm of >eukaryote cells that are concerned with respiration (generation of chemical energy by oxidation). There are on average a few hundred mitochondria in each cell. They carry their own DNA, which can become the vehicle of >cytoplasmic inheritance. The base sequence of human mitochondrial DNA has been deciphered in its entirety.

Mitosis, the process of ordinary cell division. The object is to produce two daughter cells each with exactly the same chromosome complement as the original cell (in contrast to >meiosis, the other type of cell division, which produces >gametes). The chromosomes behave in the following way:

INTERPHASE This is the non-division part of the cell cycle, but towards the end of it the chromosomes, still in elongated shape, duplicate themselves lengthwise into two >chromatids (S phase; see interphase).

PROPHASE The chromosomes begin to condense themselves by coiling up tightly. The nuclear membrane disappears. The fibrous structure known as the >spindle is formed, and each chromosome attaches itself to a spindle fibre by its >centromere.

METAPHASE All the chromosomes move towards the equator of the spindle. This is the time at which the chromosomes are at their most condensed and easiest to observe.

ANAPHASE The centromeres split, allowing the two sister chromatids to move to opposite poles of the spindle. Each chromatid will consitute a chromosome in one of the new cells.

TELOPHASE. Nuclear membranes form round each of the two groups of chromosomes. The spindle disappears. The chromosomes elongate, and get on with whatever task of protein synthesis is required in the new cells.

MLD, 'mean lethal dose', the amount of whole-body radiation that, given in a single dose, will kill 50 per cent of the individuals receiving it. For humans, the MLD is believed to be about 450 rem.

MN blood group system, a series of genetically determined antigens found on human red blood cells. The second blood group system to be discovered, it was found in 1927 by the blood-group pioneer Landsteiner and his colleague Levine, who had discovered the ABO system in 1900. The M and N antigens are so alike that each is recognized as 'self by the body's immune system whether it is the individual's own type or not. It therefore is not necessary for blood transfsions to match the MN type of donor and recipient. The genes for M and N are about equally common in Europe, so that the population consists of approximately 25 per cent group M (homozygotes), 50 per cent group MN (heterozygotes) an Australian Aborigines have frequencies of M of 75-95 per cent.

Mode, the value in a distribution that occurs most often. See also mean; median; normal distribution.

Molecular clock. See molecular evolution.

Molecular evolution, evolutionary change of particular molecules. There are many proteins that occur in a similar form in many different living things, and the evolutionary path of such a molecule from its early form to its later ones can be traced. In cases where the DNA itself has been sequenced for a gene that occurs in a wide range of creatures, the number of mutations (base substitutions, deletions, insertions) can be directly compared. If only the protein sequence is available, the number of mutations can be deduced, allowing for the fact that a proportion of mutations are silent (i.e. the same amino acid is coded for even after a base change). One must also allow for the fact sometimes a second mutation happens on the site of an earlier one.

The 'molecular clock' model proposes that mutations happen to a molecule at a particular rate per generation, and that by comparing the number of mutational changes between two species, one can deduce the number of generation (hence years) since they were the same. A study of the alpha and beta globins found that molecular clock time agreed well with geological time in the divergene of major evolutionary groups (fourlegged creatures from fish, terrestrial creatures from amphibians, warmblooded animals from reptiles and so on). Different proteins have molecular clocks running at different rates, and some parts of the molecule will be more liable to change than others (with some sequences completely invariant). For example, cytochrome c and histone H4 are extremely conservative, there having been only four mutational changes in the latter between flowering plants and mammals. On the other hand, very rapid change occurs in the satellite DNA sequences in *Drosophila*, enabling the evolutionary divergence of the hundreds of species of this fruit fly to be charted over a time-span as short as one million years.

Monera, one of the five kingdoms of living things (see classification). It includes all the >prokaryotes.

Monoclonal antibodies, a synthetic preparation of pure antibodies. The technique of producing monoclonal antibodies was devised by the British molecular biologist Cesar >Milstein in 1975. Given that each line of >B lymphocytes only produces one type of antibody, if a particular antibody is wanted the procedure is to inject a mouse with the appropriate antigen, and wait until the clone of lymph cells in its spleen begins to produce the antibody. Some of these cells are extracted, and instead of simply being cultured (lymphocytes will not grow indefinitely in vitro), they are fused with myeloma (cancerous) lymphocyte cells that proliferate speedily. These hybrid cells (known as hybridomas) combine the useful properties of both parental cells - an ability to produce the required antibody, immortality and rapid proliferation. Antibodies can be produced in quantity by this technique, for use in tissue typing and blood typing before transplants and transfusions. Monoclonal antibody is also used in the treatment of certain cancers, and to prevent haemolytic disease of the newborn of Rhesus negative mothers.

Monod, Jacques Lucien (1910-76), French molecular biologist; coproposer of the theory of the operon. Monod was trained as a zoologist, and worked at the pasteur Institute in Paris. Immediately after the discovery of >DNA's structure, he and his colleague Francois >Jacob were stimulated to look for the mechanism that controls whether a length of DNA - a gene - is active or not. In 1961 they proposed the mechanism that they called the operon, and also predicted the existence of the RNA molecule now known as messenger RNA. For this work, Monod and Jacob shared the 1965 Nobel prize for physiology or medicine with Andre Lwoff (for his work on >bacteriophages). Monod also wrote on the philosophy of evolution, proposing in his Chance and Necessity (1970) that the origin of life was a matter of chance but that its subsequent evolutionary path has been a matter of necessity.

Monoecious, adjective used to describe a plant species which has separate male and female flowers borne on the same plant. See also dioecious; flowering plants.

Monogamy, the mating system in which one male and one female pair exclusively, either for a seasion or for life. It is rare, being found in only 3-5 per cent of mammal species and a few fish, but it is the usual state for birds, with about 90 per cent of bird species being monogamous; it is hardly ever found in other vertebrates or in invertebrates.

In most species, the male's best chance of increasing the number of his genes that are passed on to the next generation is by mating with as many females as possible. But in species in which the young need more care than can be provided by one parent, it pays the male to stay with one female and raise a brood of young successfully rather than to sire a number of broods that will die. Typically, fledgling birds need one parent to keep them warm while the other forages for food, and the male and female have a similar anatomy that makes them able to share these tasks. Another cause of monogamy among bird and other species is a habitat in which only a low population density is possible, in which case it is not feasible for a male to control more than one female as they would be too widely scattered.

In most mammalian species, care of the young can be done by the mother alone, as she feeds them with her milk. The monogamous mammals are mostly canids (members of the dog family) having large litters which require the mother to be dependent for a long time on food brought to her by a mate. The young still have to have food brought to them by their parents even after they are weaned, and canids have evolved the ability to swallow food to carry it back to the lair and regurgitate it for the young. Among the higher primates, the gorilla and the chimpanzee are not monogamous; the orang-utan and the gibbon are, probably by reason of population density. As for humans, some societies are monogamous but more are not.

Monogenic, adjective describing a trait that is controlled by a single gene. This would include most of the human clinical syndromes.

Monohybrid, an individual that is heterozygous at a single locus. The term is also used for a cross involving two different alleles at a single locus, equivalent to an >F1.

Monomorphic locus, a locus at which the most common allele has a frequency of 95 per cent or greater. (Strictly it should mean a locus at always occur even at a low rate, thus providing alternative alleles.) Studies of natural populations of Drosophila indicate that as many as 90 per cent of loci are monomorphic, and this is probably typical. See also polymorphism.

Monophyletic group, a group of species (or higher taxa) descended from a common ancestor. Humans and the anthropoid apes are a monophyletic group, all being descended from a common ape-like stock. See also polyphyletic group.

Monosomy, the condition of having only one representative of a pair of chromosomes. Monosomy can cause abnormalities, of which the best known it Turner's syndrome, monosomy for chromosome X in humans.

Monozygotic twins, identical twins, those who are genetically identical because they come from a single fertilized egg which had divided in two. It is interesting that monozygotic twins are happier about being twins than dizygotic (fraternal) twins are, with one poll revealing that 78 per cent of the former are pleased to be twins as against 42 per cent of the latter. However, the idea of an unfailing mystical bond between identical twins is not borne out by the fact that 10 per cent of them dislike being twins (the data from which these figurs come do not say whether these 10 per cent of twins are from the same pairs.) Monozygotic twins tend to 'mirror' each other; for example if one twin is right-handed the other is left-handed. Monozygotic twins are much studied by geneticists, as they offer the most convenient way of estimating the heritability of a trait. (see twin studies). See also polyembryony, twinning rate.

morgan. See centimorgan.

Morgan, Thomas Hunt (1866-1945), American geneticist. Following William Sutton's suggestion, in 1902, that chromosomes might be something to do with Mendel's 'factors', Morgan experimented on ***Drosophila melanogaster*** and found that the 'factors' are indeed located on the chromosomes and that they are arranged linearly. Morgan's 'fly room' at the California Institute of Genetics became a genetical kitchen unequalled until the great days of the Medical Research Council's Molecular Biology Unit at Cambridge. In 1911 Morgan and his colleagues (including A.H. >Sturtevant) produced the first-ever genetic map, showing the location of five X-linked genes in **Drosophila**. Morgan was a skilled writer about genetics for the lay reader, and in his book The Scientific Basis of Evolution (1932) he wrote about the unique capacity of humans to evolve by non-biological means; his ideas are very close to those now proposed by Richard Dawkins (see meme). Morgan was the first geneticist to win the Nobel prize for Physiology or Medicine (1933).

Morula (from the Latin morus, 'mulberry'), the mammalin embryo when it consists of a ball of undifferentiated cells. See embryogenesis.

Mosaic, an organism whose body contains cells of two (or more) different genetic constitutions, though all of them derive from the same gamete. This could happen as the result of a mutation in a somatic cell, so that some cells have the normal genotype but the cells in the line derived from the one with the mutation are different. A particular sort of mosaic occurs in female mammals, as the result of random inactivation of one of the two X chromosomes (see Lyon hypothesis). See also chimaera.

Mouse. See *Mus musculus*.

mRNA. See messenger RNA.

Mule, the offspring of a male ass or donkey and a female horse or pony. The reverse is a >hinny, though it is often referred to as a mule, there being no significant physical difference between them. The diploid chromosome number in the ass or donkey (Equus asinus) it is 64, so the mule has 63 chromosomes, and is infertile.

Muller, Hermann Joseph (1890-1967), American geneticist. Muller worked in T.H. >Morgan's 'fly room', concentrating from the start on mutation. In 1920 he expounded the theory that, as only genes can reproduce themselves, all other components, even down to the fine structure of cells, must ultimately be produced by the genes. From this he concluded that life itself must have started with simple self-replicating molecules, which he termed 'naked genes'. As good an experimentalist as he was a theorist, Muller discovered in 1926 that X-rays induced a high rate of mutation, and he suggested (correctly) that a mutation is a chemical event within a gene. Muller was concerned with eugenics, and was an advocate of sperm banks stocked with contributions from distinguished men. Muller won the 1946 Nobel prize for Physiology or Medicine.

Multigene family, a set of similar genes that code for different parts of a particular function. They presumably arise by duplication and then diverge slightly so that they each contribute their different products. An example in humans is the group of genes coding for the different chains of the haemoglobin protein.

Multiple alleles, the situation in which a population has more than two alleles regularly occuring at a given locus. See polymorphism.

Multiple births, in humans, twins (see dizygotic twins; monozygotic twins; twinning rates), >triplets, >quadruplets, quintuplets, and so on. These occur naturally (there being some genetic element in the tendency to have multiple births), but have increased ten-fold in Western countries as a result of treatment with a fertility drug which causes more than one egg to ripen in each monthly cycle. It is given to women whose ovulation is poor, and often results in several eggs being fertilized at once.

More importantly, it is given to women as part of in vitro fertilization treatment, so that several embryos at once can be implanted to increase the chance of pregnancy. The drawback is the multiple babies are bound to be born premature and to have a poor chance of survival. One way round this is to selectively abort surplus foetuses once the pregnancy is established, but this is very risky and can result in the loss of all the foetuses; it is in any case stressful for the mother. A better solution is never to implant more than three or four embryos, a number that can have some hope of survival. Even so, there has been criticism that the multiple babies produced artificially by in vitro fertilization treatment are taking up scarce resources in neonatal intensive-care units that are needed for naturally occurring premature babies.

Multiple sclerosis, a disease in which the nervous system degenerates, causing loss of co-ordination, paralysis and early death. Its causes are not known, but there is some genetic predisposition in that the majority of sufferers are of HLA type DR2. Environmental factors are more important, however, as shown by the very different frequency in white populations (of similar European ancestry) in different parts of the world: the disease has its highest incidence in northern Europe, especially in Britain and Scandinavia, but is uncommon in North America and rare in Australia and New Zealand.

Multistate character, a genetically determined trait that can exist in three or more discrete states. An example is the >ABO blood group system. See also binary character.

Mus musculus, the house mouse; the mammal most often used in genetical research in the laboratory. Mice are found all over the world, both in the wild and in populations supported by humans, in which case they are a serious economic pest. A mouse is about 15 cm long (including the tail), weighs 25-30 g, and is omnivorous. Gestation time is 19-20 days, and the average litter is 4-6 pups (range 1-14), which mature within about two coloured, but different colours have been bred since ancient times in China and Japan and are found in some laboratory strains (with still more varieties among the mice bred as a hobby). There are over 50 main strains of mouse whose records go back to the 1920s (a proof that inbreeding is not necessarily bad). The mouse's diploid chromosome number is 40. Many mutant genes have been identified in mice, including a number that cause the same diseases as in humans, e.g. leukaemia and diabetes. See also nude mice; tail-less; waltzing mice.

Muscular dystrophy. See Duchenne muscular dystrophy.

Mutagen, any substance that raises the frequency with which >mutations occur. It may be a form of radiation, e.g. ultraviolet light or >X-rays (whose mutagenic effect was discovered in 1926 by H.J. >Muller), or a chemical, e.g.mustard gas.

Mutation, any change occuring in the genetic message that an >allele carries. The first theoretician of mutation was H.J. >Muller, who saw that the three important factors are: (1) that Darwinian evolution requires a constant supply of variation, much of it supplied by mutation; (2) that mutations are extremely rare; and (3) that mutations are almost always deleterious. It was later proved, as evolutionist had supposed, that 'mutations do not arise in response to an organism's needs not do they, except by accident, gratify them' (the discovery was Luria and Delbruck's, the words are Sir Peter Medawar's). A mutation is likely to be deleterious because it is a random change to a message - and therefore a molecule - the already works well. There are three general types of mutation: those involving a substitution at a single base-pair (see point mutation; missense mutation; nonsense mutation); those involving the deletion or insertion of one or a few base-pairs (usually resulting in a >frameshift); and large-scale rearrangments of chromosomal material. A mutation that is a change from the normal allele to a new form (which is the case in the vast majority of mutations) is called a >forward mutation; these are occasionally reversed by >back mutations. See also mutation rate; neutral allele.

Mutation load, the total genetic burden carried by a population that results form accumulated deleterious mutations.

Mutation pressure, the effect of recurrent mutation in changing the gene frequency at a given locus. This pressure is usually very weak compared to selection pressure or the effect of genetic drift.

Mutation rate, the rate at which mutational changes in DNA occur. It is expressed as the probability that an allele will undergo mutation in a single generation, or in each gamete formed (both measures are used). It was first studied by the American geneticist H.J. >Muller, who realized that the rate of mutation in nature is very low, and discovered that it could be enormously speeded up by doses of X-rays and other mutagens. Different genes have different mutation rates: the range in *Drosophila* is from 10-5 to 10-6 per gene per generation, or 10-8 to 10-9 per base-pair per generation. Similar rates are found in humans, whose mutation rate in respect of single-gene medical disorders can be estimated (for example, >achondroplasia arises at a rate of 1.3 X 10-5 per generation). There are some mutations that themselves influence the rate of mutation of other genes around them, and it is likely that mutation rates are themselves the result of genetic control and have evolved adaptively.

Myxoviruses, a group of viruses that infect vertebrates. The singlestranded DNA genome is encased in a spiral protein coat. Among the diseases caused by myxoviruses are influenza, mumps, measles, rubella (German measles) and rabies. (Myxomatosis, the disease that is lethal to rabbits, is caused by one of the >poxviruses.) These viruses tend to undergo frequent genetic change, so that vaccines that worked against previous versions of the virus are ineffective against a new strain.

N

Nail-patella syndrome, a condition in which the nails of fingers and toes are deformed and the patellae (knee-caps) are very small; there is no serious effect on health. The syndrome is caused by a single dominant gene, but the biochemical mechanism is unknown. The interest of the condition for geneticists is that the nail-patella locus is close to that of the ABO blood group system, and some of the earliest studies on human autosomal linkage were carried out using these two sets of genes.

Natural immunity, the system of immunity that is non-specific, i.e. does not rely upon recognition of antigens. This system works by (a) production of the anti-viral interferon proteins, (b) lysozyme activity to attack bacteria, (c) ingestion of particles, and (d) the action of various proteins to agglutinate foreign matter (i.e. to make it stick together in clumps). This system is the only one used by invertebrate animals; vertebrates have evolved a much more elaborate >immune system but still use natural immunity alongside.

Natural selection, the mechanism by which evolution happens. Natural selection is simply the process which ensures that some genes are relatively more numerous in the next generation than in the present one, because of the reproductive success of the individuals carrying these genes. Spencer's catch-phrase 'survival of the fittest', used to describe the action of natural selection, is misleading in two ways: firstly, 'fittest' means 'best adapted' not 'in the best condition'; and secondly, survival in itself is not necessarily important. The crucial factor is differential fertility: two organisms might live the same length of time (i.e. both are survivors), but if one raises a hundred offspring and the other only eighty the former has been favoured by natural selection. It is also important only to consider successful reproduction, i.e. the production of offspring who themselves live to reproduce. There is no profit, evolutionarily speaking, in putting one's genes into large numbers of offspring who do not pass them on.

The term 'natural selection' has, as Darwin was aware, an unfortunate implication that nature makes choices; this is clearly not so, 'natural selection' being simply a description of what happens to gene frequencies. Natural selection is also not purposive: the outcome may be either 'good' or 'bad' for the species involved in the long run; all that is being selected for is adaptedness (as measured by reproductive success) in the present environment. (See albinism for an interesting example.)

Different types of natural selection have been identified: (a) acting in different directions (see directional selection; disruptive selection; stabilizing selection); disruptive selection; stabilizing selection); (b) favouring different reproductive strategies (see K selection; r selection); (c) acting other than upon individuals (see group selection; kin selection). See also sexual selection.

Nature versus nurture, the never-ending debate on whether >genotype is more important than environment in human development and, in particular, in determining intelligence. The term was coined by Galton, who took a great interest in the matter and wrote several books on it, one entitled English Men of Science: There Nature and Nurture. It is naive to think of any trait as having a fixed proportion contributed by nature, the rest being contributed by nurture; see the examples under >heritability. Rather, it is an axiom of genetics that all variation is the result of interaction between genotype and environment (this is not to say that one or the other may not in fact contribute 100 per cent in some circumstances). None the less, the political implications of the nature-versusnature debate in terms of the development of inellignence are so importive of the environmentalist or the hereditarian position; but Galtn was also the first to realize that the study of monozygotic (identical) twins could be informative (see twin studies). See IQ for more about that issue, and see >adoption for the evidence from adopted children. Nature versus nurture has also been debated in connection with other continuously varying human traits: temperament, behaviour (notoriously, criminality) and susceptibility to certain diseases (see obesity).

• **Lancelot Hogben**, Science for the Citizen (1938; unfortunately out of print). Contains an excellent discussion of nature versus nurture, which although written more than 50 years ago says all that needs to be said.

Neanderthal Man. See human evolution.

Negative control, the situation in which a gene is not transcribed until the repressor molecule is dissociated from the operator region of the >operon (for inducible enzymes). In the case of negative control of repressible enzymes, the gene is normally transcribed until repressed. The opposite is >positive control.

Neo-Darwinism, the synthesis of Darwinian evolution and Mendelian inheritance, as put together in the 1920s and 1930s by the great geneticists >Fisher, Sewall >Wright, and others (see pp. 26-7). Darwin had not understood the mechanism of heredity, and his theory was weakened by lack of it; Mendel had not known about evolution by natural selection. Neo-Darwinism unites the two, and is the basis of all subsequent thinking in biology. It is not to be confused with >social Darwinism.

Neoteny, the retention in adult life of structures or stages of development that were appropriate for an immature stage in the ancestral form. A striking example is the axolotl, a Mexican salamander, which reaches sexual maturity while still in the larval state. See also paedomorphosis.

Neurofibromatosis, an inherited disease in which there are tumours in the nerve cells, first in the peripheral nerves and later in the spine and brain. Onset is at about the age of 20, and death at about 50. The disease is caused by a dominant gene, and occurs in about 1 per 2500 babies in European populations.

Neurospora crassa, a fungus, known as the pink bread mould, and often used in genetic experiments. For much of its life cycle, *Neurospora* grows as filaments called >hyphae, which are of two 'mating types'. To reproduce, the hyphae (which are haploid) produce >conidia (also haploid), which are spores that fertilize the opposite mating type. The zygote so produced is diploid and undergoes mitosis to produce four new haploid cells; each of these then undergoes mitosis, and the resulting eight haploid cells (called ascospores) are arranged in a row in the ascus, like peas in a pod. They can be removed and each cultured separately to be analysed genetically. This gives a very convenient way of studying cross-over frequency. As *Neurospora* is haploid in its hyphae state, all recessive alleles are expressed; this fact helped George >Beadle and E.L. >Tatum to establish that a single mutant causes the loss of a single enzyme.

Neutral allele (or neutral mutation), an allele that is a variant from the normal but does not give any selective advantage or disadvantage to the organism that carries it. According to some theoreticians, such alleles are numerous and can spread widely within a population, as there is no selection against them; others believe that it is hardly possible for a mutation to be truly neutral in selective terms. See polymorphism.

Neutralists. See genetic drift.

Nexus hypothesis, the idea that every phenotypic trait is probably influenced by more than one gene, and that each gene has an influence on more than one phenotypic trait. It is not, as appears at first sight, a contradiction of the axiom of 'one gene, one enzyme'.

Nick, a cut in a single strand of a double-stranded DNA molecule. See also restriction endonuclease.

Nilsson-Ehle, Herman (1873-1949), Swedish geneticist. A farmer's son, Nilsson-Ehle was (as Mendel had been) interested in applying science to improving crops. Realizing the significance of Mendel's work, he proved that the economically important traits in plants - size, earliness, resistance to disease - were Mendelian, even though they were continuously varying where Mendel's traits had been discrete. His papers on this (1908-11) are classics of genetics.

Nirenberg, Marshall Warren (1927-), American molecular biologist; has worked for the US National Institutes of Health since 1957. Nirenberg was the first person to decipher a 'word' of the >genetic code. In 1961, he synthesized a molecule of messenger RNA consisting of the base U (uracil) repeated indefinitely. The resulting protein consisted of nothing but the amino acid phenylalanine over and over again, so he could deduce that the codon UUU codes for phenylalanine. For this, Nirenberg shared the 1968 Nobel prize for physiology or medicine with Har Gobind >Khorana and Robert W. Holley.

Non-disjunction, the failre of homologous chromosomes to separate during >meiosis, or of sister chromatids to separate during >mitosis. The result is that the descendant cells after this event have an uneven amount of genetic material in them: too much in one line, too little in the other. Non-disjunction can give rise to >trisomy, the condition in which there are there chromosomes in a 'pair' instead of two, as in Down's syndrome.

Non-homologous chromosomes, any chromosomes that do not have an exactly matching pair. The main example is the sex chromosomes, in which the X and Y are non-homologous.

Non-identical twins. See dizygotic twins.

Nonsense codon (also known as stop codon), one of the three codons that specify the end of a genetic message. These codons do not code for any amino acid but signal the termination of the protein being synthesized. They are UAG ('amber'), UAA ('ochre') and UGA ('opal'). See genetic code; protein synthesis.

Nonsense mutation, a mutation in which a codon that codes for an amino acid is changed to a codon that codes for'stop' (see nonsense codon). The result is a shortened protein, how much shorter depending on how far along the gene the nonsense mutation occurs. See genetic code; protein synthesis.

Normal distribution, or Gaussian distribution, the statistical distribution in which the frequency of occurrences of values decreases with their distance from the mean. When plotted as a graph, a normal distribution looks symmetrically bell-shaped. The mean (the average of all the individual values), the mode (the value that occurs most often) and the median (the half-way point between the highest and the lowest values) all coincide. The >standard deviation is a measure of how steep-sided the 'bell' is. In biological data, many traits that show continuous variation conform to a normal distribution. See also bimodal distribution.

Northern blotting, a method of identifying sequences of RNA, by a technique similar to >Southern blotting (which is used for DNA). Whereas Southern blotting was called after its inventor, Northern blotting got its name by analogy.

Nucleic acid, one of the polymeric molecules consisting of nitrogenous bases connected to a sugar-phosphate structure. They can be either double- or single-stranded. The ones that occur in nature are >DNA and >RNA, and these can be artificially joined by hybridization (see Hybrid DNA.).

Nucleolus, a spherical organelle within the cell nucleus where >ribosomes are made. There are about five or six in each somatic cell nucleus.

Nucleoside, a compound consisting of a nitrogenous base attached to a sugar. See also nucleotide.

Nucleosome, the structural unit from which chromosomes are made up. Each nucleosome consists of a block of eight molecules of the protein >histone with a length of about 145 base-pairs of DNA wound round it, plus a linker unit of DNA of a length of about 55 base-pairs, and a single histone molecule to join it to the next nuclesome.

Nucleotide, a >nucleoside (a compound consisting of a base joined to a sugar) with a phosphate group attached. Nucleotides are the units of which the nucleic acids DNA and RNA are composed.

Nucleus, the roughly spherical structure in a eukaryote cell that contains the chromosomes. It is also the place where messenge RNA is made (by transcription of the DNA in the chromosomes); these RNA molecules must pass out through the nuclear membrane to the cell's cytoplasm where the ribosomes carry out protein synthesis.

Nude mice, mice that are homozygous for the allele nu. These mice have no thymus gland, a crucial organ in the immune system as it produces T lymphocytes, the cells that recognize foreign particles or tissue. These mice are therefore very susceptible to any kind of infection, but they will accept grafts or transplants from any donor even if genetically completely unrelated. Thus they are often used in experiments concerned with transplant rejection. The fact that they are hairless is a side-effect of the same gene (see pleiotropy).

Null hypothesis, the assumption that the effect being statistically investigated does not exist. For example, if one has two samples with different means for a particular trait, the null hypothesis is that they actually come from the same population. Only if the means can be shown to be differnt at a level that is statistically significant can the null hypothesis be rejected and the samples accepted as (probably) coming from different population.

Nullisomy, the condition of having both members of a pair of chromosomes missing. Such a condition would almost always be lethal, as none of the genes located on the chromosome in question would be present.

O

Obesity, Excess of body fat. How much fat constitutes an 'excess' has not been scientifically determined though there are several rule-0f-thumb ways of estimating it. One is the Broca index, which says that the 'ideal' weight in kilograms is the person's height in centimetres minus 100 minus a further 10 percent for males or 15 per cent for females. Another is W/H2, where W is the person's weight in kilograms and H is their height in metres, the ideal ratio being 25.0 for males, 24.2 for females. (These two methods do not give the same ideal weight, as readers can describe anyone between 10 and 19 percent over the ideal weight as overweight, and anyone more than 20 per cent over as obese.

Fat is laid down when the food intake is surplus to the body's energy requirements; in practice, this is almost always because of eating too much rather than exercising too little. Obesity presents a >nature versus nurture problem almost as great as that of inheritance of >IQ. Fatness does indeed run in families, but is it in the genes or the lifestyle? Most fat people have at least one fat parent, but the fact that this may equally often be the father or the mother points towards a genetic element. It has been observed that people with a flattened-front-to-back frame, the frame shape being presumed to be predominantly inherited. It has also been found that, while mice of most strains become obese when fed on a high-fat diet, there are two strains that do not. None the less the outcome of studies on twins and on adopted children (see twin studies; adoption) has shown that the heritability of obesity is rather low. In a tiny minority of cases, obesity is due to a specific physiological defect, possibly genetic in origin. This is the case with the Lawrence Moon-Biedl syndrome in humans, where there is a defect in the hypothalamus (a Part of the brain concerned with regulating body functions), and with three other genetic syndromes. There are three different >mutations in laboratory mice that cause obesity, each via a different > inborn error of metabolism, it is likely that each of these also exists in humans.

Modern medical opinion is that it is not sensible to treat patients for obesity if the distress caused to them by dieting is disproportionate compared to the risk to their health of staying fat; this is particularly the case with late-onset (40-ish) obesity, in which the extra risk of, say, heart disease is in fact small.

Objectivity. See Quantification.

Ochre Codon, One of the three >nonsense codons in DNA that code for 'stop' and therefore signal the termination of the protein being synthesized. The ochre codon is UAA.

Oenothera, the evening primrose, a plant much studied in genetics. It has an interesting arrangement of chromosomes, which are joined end to end in one large ring rather than lying in pairs. It is no relation of the British wild primrose (***Primula vulgaris***).

Ohno's rule, a theory that, because the X chromosome has intrinsic properties of gene regulation, certain genes are always situated on it, whatever the species. An interesting example of this is the fact that the mutation that causes haemophilia is found in dogs as well as in humans, and is on the X chromosome in dogs as it is in humans. It is called after the American genticist Susumu Ohno.

Okazaki fragments, short sequences of DNA that are formed on the 'back ward-facing' strand during >replication and are joined up by ligation. They are called after their discoverer, the Japanese geneticist Reiji Okazaki.

Oligogene, a gene with a major effect on the phenotype. Many are genes that control some part of the developmental process, and are sometimes referred to as ' switch genes'.

Oligonucleotide, a sequence of >nucleotides (as in DNA or RNA) but containing only a few nucleotides. Such a sequence is more usually synthetic than naturally occuring.

Oncogene, a gene that can cause a cell to become a tumour cell. originally identified in >retroviruses, oncogenes have been recognized as part of the normal genotype of most cells. They are probably concerned with the control cell growth and division. The oncogenes found in cells are known as proto-oncogenes, to indicate the fact that they are only potentially, not actually, oncogenic (i.e. cancer-causing). They become cancerous when something interferes their normal regulation system. Viral oncogenes are proto-oncogenes which during the evolution of the virus have been picked up from the genomes of host cells and have become changed so that can cause cancer. See cancer and.

Ontogeny (from the Greek ont, part of the verb 'to be'), the sequence of development of the individual organism from fertilization to maturity. See Haeckel's law.

Oogenesis, the process by which female organisms make an ovum, or egg. In humans, the eggs are usually matured in the ovary one at a time (if more than one egg is matured simultaneously the result may be a multiple birth). A primary oocyte (immature egg) goes through the first division of >meiosis to produce secondary oocyte with a non-functional >polar body attached to it; the second meiotic division produces the mature ovum with will ever have are already present; this may be as many 500,000, vastly more than will be needed in 30 years of reproductive life. The details of oogenesis differ from one species to another, and are quite different in plants. Compare >spermatogenesis.

Opal codon, one of the three >nonsense codons in DNA that code for 'stop' and therefore signal the termination of the protein being synthesized. The opal codon is UGA.

Operator, a part of the >operon. The operator region, immediately upstream of the structural genes, is where transcription of the rest of the operon is blocked or not, according to whether the repressor is bound there or not.

Operon, a group of genes all coding for parts of a metabolic pathway (a group of enzymes that work one after another to control a metabolic process within a cell), plus their associated regulatory apparatus. The parts of the operon lie adjacent to one another in consecutive order, with only short non-coding sequences interspersed. Credit for the discovery of the operon goes to the French molecular biologists Jacques Monod Francois Jacob, who reported it in 1961 (see lac Operon). The parts of the operon are: (1) a gene for the repressor molecule (a protein); (2) a promoter region; (3) an operator region; and (4) the structural genes, in a row. RNA polymerase that will transcribe the operon binds to the promoter region; if the operon is inactive, the repressor molecule is bound to the operator region immediately adjacent, and in that case, the RNA polymerase is unable to procees so no transcription takes place.

In the presence of the inducer molecule (very often the substrate of the enzymes in question), the repressor combins with the inducer and dissociates itself from the operator region, thus allowing the RNA polymerase to proceed and transcribe the structural genes. This goes on until there is no more of the inducer in the cell, so the repressor binds on to the operator once more, blocking transcription. This is the process in operons that are negatively inducible. In positively inducible operons, there is no bound repressor, but the same function is carried out by a protein co-activator (a small molecule, analogous to the inducer) and an apo-activator (a protein, analogous to the repressor), which combine to allow RNA Polymerase to pass the operator region. In repressible operons, transcription goes on continuously until the process is shut off by the binding to the operator region of a co-repressor (small molecule) combined with an apo-repressor (protein).

Operons occur particularly where the enzymes are involved in only one metabloic process, so that they are needed at the same time and in the same quantities; in other cases, genes for related enzymes may be scattered at separate loci. Numerous Operons have been found in prokaryotes, and are assumed also to occur in eukaryotes though the evidence for this is lacking.

Organelle, any structure within a cell that has a membrane round it and has a particular function in cell metabolism. See eukaryote.

Origin of life, an event that may have taken place a number of times. There is a statistical argument in favour of life existing on other planets (see exobiology). Life may have started on this planet more than once, and come to nothing, but it is overwhelmingly likely that all life as existing today had a single origin, the evidence being that the genetic code is identical in all life forms. Darwin wrote in 1871: 'If (and oh what a big if) we could conceive in some warm little pond, with all sorts of ammonia and phosphoric salts, light, heat, electricity, etc., present, that a protein compound was chemically formed....' This was taken literally more than 80 years later by the American Physical chemist Harlod Urey; in 1953, he and his research student found that a gas consisting of water vapour, methane, ammonia and hydrogen will, if exposed to electric sparks as generated by a thunderstorm, produce various organic molecules including amino acids. later work has produced even more encouraging results, though nothing approaching a molecule as complex as a protein or a nucleic acid (i.e. DNA or RNA). An alternative theory has it that life orginated elsewhere and was brought to Earth on a meteorite. However, this does not solve the problem of how that life form was originated; it merely shifts the scene to another planet. If experimental work does succeed in producing life - which is far from impossible - it will be very interesting (to say the least) to see how it differs from the system that already exists. The first origin of life was an exception to the rule that spontaneous generation cannot occur, and so will the second one be - if it happens.

Ortet, the original organism from which a >clone is derived.

Outbreeding, a mating system in which parents are not closely related to each other, technically called >exogamy. This is the system that obtains in nearly all natural populations.

Overdominant, a dominant allele which, in the heterozygous state, causes the trait to be more strongly expressed than it does in the homozygous state. For example, in a plant with allele C for curly leaves and allele c for flat leaves, the heterozygote Cc would have curlier leaves than the homozygote CC.

The phenomenon is related to hybrid vigour.

Overlapping genes, a genes that have coding sequences in common, overlapping each other. The two Sequences are translated with a different >reading frame. This system is highly economical of space, and occurs mainly in bacteriophages which have a very small genome. it does not occur at all in eukaryotes.

Oviparous, bringing forth young in eggs. Generally this is a primitive evolutionary feature, shared by most invertebrates, fishes, reptiles and amphibians and all birds. The Monotremes (an anomalous group containing the duck-billed platypus and the spiny anteater) are the only egg laying mammals. An immature form of the adult may be a larval stage. See also ovoviviparous; viviparous.

Ovoviviparous, bringing forth live young that have been hatched from eggs inside the mother's body. Some invertebrates fish, reptiles and amphibians are ovoviparous. See also oviparous; viviparous.

Ovum, the >gamete produced by the female. The mature human ovum is about 0.15 mm in diameter. See also oogenesis.

P

P, symbol used to designate the short arm of any chromosome. Thus when a gene's locus is given as 11p, this means that it is on the short arm of chromosome 11. The long arm is called >q.

P1,P2,P3, etc., the parental generations in a cross. The parents in the first cross are P1, and their offspring are the >F1 generation. P1's parents are P2,P2'S parents are P3 and so on.

P450, a gene superfamily, consisting of large groups of genes that code for proteins (cytochromes) which metabolize drugs or other foreign chemicals. The P450 cytochromes are coded for by 20 different families of genes, ten of which are the same in all mammals. Any one mammalian species may have between 60 and 200 individual P450 genes, each making a unique cytochrome. The drug or chemical acts in many cases as the >inducer for the relevant P450 gene. In humans, polymorphisms of P450 genes are responsible for many of the examples of patients' variable responses to clinical drugs.

Pachytene, one of the stages in >meiosis (one of the two processes of cell division).

Paedomorphosis (from the Greek for 'child-shaped'), the retention in the adult of some feature that was appropriate to the juvenile of the ancestral stock. The human skull shape has been described as an example of paedomorphosis, in that the rounded cranium and small vertical face are features of an ape child. See also neoteny.

Palaeontology, the study of evolution as shown in the succession of fossils. It predates the theory of evolution by natural selection, though it did contribute to formulation (see succession, principle of). However, Cuvier, the most notable early palaeontolgist, was a creationist (see p. 14), Modern palaeontology interprets fossil evidence in the light of genetical and ecological principles.

Palindrome, a sequence of DNA that reads the same in either direction, with a central point of symmetry. Such sequences are often the target sites for regulatory molecules or for cleavage by >restriction endonucleases.

Pangenesis, a pre-Mendelian theory of heredity. It proposes that each cell or tissue of the produces a minute copy of itself (called a 'gemmule') which is sent to the ovary or testes to be assembled into a gamete and transmitted to the offspring (this is quite like Hippocrates' idea of a genetical fluid;. Darwin adopted the idea of pangenesis, in the absence of a more plausible mecanism of inheritance. It was, however, soon disproved by Galton, who carried out blood transfusions between black and white, and observed that no mixing of gemmules took place: the theory predicted that the rabbits' offspring should be speckled or dappled, but they were not. Pangenesis was essentially a refinement of >preformation, in that the gamete was supposed to be a conglomerate of minute particles from each of the tissues present in the adult body.

Panmixis, unrestricted >random mating within a closed population, there being no contribution to the gene pool by migration from outside.

Papovaviruses, a group of viruses whose genome is a ring of doublestranded DNA. One of them is SV40, a virus originally identified in green monkeys and now extensively used in virus research. The whole sequence of its genome, 5224 base-pairs long, has been decoded.

Paracentric. See inversion.

Parallelism, evolutionary change in which two or more lineages derived from a common stock retain similar features even though they are developing in isolation from one another. An example is the two groups of monkeys - the Old World monkeys (Cercopithecoidea) and the New World monkeys (Ceboidea) - which have been separated for 35 million years but are still very similar in general anatomy. See also adaptive radiation; convergence.

Paramecium, a single-celled eukaryote (one of the Protozoa), very common in fresh water. It is interesting genetically because it has two nuclei in its cell. The larger macronucleus is used when **Paramecium** reproduces by simple division, and the smaller micronucleus is only used during conjugation.

Paramutation, an allele that appears to have mutated but has in fact no change in its DNA sequence. The change in phenotype that it seems to cause is due to activity by the other allele at the same locus. The condition is extremely rare, but has been much studied in maize (*Zea mays*) by R.A. Brink.

Parasexual reproduction, a cycle of reproduction that occurs in some fungi as an alternative to their normal cycle via spores. The process involves the fusion of two haploid nuclei, a mitosis division with crossing over (not a meiosis division), and a return to the haploid state.

Parthenogenesis (from the Greek parthenos, 'young woman' or, sometimes, 'virgin', and genesis, 'creation' or 'birth'), the process of reproduction in which an organism develops from an egg without its being fertilized. Parthenogenesis can be artificially induced, by physically or chemically interfering with an ovum to stimulate it into division; the resulting individual is haploid. It also occurs frequently in nature. One example is found in all the social insects of the order Hymenoptera, such as wasps, bees and ants. Here females are diploid whereas males are haploid; any unfertilized egg develops as a male, and a fertilized one develops as a female. The colony of insects is able to adjust its own sex ratio by choosing how many eggs to fertilize. Aphids have an optional phase of parthenogenesis that they use when they want a rapid increase in population at the beginning of summer: they produce large numbers of diploid eggs that develop without fertilization (these eggs are diploid as the result of a speciam modification of meiosis in which only the first meiotic division is completed). Some species of fish, amphibians and lizards are also capable of parthenogenesis. See sex determination.

Partial dominant. See semi-dominant.

Passive immunity, immunity to a disease or toxin that is not produced by the individual's own immune response. Passive immunity can be given to a patient by injecting a substantial quantity of >antiserum. This contains antibody to the antigen in question, so that immunity starts immediately without waiting for the patient's own immune system to produce antibody. One example is the prevention of tetanus by injecting antiserum after an injury, and another is the use of anti-Rhesus antibody to prevent haemolytic disease of the new born. Passive immunity also exists in nature, as the mother passes on some immunity to her foetus through the placenta and later to the newborn baby in colostrum and breast milk.

Pasteur, Louis (1822-95), French chemist. Only a moderate student at school, Pasteur intended to be a teacher until he accidently discovered chemistry at college. His first research, in 1848, concerned the two forms of tartaric acid, which he discovered were dissymmetrical (i.e. mirror images); this was the beginning of stereochemistry. Next he became interested in fermentation (a subject of great practical importance in wine producing France), and in 1857 he proved that fermentation of a sugar solution or putrefaction of a broth both depend on the presence of microsolution or putrefaction of a broth both depend on the presence of microorganisms (yeast or bacteria). This was the final disproof of >spontaneous generation, and is Pasteur's most important contribution as far as genetics is concerned.

This work led to his invention of what came to be called 'pasteurization', heating fluids for 30 minutes to about 65°c (150°F) to sterilize them. It also led to his proposition that all diseases and infections are caused by micro-organisms. This was put to immediate practical use in his investigation into the diseases of silkworms that were ruining France's silk industry; Pasteur quickly showed that bacteria were responsible. He then extended Jenner's work on >vaccination (a word coined by Pasteur), producing in 1880 a very successful vaccine against anthrax (then a common disease), and later one against rabies. Although Pasterur's discoveries were of vast economic importance, he refused monetary reward and lived simply.

Paternity, the relationship between a male and the offspring that he has sired. Humans are not the only species who hold paternity to be important: in lions and in langurs (arboreal monkeys of southern Asia), it has been observed that if a new male takes over a harem he will kill all the infants sired by his predecessor. Paternity in the sense of a child-caring role for the father is widespread among birds: 90 per cent of species are monogamous and share the care of the offspring. For a discussion of human attitudes, see pp. 44-5. The technique of establishing whether someone is or is not the father of a particular child has greatly improved with >genetic fingerprinting.

Pauling, Linus Carl, American chemist; has taught at Stanford University, California since 1969; director of the Linus Pauling Institute of Science and Medicine from 1973. Pauling's work covers an enormously wide range, but from the point of view of genetics, the most important areas of his research are: (1) his discovery of the alpha-helix structure of some proteins (which was a clue to Crick and Watson in their search for the structure of DNA); and (2) his suggestion that the deficient haemoglobin in sickle-cell anaemia is due to a change in the amino acid sequence. Pauling believes that large doses of vitamin C are effective in preventing the common cold and other illnesses. He has won two Nobel prizes, for chemistry (1954) and, as a vigorous proponent of nuclear disarmament, for peace (1962).

Pea (***Pisum sativum***), the ordinary culinary pea, which was the subject of Mendel's historic experiments. Mendel studied seven qualitative traits of pea plants (listed here with dominant variant first): flower colour (purple or white), seed shape (round or wrinkled), cotyledon colour (yellow or green), pod shape (inflated or constricted), pod colour (green or yellow), flower position (axial or teminal) and stem length (long or short). From looking at the way these behaved in crosses, he discovered the basic machinery of inheritance. The pea has seven pairs of chromosomes, and the fact that the gene for each of Mendel's seven studied traits was on a different chromosome was either a 1-in-66 chance or was due to the fact that Mendel already knew what he was looking for.

Pearson, Karl (1857-1936), British statistician. Trained in law, Pearson soon turned to mathematics, and became professor of mathematics at University College, London during Galton's time there. His work in the theory of probability and distributions is still the foundation of statistics. He defined the standard deviation of a distribution, and devised the >chisquare test.

Pedigree, a diagram representing the genetic relationships between individuals. Usually drawn up with reference to one (or a few) specific traits, the pedigree uses conventional symbols to show the sex of the individuals (usually squares for males, circles for females); a diagonal line through their symbol shows that the person is deceased. Whether or not the individual is affected with the trait in question is shown by having their symbol solid or open, perhaps with heterozygots (proven or surmised) shaped.

Penetrance, the frequency with which a dominant allele is expressed in the phenotype of the individual carrying it. A completely penetrant allele is expressed in every case, and this is normally the position with dominant alleles having major effefcts. An allele is said to have incomplete penetrance if it is not always expressed (this will depend on the genetic background and on the environment). See also expressivity.

Pericentric. See inversion.

Persistence, the number of generations that elapse before a superseded allele is entirely replaced by a superior >mutation (see fixation).

Perutz, Max Ferdinand (1914-), Austrian-born British X-ray crystallographer and molecular biologist. From 1947 Perutz was director of the Medical Research Council's Molecular Biology Unit at Cambridge, where the structure of DNA was discovered. In 1953 Perutz showed that the structure of a protein could be deduced from comparisons of X-ray diffraction photographs, and using this method he solved the structure of haemoglobin, the first protein over to have its physical structure described. His colleague John Kendrew did the same for myoglobin (the oxygen-carrying protein in muscle), and the two shared the 1962 Nobel prize for Physiology or Medicine. Perutz's MRC Unit (later Laboratory) employed many of the most brilliant molecular biologist in the world, including >Brenner, >Crick, >Milstein, >Sanger and >Watson. He has also been chairman of the European Molecular Biology Organization (1963-9).

Phage. See bacteriophage.

Phene, a 'unit' of the bodily characteristics of an individual, thought of as being equivalent to the genetic unit, the gene. Geneticists challenge the usefulness of this term, pointing out that all aspects of an organism's phenotype are actually the result of interaction between the genotype and the environment, and therefore there cannot be a pure unit of phenotype that can be mapped on to the corresponding part of the genotype. None the less some traits such as blood groups, being uninfluenced by environment, do behave as implied by the word 'phene'.

Phenetic classification, a system of classifying species by their phenotypes (visible or otherwise measurable) rather than with reference to their ancestry. See also phylogenetic classification.

Phenetic distance, the total amount of difference in measurable traits between any two related species. This may or may not reflect how closely they are related to a common ancestor, depending on whether there has been convergence or parallelism. See also cladistic distance.

Phenocopy, a phenotype caused by environmental influences that is identical or very similar to the phenotype controlled by the genotype under consideration. For example, dietary deficiency may mimic the effects of an allele that prevents the organism from digesting a particular nutrient. The characteristics of the phenocopy are not heritable.

Phenotype, the totality of the physical characteristics of an organism. The term is used in contradistinction to the >genotype, the total of the organism's genetic information. The genotype has an influence on the phenotype, modified to a greater or lesser degree by the environment. The phenotype used to be taken to mean only those characteristics that could be directly observed, but now includes biochemical characteristics such as the ability or inability to produce particular enzymes; it also includes behavioural characteristics.

Phenotypic variance, the total amount of >variance observable in a trait. It consists partly of the genetic variance and partly of a component contributed by environmental factors.

Phenylketonuria (PKU), a human disease caused by a recessive >allele. The actual mutation is a failure to produce the liver enzyme phenylalanine hydroxylase, so that phenylalanine cannot be metabolized and accumulates in the bloodstream in concentrations up to 100 times normal (some is also excreted in the urine, hence the name of the disease). These excessive concentrations cause severe damage to the brain. A baby with phenylketonuria is protected before it is born because its mother has the normal enzyme and deals with the baby's excess phenylalanine as it passes through the placenta, but at birth, the baby must immediately be put on a diet low in phenylalanine to prevent brain damage occurring. (Some phenylalanine is essential, as it is one of the amino acids necessary for protein synthesis.) All babies born in Britain are routinely screened for this disorder, which occurs in about 1 in 10,000 births. A prenatal test is now available for mothers who have already had one phenylketonuric baby.

Phenylthiocarbamide. See taste blindness.

Pheromones, a group of chemicals that act as messengers to alter the behviour of other members of the same species, in a way analogous to hormones altering the activity of other tissues within the body (hence the pheromone's altering name 'sociohormone'). Pheromones are certainly the oldest form of intra-specific signalling, and the chemical communication system has a number of advantages over the later auditory and visual ones: it is bioenergetically very economical; its signals can be transmitted in the dark and around obstacles, in air or in water; and it has by far the greatest range and duration. Pheromones have been found in every major animal group, including mammals, and though none has been identified in humans, their preence is suspected. They are particularly important in social insects, which may have up to a dozen different pheromone-producing glands, each for a different message. Sex-attractant pheromones are also very widespred, especially the ones put out by females, and can often operate over a distance of several kilometres. Territory and trail markers are common too. Odours to indicate dominance status are found in mammals. There has to be a simultaneous evolution of both the production and the reception apparatus within the species for a pheromone system to work. See allomones.

Philadelphia chromosome, an abnormal chromosome found in the cancerous white blood calls of those suffering from chronic myeloid >leukaemia. In these cells, there is a translocation of part of chromosome 22 to chromosome 9; all the rest of the cells are normal. It seems that what happens is a genetic accident within a single white blood cell; that cell is transformed into a cancer cell which then proliferates out of control and oncogene.

Photoperiodism, biological activity that is influenced by changing day length. Many plants, especially flowering ones, are photoperiodic, being generally sensitive to the length of the dark period rather than the light (the two can be experimentally manipulated to be independent). Bird migrations are mostly also photoperodic. The system probably evolved as a more reliable way of timing hazardous operations such as flowering and migrating than by temperature: a sudden spell of unseasonable weather could otherwise deceive individuals into acting at the wrong time.

Photoreactivation. See DNA repair.

Phyletic evolution, evolutionary change in which one species transforms over time into another species, without any branching of the ancestral tree; there are never two different species at the same time. An example is the emergence of *Homo sapiens* from the earlier species *Homo erectus*. See also clacogenic evolution.

Phyletic gradualism. See gradualism.

Phylogenetic classification >a system of classifying species and higher taxa according to their ancestral relationships (in contrast to >phenetic classification, i.e. by observable characteristics). It has been noticed that classifications made by peoples without knowledge of evolution, e.g. Australian Aborigines, are very much closer to the phylogenetic rather than the phenetic system, contrary to expectation.

Phylogeny, the relationship between a specie and its evolutionary predecessors. See Haeckel's law: 'Ontogeny recapitulates phylogeny.'

Pigmentation, any colouring in the tissues, but particularly the external tissues such as skin and hair. The molecule that causes the black or brown colour in skin and hair is >melanin, and the concentration of this in the cells determines the actual colour seen.

Although it is obvious that populations in the tropics (or whose ancestors came from the tropics) have strong pigmentation, the evolutionary reason for this is a matter for guesswork. It is said that the original human colour was black (though there is absolutely no evidence) and that this colour is advantageous because of the protection it gives against >ultraviolet light, which causes skin cancer; but was the risk of skin cancer ever large enough to be a selective agent? Or it has been suggested that vitamin D synthesis is involved; the vitamin is synthesized in the skin, and without the shield of pigment, an unhealthy excess would be produced under the tropical sun. Either or both or none of these may have been factors. The genes responsible for skin pigmentation have not been identified, but analysis of the amount of variation suggests that alleles at only three or four loci are involved.

As for hair pigmentation, there is no theory as to why the vast majority of humans, living in environments ranging from the tropics to the Arctic circle, should have black hair, while blonde, brown and red hair colours have evolved in Europe.

Plaque, a small patch within a culture of bacteria (a 'bacterial lawn') where there are no bacteria growing. This is because the cells are being lysed (see lysis; bacteriophage) or because they are sensitive to an antibiotic being tested.

Plasmid, an extrachromosomal genetic element, consisting of a small ring of double-stranded DNA. Plasmids replicate themselves independently of the chromosomes, and segregate either regularly or irregularly during cell division. They occur in widely varying numbers, from one per cell to a hundred or more. Most commonly found in bacteria, they often carry genes for resistance to antibiotics (see R plasmid). Bacterial plasmids are used in >gene cloning. Plasmids are uncommon in eukaryotes, but do occur in the mitochondria of yeast and some plants.

Plasmodium falciparum, the protozoan parasite that causes >malaria.

Pleiotropy, the condition in which a single mutant allele causes a number of different effects. These effects may be very diverse in their end result, but are presumably due to a 'pedigree of causes' in which the primary effect of the mutation, especially if expressed early in embryonic life, leads to a chain of adverse consequences. Many of the severe human disorders are examples of pleiotropy, e.g. Marfan's syndrome, in which the primary defect in collagen synthesis leads not only to skeletal deformities but to heart and eye defects.

Ploidy, the number of multiples of the basic set of chromosomes that ar present in a cell or organism. The basic set, with one copy of each chromosome, is called >haploid and is found in gametes and some simple plants. The set with two copies of each chromosome is called >diploid and is found in the cells of almost all organisms. Many specie of plants tend to >polyploidy, when there are more than two times the haploid number. The word-ploid can be attached to a number; for example, 16-ploid means having 16 times the haploid number of chromosomes. See also aneuploidy.

Plus and minus strands, designations for the strands of DNA and RNA in their role in transcription and translation. The coding strand of DNA is a 'minus', and the messenger RNA taken from it is a 'plus'strand.

Point mutation, a mutation involving the substitution of just one basepair for another. The term is also used to mean a mutation giving rise to a new allele, in contradistinction to a genetic event involving a change in chromosomal structure.

Poisson distribution, a distribution that represents the probability of occurrence of independent events. The frequency of cross-overs is fairly close to a Poisson distribution, except that one cross-over is fairly close to a Poisson distribution, except that one cross-over interferes somewhat with the occurrence of another, so they are not truly independent events. The Poisson distribution was described by the French mathematician Simeon Poisson (1781-1840), professor of pure mathematics at the Ecole Polytechnique in Paris.

Polar body, a non-functional by-product of meiosis in vertebrate females. When the primary oocyte (immature egg) goes through the first meiotic division, the chromosomes are evenly divided into the two new cells but the cytoplasm is not: one cell is large, and will become the secondary oocyte, but the other has virtually no cytoplasm and becomes a mere bump on the surface of the oocyte. The secondary oocyte then goes through the second meiotic division and again produces one cell (now the ovum) with all the cytoplasm and a polar body which joins the first on the cell wall of the ovum. In some species the first polar body goes through a second meiotic division itself so that in the end there polar bodies. Each polar body has a haploid set of chromosomes, and is genetically different from the ovum and from the other polar bodies.

Poly-A tail, the sequence of adenine (A) bases at the >3' end of a molecule of >messenger RNA, usually consisting of 100-200 As. Its function is thought to be stabilization of the RNA molecule. See terminal transferase.

Polycistronic RNA, a molecule of messenger RNA that includes the transcription product of more than one gene. These are produced when an operon is being transcribed. In other cases, they occur where structural genes are close to each other with little intervening matter, though this kind of polycistronic RNA may be cut into separate messages for translation (see protein synthesis).

Polycystic kidney disease, a disease in which the kidney develops multiple cysts (abnormal, fluid-filled swellings), with probable total kidney failure in middle age. Polycystic kidney disease is inherited as a dominant, and is quite common in European populations, affecting 1 per 1250 babies; the majority of patients who need treatment by kidney dialysis have this diseas. The gene has been treatment by kidney dialysis have this disease. The gene has been located to chromosome 16, where it is close to the gene for alpha-globulin (one of the components of haemoglobin). This genetic marker can now be used (in some families) for predicting whether a foetus has the polycystic gene.

Polyembryony, the production of more than one embryo from a single fertilized egg. The commonest examples is >monozygotic (identical) twins, but another is provided by the nine-banded armadillo which has litters of four genetically identical young all derived from a single fertilized egg.

Polygenes, a group of genes, not separately identifiable, that are involved in the control of a >continuously varying trait. An individual polygene is a Mendelian allele like any other. Each polygene may have a simple additive effect on the phenotype, or there may be dominance or other interaction between the polygenes; there will also be interaction with the environment. The number of polygenes involved in any one trait can be as small as two, or as large as 100 or so, though in most cases it is probably under ten.

Polymerase, any of the enzymes that construct a polymeric molecule of nucleic acid - i.e. DNA or RNA - by joining nucleotides together. DNA polymerase makes a new strand by using a single strand of DNA as a template (see replication). There are different RNA polymerases for building messenger, ribosomal and transfer RNA. See also reverse transcriptase.

Polymerase chain reaction (PCR), a technique that allows a DNA fragment to be analysed without having to be cloned first. Two primers have to be prepared, which are oligonucleotides (very short synthetic DNA sequences, about 30 bases long) that are copies of sequences known to occur in the region of the DNA to be studied. The original DNA in the genome is subjected to melting to reduce it to single-strand state, in the presence of large quantities of the two primers; the primers find their complementary sequences on the genomic DNA and hybridize to them. A polymerase called Taq, which can work at high temperatures, is added; it synthesizes a strand of DNA, starting form the end of the annealed primer and using the genomic DNA as a template. The resulting double-stranded DNA is melted again, and the process repeated. The DNA in the region between the two primers is doubled each time the cycle is repeated: after one cycle, there isquantity; after two cycles, four times; three cycles, eight times and so on. As each cycle takes only about five minutes, it is possible to produce large quantities of DNA within a few hours. This can then be analysed using >restriction enzymes. A DNA sample form only a single cell or a few cells is enough to carry out PCR amplification, and results can be obtained within 24 hours.

Polymorphism, 'The occurrence together in the same locality of two or more discontinuous forms of a species in such proportions that the rarest of them cannot be maintained merely by recurrent mutation.' This definition by the British geneticist E.B. Ford is one of the classic texts of genetics. By 'discontinuous forms' Ford did not mean types that would be visually distinct, but rather types having different genotypes at a particular locus. For example, the >ABO blood group system is a polymorphism, and people with blood groups A, B, AB or O are the discontinuous forms. In higher eukaryotes, about 10 per cent of loci are polymorphic. This may be a stable situation in a population, maintained because the heterozygote has superior fitness compared to the two homozygotes (see heterozygous advantage). More usually the polymorphism will be transient, the population being part way along the course of having a new mutation replace the old allele. Some geneticists believe that many polymorphisms are maintained merely at random, the mutant allele being neither better nor worse than the original so that their relative frequencies fluctuate by chance (see genetic drift).

Polypeptide, a molecule that is a chain of up to about 50 amino acids (above that it would be referred to as a >protein). It may be an independent molecule in its own right (e.g. a >hormone) or it may form part of a protein.

Polypheny, the occurrence in a population of two or more distinct phenotypes not due to underlying genetic differences. The phenotypic variability is due toenvironmental effects, particularly ones happening during the individuals' development (see epigenetic landscape). An example is the garden hydrangea, whose flowers are either pink or blue according to whether the soil is alkaline or acid.

Polyphyletic group, a group of taxa who have different evolutionary ancestors. Such groups occur when there has been >convergence in their evolutionary history. See also monophyletic group.

Polyploidy, the condition of having a chromosome complement of more than twice the haploid number. The most usual forms are tetraploidy (four times the haploid number) and higher evern numbers; >triplody is unusual. Polyploidy is common in higher plants and occur in some lower animals, but is rare in vertebrates. Tetraploidy can be maintained as a stable condition in self-fertilizing plants; they produce diploid gametes which, on fertilization, give tetraploid offspring. The same is true for octaploids (eight times haploid) with tetraploid gametes and so on. This process has been important in the evolution of plants, being a short-cut to new species (the tetraploid plants cannot make fertile crosses with the original diploid variety). Occasionally one particular cell line in the body of a diploid organism can be polyploid, as with the >silk moth.

Polyposis coli, an inherited disease in which hundreds or even thousands of polyps (small soft growths) develop in the colon. It is caused by a dominant gene, whose mode of action is unknown. There is a high probability of the disease progressing to cancer of the colon, and in many cases removal of the colon is carried out as a preventive measure. Polyposis coli affects about one person in 10,000.

Polysome, a group of two or more ribosomes with an mRNA molecule, actively engaged in protein synthesis.

Polysomy, the condition of having more than two copies of a particular chromosome. The most common form is >trisomy.

Polyspermy, entry of more than one sperm into the ovum. Only one of the sperms joins its pronucleus to that of the ovum, while the other is absorbed into the cytoplasm. It happens only rarely, and has no genetic effect, as only one of the sperm is involved in fertilization.

Polytene chromosome, a structure consisting of a homologous pair of chromosomes fused together and with up to 1000 or so >chromatids in parallel. The DNA replicates many times to make new chromatids but without any separation; the chromatids lie alongside one another to form a wide ribbon. Polytene chromosomes are so large (up to 10 pm wide) that one can easily observe the bands of heterochromatin (condensed nonactive DNA) alternating with paler interbands where genes are being transcribed (these sometimes expand into a >puff). Sites of deletions, inversions and translocations can also be seen, where one side of the polytene chromosome does not match the other. Polytene chromosomes occur in the salivary glands of *Drosophila* and other insects.

Polytypy, the existence within a species of a number of types. These are not distinct enough from each other to be separate species but are recognizably different, and may be classified as subspecies or variant forms. The types are often geographically separated. An example of polytypy is the horse *Equus caballus*, which even in its original wild forms varied between Shetland ponies and the swift horses of the Arabian peninsula.

Population, a group of individuals forming an interbreeding community. Some degree of isolation from other populations is implied.

Population cage. See *Drosophila*.

Population genetics, the study of how genes behave in populations during evolution. The mathematical theory was laid down in the 1930s, the key thinkers being Sir Ronald >Fisher, Sewall >Wright, Godfrey Hardy and W. Weinberg (see Hardy-Weinberg law). The key factors in determining change in >gene frequencies were found to be >mutation, >natural selection, >migration and >genetic drift. Understanding of these factors was the foundation for practical studies in ecological genetics, such as the example of how a species of moth evolved black coloration (see melanism) in an area heavily polluted by soot emissions from factory chimneys. But there were no new theories in the field of population genetics until it was suggested, in 1957 by the British geneticist C.H. Waddington, that evolution processes could perhaps be analysed using >game theory. See also ecology.

Porphyria, a human existing in various forms, some of them genetically determined. The sufferer produces excess amounts of haem, the compound that gives red blood cells their colour and enables them to carry oxygen; this excess is metabolized into porphyrins which can be detected in the urine. There is also a variable amount of mental disturbance, sometimes taking the form of periods of hallucinations interspersed with complete lucidity. The most famous example of a porphyria sufferer is George III (1738-1820), king of Great Britain from 1760, who may have had acute intermittent porphyria, which would explain the insanity that eventually made him unfit to rule; this form is inherited as a single dominant >allele. Another form, also a dominant, is quite common in Sweden. A third, again dominant, is found among white South Africans at a frequency of 1 in 300, all of whom can trace their descent from a single Dutch couple who went to the Cape in 1688. There is no treatment for porphyria, but sufferers can avoid things that are known to provoke attacks, such as barbiturates. Mortality can be as high as 20 per cent.

Position effect, a change in the expression of a gene as a result of its changing position within the genome (or of another gene normally adjacent to it moving away). The splitting up of an operon will cause a change of expression, but in eukaryote genes, position effects are not common, and most genes are expressed in a similar way regardless of their position.

Positive control, a form of gene regulation in which transcription cannot occur until an activator molecule binds to the DNA in order to allow RNA polymerase to proceed with transcribing the structural gene or operon. The opposite is >negative control.

Posthumous conception, one of the possibly desirable results of keeping sperm or eggs frozen for future use. Storing sperm is usually done for men about to have a vasectomy, so that they can still father children if they later want to. Posthumous conception using a dead man's sperm would be technically simple, but it is hard to imagine anyone setting up the much more complicated business of in vitro fertilization and surrogacy that would be involved in posthumous conception using a dead woman's egg. Those who worry about telling a child about its 'real' (i.e. genetical) parent would find either of these scenarios difficult. See also sperm banks.

Poxviruses, a group of rather large viruses with a genome consisting of double-stranded DNA. It includes the viruses that cause cowpox and myxomatosis; the virus that caused smallpox is now extinct in the wild but is maintained as a laboratory stock in only a handful of centres for scientific interest. Vaccination has been effective against poxviruses, as they do not evolve new unrecognizable strains.

Pre-adaption, the situation in which a mutation gives an organism a phenotypic character which happens to be adaptive to an environmental change that occurs later. See adaptation.

Preformation, the idea that the human being is already fully formed at the moment of conception. Some performationists held that the miniature human homunculus was to be found inside the sperm; others believed it was in the ovum. The latter (called 'ovists') went so far as to say that each female contained in her body the fully formed bodies of all her potential sons and daughters, and that her daughters within her body had all their offspring within their miniature bodies, and so on. If this theory were true, Eve would have held within her the actual bodies of all the humans that would ever be. Improved microscopy put the preformationists into retreat in the first part of the

nineteenth century, and the researches of von Baer and others gave rise to the theory of >epigenesis, the progressive elaboration of the organism form the single-celled zygote. See also germ plasm theory; pangenesis.

Pre-implatation diagnosis, the inspection of an embryo fertilized in vitro (see in vitro fertilization) before it is implanted into the uterus, to check for the presence of genetic defects It is also possible that this could be done in the case of a naturally conceived embryo, if it were removed before it had implanted, and then put back. It is necessary to remove one cell from the embryo (this does not damage the development of the embryo if it is done early enough) and to amplify the DNA using the >polymerase chain-reaction technique. The DNA can then be inspected for certain defects by using restriction enzymes to identify mutations that cause loss of restriction sites or by using allele-specific oligonucleotides. Linkage to restriction fragment length polymorphism alleles can show the probability of the embryo having a defective gene, if the family pattern of linkage is known. If these tests are to be carried out before the embryo is implanted, they must be done fast, so they must focus on defects that are suspected of being present. An alternative approach would be to implant one while keeping the other for more time-consuming tests, including the cellculture tests required for detecting chromosome abnormalities. This is ethically controversial. In the case of X-linked conditions, preimplantation diagnosis is used to ascertain the sex of the embryo, so that it can be destroyed if male; this too is ethically controversial, as only half the male offspring of a carrier inherit the condition.

Prenatal diagnosis, the various techniques that are used to discover whether an unborn baby has a genetic disorder. The simplest is ultrasound inspection, which can detect defects that cause substantial anatomical abnormality. >Amniocentesis is very useful in that it involves little risk to the foetus and can yield information on a wide range of genetic disorders. The drawback is that the tissue culture procedures take two or three weeks, with the cytological tests for chromosome abnormalities (including such relatively common conditions as Down's syndrome) taking longest; this means that any abortion following diagnosis will necessarily be a late one. The sampling of foetal blood can be used for diagnosing blood conditions (including thalassaemia and sickle-cell anaemia), but there is a significant risk to the foetus, and it cannot be carried out until about the same time as amniocentesis (12 weeks plus). A newer technique is >chorionic villus sampling. Samples of the foetus's >DNA obtained in this way can be used for direct diagnosis (see polymerase chain reaction) of a number of genetic disorders that can be identified because they involve the loss of a cutting site for a restriction enzyme, or for indirect prediction that the foetus carrier a mutant gene because it is known to be linked to a marker (see restriction fragment length polymorphisms).

Pribnow box, a sequence found in prokaryote genes. Part of the >promoter region, it occurs precisely 10 base-pairs upstream from the point at which RNA transcription is to start. Eukaryotes have a >TATA box. It is called after its discoverer, the American biochemist David Pribnow.

Primary sexual characters, the reproductive organs, both internal and external. In some animals, including humans, there are also >secondary sexual characters.

Primates, the group of mammals comprising humans, apes, monkeys and prosimians (lemurs, tarsiers, etc.). The term was devised by Linnaeus, who certainly intended it to have a sense of 'first and highest', witness the fact that he also coined the words 'secundates' for mammals other than primates and 'tertiates' for all other animals.

Primitive streak, one of the layers of cells in the very early stages of >embryogenesis; from it the skeleton, muscles, heart and blood system and several other tissues are derived. The time of its appearance (at about 14 days after fertilization) has become, in the Human Fertilization and Embryology Act 1990 in Britain, the limit on the time during which embryos can be experimented upon.

Primrose (***Primula vulgaris***), a plant studied by Darwin for its mechanism of >self-sterility. The evening primrose is >*Oenothera*.

Probability, a measure of how confident one can be that a certain event will occur. It ranges from P = 1 (absolute certainly that it will occur) to P = 0 (absolute impossibility). An estimate of probability is often used to indicate the significance of statistical results. For example, if two samples have different mean values for some trait, it may be that the probability of their having come from the same population, with the difference in means arising only by chance, is P = 0.001, i.e. one in 1000.

Probe, a synthetic DNA or RNA sequence used to identify fragments in a sample (see Southern blotting). Synthetic DNA sequences can be made up to about 20 base-pairs long; this is enough to mimic a gene for a particular protein in some circumstances, if the sequence of the protein is known but the gene has not been isolated.

Progeny, a group of individuals all having the same two parents. The word can also be used to mean all the offspring of one individual, regardless of different mates.

Progeny testing, a technique used in the breeding of cattle and other livestock to improve the efficiency of selection. Assuming that the environments are roughly comparable, if the progeny of Cow M has on average a higher milk yield than that of Cow N, one can conclude that Cow M's genotype for milk yield is superior. This is a more accurate measure than simply testing each cow.

Progeria, a very rare inherited disease whose symptoms are an immensely speeded-up ageing process - wrinkling of the skin, loss of hair, bone shrinkage. By the age of ten or so, sufferers look as if they are 70. Death from heart disease usually occurs in the early teens. The gene responsible is recessive; its mode of action is unknown but is suspected of being concerned with DNA repair.

Prokaryote, any organism that has its genome in the form of a single DNA molecule (comparble to a chromosome), sometimes in the form of a ring. A prokaryote also has a simple cell structure, with no separate nucleus. The prokaryotes comprise bacteria and blue-green algae, and are a simpler form of life than the later >eukaryotes (see classification).

Promoter, the sequence of DNA immediately upstream from the coding sequence of a structural gene, where RNA polymerase binds on to begin transcription. In operons in prokaryote cells, the promoter is followed by the operator sequence (where the repressor or activator acts). In eukaryotes, there are particular promoter sequences for different genes, determining which of the various types of RNA polymerase will bind to the promoter sequences contain marker sequences at a fixed distance from the beginning of the coding sequence: these are the Pribnow box in prokaryotes and the TATA box in eukaryotes. Promoters are either 'strong' or 'weak' according to the consensus between their sequences and those of the relevant RNA polymerase type.

Pronucleus, the nucleus of a gamete (sperm or ovum, in animals). It has one chromosome of each pair, so that at fertilization the new individual has a complete set of chromosomes in pairs.

Proofreading, the process in which DNA polymerase checks during replication that the correct base has been put in place, and removes any incorrect one.

Prophage, a >bacteriophage that has entered a bacterium and has become integrated into its host's genome but is quiescent. The bacteriophage's nucleic acid is replicated every time the host cell divides, but the genes are not transcribed to produce more phages until specifically induced by some genetic event or by radiation.

Prophase, one of the phases of the two types of cell division, >meiosis and >mitosis (see diagrams).

Proposita, propositus, in medical genetics, the person in whom an inherited condition was first diagnosed. The family pedigree around the proposita (female) or propositus (male) is then traced to ascertain the pattern of inheritance and to identify other persons who may have the same gene. The proposita/propositus can be a member of any of the living generations of a family. See genetic counselling.

Protein, a molecule that consists of amino acids strung together linearly. Typically a protein has two or more separate polypeptide chains; the sequence of amino acids (the primary structure) determines the shape that the chains form when folded together (the secondary structure) which is critical for the protein's function. Proteins can be (a) >enzymes, (b) >antibodies, (c) >hormones (d) structural constituents, e.g. collagen, (e) transport molecules, e.g. >haemoglobin, (f) nutritional, e.g. casein, or (g) contractile molecules in muscle. See protein synthesis.

Protein synthesis, the manufacture of proteins of all sorts from the instructions in the genes. The central dogma of molecular biology states that DNA is transcribed into RNA which is translated into protein. The DNA is situated in the chromosomes, in the nucleus of each cell.

When the protein synthesis is to start (a complicated matter in itself; see gene regulation), an enzyme called RNA polymerase begins to build a molecule of messenger RNA (mRNA for short) according to the code in the DNA: this works in the same way as the DNA replicating itself, i.e. each base in the DNA must be matched with the complementary base in the RNA. The mRNA goes out of the nucleus into the cell's cytoplasm; among the many things floating around in this crowded space are (a) molecules of transfer RNA (tRNA) and (b) ribosomes. The tRNA molecules consist of only three bases (i.e. a single triplet); these three bases form an anticodon, and each of these matches on to a codon in the mRNA. Thus if the mRNA reads GCA-AAG-AAA, it will attract to itself three tRNA molecules having the anticodons CGU-UUC-UUU. Each of these has the appropriate amino acid (as specified in the >genetic code) attached to its other end.

Actual assembly takes place on the ribosomes, cottage-loaf-shaped bodies made of proteins and RNA (rRNA or ribosomal RNA), which may be free-floating or embedded in a network of membrances called the endoplasmic reticulum. The linear molecule of mRNA is gripped in the groove of the ribosome, the starting point always being the codon AUG in the mRNA. The ribosome has two slots alongside each other, and a tRNA molecule with the anticodon UAC (complementary to the AUG in the mRNA) moves into the first one; then the tRNA appropriate for the next mRNA codon moves into the second slot, and the ribosome joins their amino acids are joined, and so on until the ribosome comes to a 'stop' codon in the mRNA. This is a codon that does not have a corresponding amino acid, so no tRNA fits with it and the chain comes to an end. It has already been folding itself into shape at the front end while the back was still being synthesized.

All the RNA components in this process get recycled. The tRNA molecules float off the ribosome empty, having had their amino acids built into the protein, so they go off and pick up a new amino acid. The mRNA, once used for a new mRNA, molecule.

Protoctista, one of the five kingdoms of living things (see classification). It includes all the single-celled >eukaryotes.

Proto-oncogene. See oncogene,

Protozoa, a group of single-celled organisms that are not plants, forming part of the kingdom Protoctista (see classification). The amoeba, >Paramecium and >Chlamydomonas are the best-known examples.

Provirus, the genome of a virus when it is incorporated into the genome of a host eukaryotic cell but is quiescent. The viral genome is replicated during cell division but none of the viral genes is expressed. It is a characteristic of >retroviruses that they become transmitted as proviruses. See also prophage.

Pseudodominance, the phenomenon in which an allele that is not normally dominant behaves as if it were because the corresponding allele on the other chromosome has been removed by a deletion.

Pseudogene, a sequence of >DNA that resembles a normal gene but cannot be transcribed (see transcription) because of the number of nonsese mutations in it. They occur only in eukaryotes, and are oftengrouped together in a single region with many repetitions or nearly identical copies.

Psoriasis, a disease in which there is overproliferation of skin cells, leading to irritating rough patches; there is no direct effect on general health. Psoriasis has a genetic component, though there is no straightforward Mendelian pattern of inheritance. There is some association with the immune system, as some HLA types are more susceptible than others. Psoriasis affects 1-2 per cent of white populations.

Pteridophytes, a group of plants including the ferns, club mosses and horsetails that reproduce without seeds. They have alternation of generations, as seed-bearing plants do.

Puff, part of a >polytene chromosome where one of the bands is greatly expanded. The puff consists of loops of DNA that has unwound itself in order to be transcribed. It is believed that puffs are the sites of transcription of cell-specific genes (seé luxury gene), since the pattern of puffing is specific to the organ concerned, within the same organism. The largest puffs are known as Balbiani rings.

Punctuated evolution, evolutionary change which, over the long term, consists of periods of relative stability punctuated by bursts of rapid species formation ever short periods. This is not what was originally envisaged by Darwin and Wallace, but is perfectly consistent with their theory of evolution by natural selection.

Punnett square, a technique for predicting the outcome of a cross, invented by the British geneticist R.C.Punnett in 1906. A table is drawn up with a column or row for each of the possible parental gametes; the genotypes that will occur among the offspring are then read off from the squares. The method also shows what proportions of each genotype there will be.

Purine nucleoside phosphorylase (PNP), an enzyme involved in the immune system. Lack of it (inherited as a recessive) causes one form of >severe combined immune deficiency.

Purines, one of the two types of nitrogenous bases that are the important constituents of DNA and RNA. The commonest (and the only ones occurring in the nucleic acids) are adenine and guanine. See also pyrimidines.

Pyloric stenosis, an inherited disorder in which the pylorus (the muscular ring at the lower end of the stomach) is enlarged and consequently the outlet to the small intestine is narrowed. A baby with this disorder is normal at birth but begins to vomit at about two months old; without surgical intervention to clear the enlargement the baby dies. The inheritance is polygenic, and for some unknown reason boys are about four times more likely to be affected than girls. It is a common disorder, with an overall frequency of 1 in 200-300 births.

Pyrimidines, one of the two types of nitrogenous bases that are the important constituents of DNA and RNA. The commonest (and the only ones occurring in the nucleic acids) are cytosine, thymine (in DNA) and uracil (in RNA). See also purines.

Q

q, symbol used to designate the long arm of a chromosome. Thus if a gene's locus is given as 22q, this means that it is on the long arm of chromosome 22. The short arm is called >p.

Quadruplets, four babies born together. The frequency with which quadruplets are born is the cube of the frequency of twins (with triplets being the square); their is no theoretical basis for why this should be so (see twinning rate). This frequency will no longer hold because of the high frequency of >multiple births following the use of >fertility drugs. Human quadruplets may be identical or non-identical. The word would not be used of animals having four young at once, except perhaps in the case of nine-banded armadillos, which have four identical pups derived from a single fertilized egg.

Quantification, the application of measurement to experimental results. Sir Francis >Galton was the pioneer of quantification in genetical studies. Quantification is not equivalent to objectivity: no matter how good one's measuring apparatus or one's technique in using it, one has to decide what to measure. A brilliant hypothetical example of this was proposed by the British mathematician Alan Turing (whose theoretical work led to the development of computers while at the same time defining their ultimate limitations). Suppose that the photograph taken of an apparent dead-heat dead-heat in a horse race shows that horse A has its muzzle first across the line, but horse B has a stream of saliva coming out of its mouth, which crossed the line before the muzzle of horse A. Which is the winner? Quantification in this case in measuring the presence or absence of a horse at the line at a given time, and the subjective element is deciding whether or not saliva is part of a horse. (In deciding this, one must remember that saliva is so identifiable as belonging to one individual that it has often been used as forensic evidence.)

Quantitative trait. See continuous variation.

Quantum evolution, evolutionary change in which new species are formed very rapidly. The circumstance in which this is likely to happens is the colonization of new habitats by small populations, in which both founder effect and genetic drift contribute to genetic change.

R

Race, an informal classification of people, supposed to be based upon biological differences. It need hardly be said that the political application of this idea was and continues to be normally outrageous. The genetical perspective on the matter is this: there are difference in the frequencies of genes between one human population and another, as natural selection has adapted them to different environments. There are no cases of hard-and-fast divisions between populations on the basis of these genefrequency differences, and attempts to classify human groups according to the available data have resulted in widely different interpretations: for example, in 1962 the geneticist Theodosiuns >Dobzhansky estimated that there were 14 identifiable groups; the year before, S.M. Garn (an American anthropologist) had said that there were 32. So much for the idea that a 'race' is a biological entity.

To get back to the matter of gene frequency, the fact is that the genes that cause the difficulty are the ones controlling a minor aspect of a person's appearance: skin colour. These genes have not been separately identified, but it is estimated that alleles at only three or four loci need be involved (see pigmentation). Out of a genome of 100,000 genes or so this is an infinitesimal difference; but it is true that, at these few loci, the genes for dark colour occur at a much higher frequency in populations who themselves live or whose ancestors lived in the tropics.

Colour prejudice is irrational enough, and it would be just as irrational to discriminate against Amerindian populations on the grounds that they have practically no individuals with blood group B or Rhesus allele r. Again these are only a very few genes out of the whole human genome - but invisible ones.

Radiation, one of the causes of raised mutation rates and of increased incidence of cancers. There is an appreciable amount of naturally occuring radiation, from cosmic rays and from minerals in rocks. Artificial radiation comes from such sources as fall-out from nuclear weapons, the by-products of nuclear energy, and medical and dental uses (e.g. >X-rays). There is no safe level of radiation: the amount of genetic damage is linearly related to the radiation dose, starting with the smallest doses.

Genetic damage, in the sense of mutations arising in the gametes, is the result of the sex organs receiving radiation; in humans, males are more at risk because the testicles are external, while the ovaries of females have some protection by being internal. Males may also be more at risk because >spermatogenesis goes on all the time, whereas females store their ova already partly matured. There may also be a cultural factor, in that men have been more exposed to radiation hazards in the belief that they are less vulnerable than women (see leukaemia).

Radiation causes cancers (including leukaemias) by damaging the chromosomes in the cells of the body. The data from the survivors of the atom bombs dropped on Japan in 1945 show that the effect or radiation is indeed linear: the greater the dose, the higher the probability of the person developing cancer. Radiation received is measured in rem (roentgen-equivalent-man). The amount that the average person receives from background radiation is about 0.1 rem per year, and on average, medical X-rays add another 0.1 rem annually. The LD50 (i.e. the dose that would kill 50 per cent of a population) is believed to be about 450 rem for humans, although some survivors from the 1945 bombs received about 500 rem.

Ramet, an organism raised as a member of a >clone but capable of independent existence. See also genet; ortet.

Random mating, a system of mating in which neither genotype nor phenotype influence the choice of mate. There is no systematic inbreeding or outbreeding or assortative mating. Random mating is used as the presumed mating system in many calculations in population genetics.

Reading frame, the mode of reading of a sequence of DNA or RNA, depending on the exact point of starting. For example, the sequence AGGUCGUCC could be read as AGG-UCG-UCC, but with a different reading frame, it could become GUC-GUC-C and so on. >Mutation that add or subtract one or two bases alter the reading frame (see frameshift mutation). See also overlapping genes.

Recapitulation. See Haeckel's law.

Recessive, an allele that does not express its characteristic unless it is present in double dose, i.e. in the >homozygote. The majority of mutations are recessive, because generally they code for a defective protein which is compensated for by the normal protein from the other allele. Note that recessiveness is relative: allele a1 can be recessive when associated with allele a2 but dominant in relation to allele a3 (see allelic series). See also X linkage, a special case of recessiveness.

Reciprocal cross, two crosses, in one of which the parent from strain A is the female and from strain B the male, and in the other the reverse. The resulting offspring will be identical unless there is X-linkage or maternal inheritance.

Recombinant DNA, any artificially created DNA that consists of a sequence from one genetic source spliced to a sequence from another. It is now almost always DNA from two different species, but the first recombinant DNA (made in 1973 by the American biochemist Herbert Boyer) was from two different >Escherichia coli >plasmids. (See splicing.) Recombinant DNA technology is a major part of the repertoire of >genetic engineering.

Recombination, the formation of new combinations of alleles on one chromosome, following >corssing-over at >meiosis. Recombination is second only to mutation as a source of new variation for natural selection to work upon. It also enhances the effects of mutation: a newly arisen mutant allele may not be beneficial in its original setting, and may only become so when, by recombination, it finds itself acting in concert with other genes in a new and superior combination.

Suppose that there are two strains of goats: one has long hair (genotype HH) and long legs (LL), the other short hair (hh) and short legs (ll), the loci being on the same chromosome (see linkage). All the first-cross (see F1) goats are Hh Ll - that is, long-haired and long-legged - with one chromosome carrying H L and the other hl. With no recombination, the F2-cross goats are Hh Ll-that is, long-haired and long-legged or short-haired and short-legged types seen in the original strains. But if in a few cases during meiosis in the F1 animals, the homologous chromosomes have exchanged sequences, the result will be a corresponding number of animals with recombined genotypes. The recombined chromosomal groupings are Hl and hL. They will be visible in the different phenotypes in crosses with the double recessive hl, which will result in individuals with genotypes Hh ll - long - legged and short-legged-and hh Ll - short-haired and long-legged.

Note that the original chromosomal grouping of HL and hl has also been perpetuated; recombination provides an illustration of the combination of innovation and conservation that is often found in genetics.

Red queen hypothesis, the idea that species inhabiting the same area are in a perpetual evolutionary race with one another, because an improvement in one species' adaptedness to its environment must be at the expense of another species.

Reduction division. See meiosis.

Redundant code (also called 'degenerate code'), the >genetic code of DNA and RNA, described as 'redundant' because each amino acid has more than one codon to specify it.

Regeneration, the regrowth of an organ or tissue that has been lost by accident or disease. The ability to regenerate is, perhaps paradoxically, a primitive character in evolutionary history, becoming progressively lost as organisms become more complex. Many invertebrates can regenerate an entire new individual from a severed fragment; crustacea and reptiles can regenerate legs or tails. Mammals can only regenerate damaged skin, bone and nerve fibres and those only to a limited extent. They also show the related phenomenon of compensatory hypertrophy: if one kidney is destroyed, the other grows to nearly double the size to compensate (the same happens if part of the liver or pancreas is lost). Plants regenerate extremely readily, with whole new plants able to grown from cells taken from roots, stems or leaves in most species.

Regression line, a line drawn over plotted points on a graph where there are two variables. One is the dependent variable, and the other is the independent variable; the regression line shows how much change in the independent variable. It is a measure of statistical association only, and does not imply that the two variables are causally related.

Regulator gene, a gene whose product is concerned with controlling the rate of synthesis of the products of structural genes. The structural genes may be either adjacent (cis-) or somewhere else (trans-) in the genome. See lac operon for an example of a regulator gene as part of an operon.

Rejection, the adverse reaction of a recipient's immune system to a graft or transplant from an unrelated donor, which it recognizes as foreign tissue. The cells of the grafted or transplanted tissue are attacked, and become non-functional. The more closely the histocompatibility antigens of donor and recipient are matched, the less likely rejection is to happen. It can be controlled to some extent by immuno-suppressive drugs.

Relationship, coefficient of, the proportion of genes that are shared (on average) by relatives. It depends the number of steps between them: parent-child or brother-brother is one step, and the number of genes in common is a half. For uncle-nephew or grandparent-grandchild, both two steps, it is a quarter; first cousins, three steps, one eighth and so on. The number of genes in common is halved for each step. These are average numbers with the actual depending on the distribution of the chromosomes during meiosis (except in the case of parent-to-child, which is necessarily exactly half).

Relict population, a remainder, usually in a geographically isolated position, of a formerly widespread species.

rem, roentgen-equivalent-man, a measure of the amount of radiation absorbed by the body. There is no safe level below which genetic damage does not occur, or where there would be no risk of cancer (see pp. 33-4, and roentgen).

Reoviruses, a group of viruses with a double-stranded RNA genome. They have been found in human lung and intestine tissue but have not been associated with any disease; they are therefore known as 'orphan' viruses. (The name reo is from 'respiratory and enteric orphan')

Repair synthesis, the process in which a section of DNA that is damaged is removed by enzymes and replaced with a newly synthesized sequence. See also >DNA repair.

Repetitive DNA, any sequence of DNA that occurs more than once in the genome. Repetitive DNA can be moderately repetitive, when it consists of many similar copies of sequences coding for regulatory functions and for transfer and ribosomal RNA; it also includes multigene families. Highly repetitive DNA is different in that it does not seem to have a genetic function. Higher organisms have quite lare amounts of highly repetitive DNA: in humans, there is one sequence that is repeated about 300,000 times (see Alu sequence) and guinea pigs have a six-base-pair sequence repeated several million times. See also selfish DNA.

Replication, the process by which a new complete molecule of DNA is made, by one single strand being used as a template for the assembly of another single strand. This is semi-conservative replication, as opposed to conservative replication, in which the whole original double-stranded molecule would act as a template for a wholly newly synthesized doublestraned one. In 195'' Meselson proved that DNA uses the former method, as Crick and Watson had earlier suggested.

In simple terms, replication begins when the two strands of DNA part in what is called a 'replication fork', and DNA polymerase begins to assemble a corresponding new strand alongside each of the original ones. DNA polymerase reads the parent strand from the 3' end going towards the 5' end. Because of the antiparallel arrangement of the two strands, only one is lying the right way round for the enzyme to work in this way; this is known as the 'leading strand'. On the other, the 'lagging strand', the enzyme still works from the 3' end, but to do so, it has to jump ahead and work back in short bursts. The resulting short fragments of newly synthesized DNA (which are called >Okazaki fragments and are about 1000 bases long) are quickly ligated together to form a continuous strand. Note that in reading from the 3' end of the parent strand, the enzyme is building the new strand's 5'end first and working towards its 3' end.

In eukaryotes, chromosomes are replicated by DNA synthesis taking place in both directions from each of a number of points. Once replication has been initiated at one point, two replication forks travel away from each other (at a speed of about 1 pm per minute). Each stretch between the points is known as a replicon, and these eventually join together to form the complete new chromosome. In prokaryotes, with a circular chromosome, DNA replication begins at one point and proceeds in both directions, meeting on the opposite side (or sometimes by >rolling circle replication).

Replicon, a length of DNA between points of initiation of replication. It is not known how these points are recognized by the enzymes involved. In eukaryotes, a replicon is typically about 40,000 base-pairs long.

Repressible enzyme, an enzyme that is produced all the time, except when its gene is switched off by the binding on of the >repressor molecule to the operator. The opposite is an inducible enzyme. See gene regulation; operon.

Repressor, the protein that acts in an >operon to prevent transcription of the structural genes. The repressor is coded for by a sequence at the upstream end of the operon; only a very few molecules are made (perhaps only ten molecules per cell of the lac operon's repressor) as there is only need of enough to be sure that the operon will be turned off by one repressor molecule when required. The scarcity of repressor molecules in the cells of bacteria was the reason why it took a long time for them to be discovered and identified as proteins. The repressor is normally bound to the promoter region of the operon, but when the inducer molecule is present, it combines with the repressor so as to make the repressor dissociate from the operator, allowing transcription of the structural genes downstream to proceed.

Reproduction, the process by which a living thing creates one or more individuals similar to itself. (Viruses are not able to reproduce by themselves, they can only do so by incorporating their DNA into a host cell; they are therefore usually not considered to be living things.) The simplest method of reproduction is equal cell division, in which the parent cell divides to form two identical cells. There is no difference in the genetic content of these two cells (barring mutation). This method is used by bacteria and other very simple forms of life. A variant of this is unequal budding, which is essentially the same except that the parent cell continues its independent existence after extruding a small cell as a bud; this happens in yeasts. Again there is no genetic difference between the cells involved.

The earliest system to incorporate any exchange of genetic material during reproduction is >conjugation, the process in which bacteria transmit DNA to one another on >plasmids. Sexual reproduction can be defined as reproduction by the fusion of a male and a female >gamete. It involves exchange and reassortment of genetic material, and thus provides much more genetic diversity for natural selection to work upon. It is the system used by plants and animals (note that it does not imply an act of sexual intercourse: sexual reproduction involving wind-borne pollen fertilizing a distant flower is still sexual reproduction).

'Reproduction is by far the most important biological function that any living species has to perform. The only alternative to satisfactory performance of this function is extinction.' This quote is from the American geneticist Theodosius Dobzhansky. See also meiosis; recombination; vegetative propagation.

Repulsion. See trans configuration.

Resistance, the ability of some bacteria to resist the effects of antibiotics because of their genetic constitution. Resistance is passed from one bacterium to another during >conjugation, when a plasmid (a nonchromosomal DNA element) carrying the protective gene is transferred. Usually one plasmid codes for resistance to a particular antibiotic, and to be resistant to more than one, a bacterium must have a number of R plasmids code for more than one types of resistance. Plasmids can be transferred between species or even genera of bacteria. The problem from the point of view of public health is that the more extreme the measures taken to eradicate harmful bacteria with antibiotics, the stronger the pressure of selection to produce resistant strains.

Restriction endonuclease, an enzyme that cuts a DNA molecule but only at the site of a particular sequence (usually 4-8 bases) that it recognizes. Since the first one was discovered in 1968 by the Swiss molecular biologist Werner Arber, over 350 different restriction endonucleases have been identified, and they have proved to be extremely useful tools in DNA research and >genetic engineering. Some cut the DNA straight across, leaving a 'blunt end', but others cut so as to leave a few unpaired bases on one strand (a 'sticky end', but others cut so as to leave a few unpaired bases on one strand (a 'sticky end'). Their function is thought to be the protection of the cell from any incoming foreign DNA (which is cut up by the enzymes). See restriction map; restriction fragment length polymorphisms.

Restriction fragment length polymorphisms (RFLPs), a number of different polymorphisms in humans, detectable by the use of >restriction endonucleases (specific types of enzymes). If the total DNA in an individual's genome is digested with a restriction enzyme, it ends up as a great many fragments; these can be sorted by Southern blotting, using a probe from the part of the genome that is being investigated. This probe will find a small number of fragments of defined length. For example, when the endonuclease Hind III is used to test for >sickle-cell mutation has knocked out the DNA sequence at two of the sites where Hindi III can act). Because not all people have the same fragments, this is said to be a polymorphism. Sometimes instead of a fragment being missing it will be of an atypical length, when a mutation has inserted or deleted some DNA between the cutting sites. If these fragment-length polymorphic loci are very closely linked to a gene causing a genetic disorder, they can be used for prenatal diagnosis.

Restriction map, a diagram of the sites on a DNA molecule, in terms of where particular >restriction endonucleases have their cutting sites. By digesting the same DNA molecule with different endonucleases (all of which have different sequences that they recognize as restriction sites, i.e. the places where they cut) one can obtain a picture of the molecule as a whole. A large fragment is digested by one endonuclease and a number of smaller fragment obtained; if this process is repeated the overlapping resulting fragment can be built up into a picture of what order the fragments are in the complete molecule. The practical complication of this procedure can be judged from the fact that even so small a genetic entity as a plasmid may have about 500 restriction sites.

Retardation, the appearance of a structure in embryogenesis later than it did in ancestral forms. The opposite is >acceleration.

Retinitis pigmentosa, an inherited condition in which pigment forms in clumps in the retina at the back of the eye; these restrict the visual field and eventually cause total blindness (unless treated). There are at least four separate loci at which genes causing retinitis pigmentosa are found: at one the mutant allele is dominant, and at another it is recessive, and at a third there are two mutant alleles, one dominant and the other recessive to the normal allele. The fourth is an X-linked recessive. The overall frequency of the disorder is about 1 in 5000.

Retinoblastoma, a form of cancer that is inherited as an autosomal dominant. Children who have the retinoblastoma gene develop tumours in both eyes in early childhood unless treated (both surgery and irradiation are effective). The gene has been located to chromosome 13, and appears not to be a mutant allele but a very small deletion of chromosomal material. This deletion happens at the site of a tumour suppressor gene; when this is knocked out, the cancer can arise. Another form of retinoblastoma is inherited as an >X-linked recessive. The autosomal dominant kind occurs in about 1 in 30,000 births.

Retroviruses, a group of viruses with a single-stranded RNA genome. Their peculiarity is that they possess the enzyme >reverse transcriptase, by means of which their RNA is back-transcribed into DNA, which is then inserted into a host cell's genome. Retroviruses are used in >genetic engineering, their reverse transcriptase include a gene that is an >oncogene, i.e. it can induce cancer. Examples are Rous sarcoma virus, several viruses causing leukaemia and other cancers in mice, and HIV (human immunodeficiency virus) which is responsible for AIDS.

Reverse genetics, term used to describe the process in human genetics in which gene mapping (often with restriction fragment length polymorphisms as markers) is used to locate a gene that causes a clinical condition. The gene can then be cloned and its product identified. This is the 'reverse' of the traditional pattern in medical genetics, when the biochemical nature of the disease would be known but the location of the gene unknown. The molecules involved in cystic fibrosis and Duchenne muscular dystrophy were both found by means of reverse genetics, and it is hoped that this is a stage towards a cure for these diseases.

Reverse mutation. See back mutation.

Reverse transcriptase, an enzyme that can make a strand of DNA using RNA as a template. This is a contravention of the >central dogma of molecular biology, which states that the flow of information is always from the DNA outwards, not the other way round. Reverse transcriptase occurs in >retroviruses, enabling them to get their RNA genomes copied into the DNA genomes of the host bacteria. It may also be responsible for some gene copies within eukaryote genomes: a copy that lacks the introns of the original may well have been made by reverse transcription from messenger RNA (see Alu sequence). Reverse transcriptase is widely used in genetic engineering. It was discovered independently in 1970 by both David >Baltimore and Howard Temin.

Rhesus blood group system, a set of antigens on the red blood cells in humans (and other primates - the name comes from the Rhesus monkey, in whose blood the antigen was first isolated). The Rhesus system was discovered in 1940 by Landsteiner, the blood-group pioneer who was also the discoverer of the first known system, ABO, in 1990. There are five Rhesus antigens: C,D,E, c and e, of which D is the clinically important one. People with D are said to be 'Rhesus-positive', and those without it are 'Rhesus-negative'. According to one theory, the genetic control is at three closely linked loci, with the D locus having two alleles: D (dominant) and d (recessive): Another theory has a single locus with multiple alleles. The gene frequencies vary a great deal between populations: Rhesus-negative people comprise about 17 per cent of European populations but only about 3 per cent of people of African descent, and are extremely rare among people of Chinese descent, Rhesus incompatibility is responsible for >haemolytic disease of the newborn.

Rheumatoid arthritis, an auto-immune disease in which there is probably an element of genetic susceptibility. The joints (usually starting with fingers and toes) are inflamed and become stiff and ultimately unusable; this is due to the immune system attacking the connective tissue. Women are three times as likely to be affected as men, and some 15 per cent of women over the age of 65 have this disease. There is some relationship to the >HLA system, in that four out of five rheumatoid arthritis sufferers have HLA type DR4.

Ribosomal RNA (abbrevation: rRNA), the form of >RNA that is found in >ribosomes. In eukaryotes, there are four types (of different lengths), the genes for which occur in many copies, making up part of the repetitive DNA in the genome. Sometimes these RNA genes are also the subject of >gene amplification. The rRNA is synthesized as a continuous long molecule which has spacer RNA (coded for by >spacer DNA) between the sequences of actural rRNA. Prokaryotes have three types of rRNA, and only sever copies of the genes for each.

Ribosomes, >organelles found in large numbers in the cytoplasm of the cells of both prokaryotes and eukaryotes. They are cottage-loaf shaped, made of >ribosomal RNA and protein, and their function is to bring together the amino acids during >protein synthesis.

Rickets, a disease in which a deficiency of vitamin D leads to failure to absorb calcium and consequently to stunted r softened bones. The vitamin is found in yeast, egg yolks and fish liver, and can also be synthesized in the body under the stimulus of the ultraviolet rays in sunlight (see pigmentation). In its most common form, rickets is a nutritional disease, the lack of vitamin D being simply due to a poor diet. Other cases are due to a lack of vitamin D being simply due to a poor diet. Other cases are due to a lack of vitamin D being simply due to a poor diet. Other cases are due to a lack of exposure to sunlight. There is also an inherited form known as vitamin-D-resistant rickets, in which the inability to absorb calcium into the bones is not involved with vitamin D but rather with phosphate metabolism. This condition is an X-linked dominant, with heterozygous females showing much variation in how severely they are affected.

Ring chromosome, an abnormal chromosome that has the ends of its arms joined to each other to form a ring.

RNA (ribonucleic acid), a molecule chemically similar to DNA. The sugar that connects the bases is ribose (rather than deoxyribose in DNA), and in almost all cases, the molecule consists of a single strand. The base-pairs in RNA are guanine(G)-cytosine(C) and adenine(A) - uracil(U) (where DNA has thymine); these carry the genetic code. RNA has a central role in >protein synthesis, at several different stages. First comes heterogenous nuclear RNA (hnRNA), which in eukaryotes is the transcriber of the DNA code; then >messenger RNA (mRNA), which carries the coded message to the >ribosomes for synthesis; next >ribosomal RNA (rRNA), a constituent of the ribosomes; and finally >transfer RNA (tRNA), which assembles the amino acids. Each is synthesized by a particular >RNA polymerase.

RNA polymerases, the various enzymes that produce the different types of RNA, by copying from a DNA strand (the >sense strand); this is the process of >transcription, and is the basis of >protein synthesis. Prokaryotes have only one RNA polymerase, but in eukaryotes there are different ones for the different RNAs: Type I synthesizes the three largest ribosomal RNAs; Type II synthesizes heterogeneous nuclear RNA and messenger RNA; Type III synthesizes transfer RNA and the smallest ribosomal RNA; and Type IV is found only in >mitochondria, which have some independent DNA.

Robertsonian translocation, a type of chromosomal translocation; also known as whole-arm fusion. The long arm of an acrocentric chromosome breaks off very close to the centromere, and becomes joined to another (non-homologous) acrocentric which has also broken close to the centromere (but on the short-arms. The latter is often lost in subsequent divisions, but as it may well contain little genetic information there is usually no ill effect. Whole-arm fusion probably accounts for the different karyotypes of related species, when some have all acrocentrics and others have the same amount of genetic material but with some metacentrics as well as acrocentrics.

roentgen, a unit of radiation, the amount that results in 2.083 x 109 ion pairs per cm3 of air. It is called after the German physicist W.K.Roentgen who discovered X-rays in 1895. See also rem.

Rolling circle replication, a mode of DNA replication occurring in circular DNA molecules (e.g. those of bacteriophages). A nick is made in one strand, and new nucleotides are added to the 3' side of the opening, displacing the original second strand, which falls away. As this unpeels, it also gets another complementary strand by replication, and the result is a new double-stranded length of DNA much longer than the original DNA circle.

Rous sarcoma virus, a >retrovirus that causes tumours in chickens. Identified in 1910, it was the first virus to be associated with a form of cancer.

rRNA. See ribosomal RNA.

Rubella or German measles, a disease causing serious >congenital defects in the foetus if the mother becomes infected with it while pregnant. The risks of the foetus being damage are: infection in first four weeks of pregnancy, 50 per cent; at 5 to 8 weeks, 35 per cent; at 9 to 12 weeks, 15 per cent; thereafter negligible. Vaccination against rubella is effective.

S

S1, S2, S3, etc., generations in >self-fertilizing plants, where every generation is self-fertilized.

Saccharomyces cerevisiae, common baking or brewing yeast, one of the most important organisms in human history. It reproduces by 'budding', that is, a small outgrowth appears on the cell wall and grows until it is nearly equal in size to the original cell, when a wall forms to separate off the new cell. Saccharomyces (which is a fungus) has also been much used in genetic research, and is one of the organisms used in commercial >gene cloning.

Salivary-gland chromosome. See polytene chromosome.

Saltation, a sudden evolutionary change, probably due to a major mutation. See punctuated evolution.

Samesense mutation, a >mutation in which there is a change in the sequence of base-pairs in the DNA but no change in the amino acid coded for (see genetic code). This is most likely to occur in the third base of a codon (see wobble).

Sanger, Frederick (1918-), British biochemist; worked at the Medical Research Council Laboratory of Molecular Biology at Cambridge from 1951 until his retirement in 1983. Sanger devised a method for analysing the amino acid sequence of a protein chain; it involved first cutting up the protein into short sequences with acids or enzymes, then labelling the 'free amino' end of each sequence. He used this technique to work out the sequence of insulin, which has 51 amino acids arranged in two chains. For that he won his first Nobel prize for chemistry in 1958. His second came in 1980, for his work on >nucleic acid sequencing: he read the whole genome of the bacteriophage Phi X 174. He has also sequenced the entire Epstein-Barr virus (see herpes viruses).

Satellite chromosome, a part of a chromosome that is joined to the end of the main chromosome by a very fine thread of non-condensed chromatin.

Satellite DNA, the DNA from a eukaryote that, when centrifuged, separates out as a small fraction distinct from the bulk of the DNA in the genome. The satellite fraction contains mitochondrial and highly >repetitive DNA. It is not clear why the amount of satellite DNA should vary widely between closely related species.

Schizophrenia, a severe mental illness in which the person's connection with reality is disturbed and their reason and emotions become disjointed from one another. Schizophrenia takes many forms and there is no overall agreement among psychiatrists as to what is or is not schizophrenia. There is like wise no single explanation of its cause(s), although a genetic element seems to be involved. The best evidence for this is that if one identical twin has schizophrenia, in 60 per cent of cases the other twin will develop it too, while the equivalent figure for non-identical twins is 10 per cent. About one person in 100 in Britain suffers from schizophrenia. Some recover from it and some do not, a fact which has been interpreted as showing that there are several separate diseases with similar symptoms but quite different causations.

Scion, a shoot of one plant that is grafted on to the stock of another. The two plants are genetically different and need not even be of the same species. The word's genetical use is therefore at odds with its literry use, when it means 'heir' or 'descendant'.

Secondary sexual characters, any charaters shown by one sex and not the other, except the reproductive organs themselves (the >primary sexual characters). In humans, the beard is a secondary sexual character. There has been discussion as to whether the female breasts are a primary or secondary sexual character: during breastfeeding they enlarge and are part of the necessary reproductive apparatus, but at other times they could (as in other mammals) return to their former size. See sexual dimorphism; sexual selection.

Segregation, the separation of the two members of a pair of alleles during the formation of gametes. It is Mendel's first law. He proved that for every trait the individual has inherited one allele (a word Mendel himself did not use) from each of its two parents, and that when this individual comes to make gametes equally frequently. The two alleles are passed on not only independently but also unchanged; there is no blending.

'Segregation' is also used to refer to the separation of the two homologous chromosomes in a pair during >meiosis. See also independent assortment (Mendel's second law).

Segregational load, the part of the> genetic load that consists of the less fit homozygotes where the heterozygote is the fittest genotype. See heterozygous advantage; fitness.

Selection, coefficient of, a measure of the extent to which natural selection is acting against a given gentotype, i.e. reducing its relative contribution to the next generation. Denoted by s, the coefficient varies from 0 to 1: if it s = 0, the genotype is not selected against at all; if s = 1, selection is total, i.e. the genotype makes no contribution to the next generation (a lethal or sterility gene). See also fitness.

Selection differential, the difference between the phenotypic mean of the population and of the individuals chosen (in artificial selection) to be the parents of the next generation.

Selection limit, the point at which further change in the genetic composition of a strain (e.g. improvement in farm animals) is impossible because there is no further genetic variability available. There can be no dogs bred to be 3 inches high because there are no alleles in the gene pool of canis familiaris which have that effect. Another reason for selection reaching a limit is that improvement in one characterisis often counterbalanced by a disimprovement in another: hens may be bred to lay more eggs per unit time, but beyond a certain point this is only achieved by reducing the size of each egg, so that the total egg output ceases to improve. See also heritability.

Selection pressure, the intensity with which natural selection is acting upon a population to change the gene frequencies from one generation to the next.

Selective advantage, the increase in >fitness of one genotype compared to others in the same population.

Selective reduction, induced abortion of one or more foetuses out of a multiple pregnancy. The technique consists of killing the foetus with a needle, using an ultrasound scanner to guide it. It was developed for the few cases where one twin has a defect, such as spina bifida or anencephaly (failure of the braib grow), that can be seen by ultrasound. It has also been used for eliminating some of the foetuses in a pregnancy with a large number (quads or more) implanted after in uitro fertilization. It carries a high risk that all the foetuses will be lost.

Self-fertilization, the process by which a female gamete is fertilized by a male gamete produced by the same individual. A primitive system, it is frequently found in flowering plants and in some lower animals that are >hermaphrodites (e.g. worms, slugs and snails). In evolutionary terms it is a disadvantageous strategy, as it leads to high levels of homozygosity within a population and consequently a less flexible genetic response to environmental change; there fore various mechanisms have evolved to prevent it (See self-sterility).

Selfish DNA (also called '**junk DNA**'), the DNA within the genome that has no apparent genetic function. In vertebrates, this can amount to as much as 90 per cent of the genome, and there has been much speculation as to why it is there. Crick and others have suggested the term 'selfish DNA', to denote that it is present in the genome for its own benefit and not which a virus inserts its own genome into that of a host cell. But it is not certain that 'selfish' DNA has no function: it may have arrived in the genome for its own sake, but perhaps the recipient organism has found a way to make it useful. Selection would tend to pick up on any usefulness of selfish DNA, for otherwise it is a great burden on the cell to have so much replicating to do for no purpose.

Self-sterility, a feature of certain flowering plants, apparently evolved to avoid > self-fertilization which increases the average homozygosity in the population. The mechanism, as studied by Darwin in the common primprose, consists of these plants having two different shapes of flowers. A bee visiting the 'pin' flower type gets pollen stuck on the middle of its proboscis from the anthers half-way down the flower tube; which has the stigms half-way down the tube. At the same time, the placing of the anthers at the top of the tube in the thrum corresponds with the placing of the stigms in the pin. Since a single plant only has the one kind of flower, it is self-sterile. Several varieties of commercially grown fruit trees are self-sterile because they have all the same kind of flower, they need a 'pollinator' tree nearby with the opposite-shape flowers.

Semen, the fluid that contains >sperm, produced in the testes of male animals. The fluid also contains a high concentration of the sugar fructose, probably as 'fuel' for the sperm. In humans, the volume of semen per ejaculation is 2-4 ml, which contains about 200 million sperm.

Semi-conservative replication. See replication.

Semi-dominant, an allele that, when present in the heterozygous state, has an effect intermediate between that of the two homozygotes. A classic example (and one that Mendel did not find in his peas) is where genotype RR has red flowers, rr has white flowers and Rr has pink flowers: R is said to be semi-dominant, or partially dominant. See also semi-lethal.

Semi-lethal (also known as semi-dominant lethal), an allele that in the heterozygous state, has a deleterious effect but is even worse, i.e. lethal, when homozygous. Very many alleles causing human disorders are loosely described as dominant when they are, in fact, semi-lethal. For example, achondroplasia (a form dwarfism in which the person has very short limbs but is otherwise healthy) does indeed behave as a dominant in that the aa homozygote is normal and Aa heterozygote has achondroplasis; but it can be seen from the offspring of marriages between achondroplasics that the homozygote AA is very severely deformed and dies not longer after birth.

A semi-lethal that kills the individual before birth gives an apparently non-Mendelian ration. For example, if the alleles are H and h, the cross Hh x Hh gives 25 per cent hh, 50 per cent HH. Since the latter die before birth, the ratio appears to be 1:2, though it is fact the Mendelian 1:2:1 with the last group missing.

Senescence. See ageing.

Sense strand, the DNA strand that serve as the template for RNA and thence protein synthesis. It is also known as the'anticoding' stand, the messenger RNA transcribed from it being the 'coding'strand. See genetic code.

Severe combined immune deficiency (SCID), an inherited disease in which the body has no immune response (see immune system).

SCID sufferes are unable to withstand any infection by viruses or bacteria; these are the 'bubble babies' who have to kept entirely enclosed from birth. The genetic cause is a recessive mutant which fails to produce the enzyme adenosine deaminase; as a result the bone marrow transpalnt if a tissue-matched donor can be found. About 1 in 25,000 babies is born with SCID, and each has only a 50 per cent chance of surviving more than six months. It is hoped that gene therapy may produce a cure for SCID.

Sewall Wright effect, the proper name for >genetic drift, named after its discoverer, the American geneticist Sewall >wright.

Sex chromosomes, those that determine the sex of an individual (See sex determination). In humans, as in many other creatures, there are two sex chromosomes, X and Y. See also chromosome abnormalities; gamete.

Sex determination, the various mechanisms that decide whether an individual is male or female. Some plants and animals do not have separate sexes; they are >monoecious (plants) or >hermaphrodite (animals). Species that do have separate sexes are >dioecious (plants) or bisexual (animals). In animals (in the broadest sense, i.e. members of the animal kingdom) the sex is determined by the sex chromosomes. Generally, the female has two X chromosomes (XX), and the male has an X and a Y chromosome (XY). In this arrangement, it is the activity of the Y chromosome that causes maleness; the extra X in the female is not responsible for causing femaleness (and indeed probably has no effect at all - See Lyon hypothesis). This can be seen from the sex of humans with the abnormal sex chromosome types XXY, who are male, while those with just one X are female (See Klinefelter's syndrome; Turner's syndrome).

In most higher plants and in the majority of animal groups, XY is the female and XY the male. The sex with the same two sex chromosomes, is called >homogametic (from the Greek homos, 'the same', because all the gametes are made the same) and the sex with two different sorts of gametes are made). The female makes only X-bearing ova, the male makes both X-bearing and Y-bearing sperm, and is therefore the parent responsible for the sex of the offspring. In some species, a similar arrangement has females that are XX and males that are XO (i.e. with only one X and no other sex chromosome): this works the same way, with males producing either X-bearing or 'blank' sperm. In some animals, it is the female that is heterogametic; in these species the sex chromosomes are called W and Z, so the female is WZ and the male is ZZ. In these groups, which include the butterflies and moths, all birds and most fish, amphibians and reptiles, the sex of the offspring is determined by the mother who produces W or Z bearing ova.

In the Hymenoptera (the social insects such as ants, wasps and bees) the females are diploid and lay haploid eggs. If the eggs are fertilized they become diploid and are therefore female, but if they are not fertilized they remain haploid and develop into males (the males produce their sperm by mitosis, not meiosis). See social insects; parthenogenesis.)

In some organisms sex is determined not by chromosomes but by the environment. Some turtles do not have sex chromosomes, and develop into one sex or the other according to the temperature at which the eggs are incubated. An extreme case is the Bonellia marine worm: the larva settles in the mud and if it has enough space round it develops into a female, after which any other larva that arrives is turned into a male by the influence of her scretions. Maleness to this marine worm involves the loss of all organs except the reproductive one, and the male becomes a parasite inside the female's reproductive tract. Some fish, e.g. wrasses, have sex reversal, with individuals able to change from one sex to the other if he population's sex ratio becomes unbalanced. A special example of this is the Pacific coral-reef fish Labroides in which a male fish has a harem of individuals who are kept female by being dominated; when the male dies, the most senior female changes sex and becomes the new male.

Sex hormones, those that control the reproductive cycle in animals. The female steroid hormones are: the oestrogens, produced by the ovary and the placenta (these are used in hormone replacement therapy in women who have reached the menopause); and the progestrogens, produced by the placenta during pregnancy. Males produce androgens, also steroids, of which the most important is testosterone, produced in the testes and important in controlling spermatogenesis. The androgens also have an anabolic (growth-stimulating) effect; synthetic anabolic androgens have been developed which give the extra growth (especially in muscles) without too much of the masculinizing effect. Both sexes produce gonadotrophic hormones in their pituitary glands, which act upon the ovary and the testes. One of these, follicle-stimulating hormone (FSH), stimulates the ovary to produce ripe eggs; it is used to treat some infertile women, and can cause multiple births.

Sex-limited trait, any trait that is only expressed in one sex. An example is milk yield in cows, which can obviously only be a trait of the female even though the genes for it are carried on the autosomes and are therefore present in both sexes. A more subtle example is male-pattern baldness in humans, which only affects males because it depends on the influence of the male sex hormones. See also X linkage

Sex linkage. See X linkage.

Sex ratio, the proportion of males to females in a population. Although in most species of animals, it seems obvious that the ratio is 1:1, the position is more complicated. Males being some what more vulnerable during embryogensis, many species have an excess of males at conception to allow for wastage. In humans there is excess mortality of boys during adolescence, and to ensure an approximately equal sex ration taken at age 70+ (in Britain) is about 74:100 (and by age 80 it is 43:100). Some species can adjust the sex ration to suit conditions (See sex determination), but in general it can be shown that the sex ratio will evolve towards being equal at maturity. The following argument was put forward by Fischer: if males are rarer, each male will have a better chance of finding a mate, so it will pay parents to produce extra males; but this advantage will cease when males are no longer rare. The same argument applies to females, so the sex ratio stabilizes.

Sex reversal, alteration of an individual's sex. It occurs in nature, especially with some species of fish (See sex determination), and can also be artifically induced in many species by surgical and chemical intervention.

Sexual dimorphism, the existence of physical differences between the two sexes, other than the differences in the reproductive organs. Some species, e.g. mosts nakes, have almost no sexual dimorphism, but others, including many birds, take it to spectular extremes. Most mammals have at least a difference in size of skeleton and musculature. In humans, sexual dimorphism has decreased during our evolution from an ape-like ancestor. Darwin proposed, and it is now accepted, that sexual dimorphism evolves by means of >sexual selection.

Sexual selection, a form of selection under which an individual has more offspring than average because of being able to acquire more mates. It was first described by Darwin, who though it both important and paradoxical. For him the puzzle was that by means of sexual selection a trait could develop that was not actually 'useful' to the species at all. The classic example is the peacock's tail, a spectacular and successful piece of sexual display but not a help to the bird in the business of getting its food or escaping predators. However, according to Darwin's correct view that what counts is the number of surviving offspring, a bird that gains more mates is more evolutionarily successful than another, regardless of their general competence in the environment. Darwin identified two components of sexual selection, and called them 'the power to charm the females' and 'the power to conquer other males in battle'. The latter leads to the evolution of offensive weapons in males, to be used not for hunting or defence but against other males of their own species (again a situation that does not on the face of it seem to be for the 'good' of the species).

Shuttle vector, a genetically engineered hybrid vector that can function in two different hosts. A plasmid that consists of DNA from both ***Escherichia coli*** and yeast is used for gene cloning with eukaryote genes. The gene us cloned in ***E. coli***, then the plasmid is isolated and transferred to yeast which, being a eukaryote (albeit a primitive One), is more easily able to express eukaryote genes.

Sibling (sib), a brother or sister. Siblings have the same two parents; half-siblings have only one parent in common.

Sickle-cell anaemia, a disease in which the red blood cells are malformed and fragile and break up faster than they can be formed; this produces severe anaemia, which often proves fatal. Sickle-cell anaemia provides the classic example of Mendelian inheritance in humans, as the whole story is known - from the mutant base-change in the DNA to the evolutionary forces that keep the gene at a high frequency in many populations (See polymorphism).

Red blood cells contain a protein called haemoglobin, whose normal form is haemoglobin A (for simplification, the normal allele can be called HbA). There are several abnormal haemoglobins, the one involved in sickle-cell anaemia being haemoglobin S (allele Hbs). Homozygous people Hbs/Hbs) produce only haemoglobin S, which tends to form semicrystalline lumps in the red blood cells, giving them a distorted shape (like a sickle, hence the name). Heterozygotes (HbA/Hbs) produce both kinds of haemoglobin, and although they sometimes have mild anaemia they are generally healthy; they are said to have sickle-cell trait.

Analysis of the structure of the haemoglobins showed that the difference between the A and S forms is just one amino acid out of a chain that is 170 amino acids long, and this in turn is due to the >codon in the DNA changing from G-A-A to G-U-A. Thus the change of a single base in the DNA can make the difference between a person being healthy or having fatal anaemia.

It was discovered that people who are HbA/Hbs heterozygotes were significantly less susceptible to malaria than people only the normal haemoglobin A. It seems that the Plasmodium parasite finds the sickletrait red blood cells more difficult to colonize and this provides the explanation for why a gene causing such an ill-effect as sickle-cell anaemia should occur at frequencies as high as 40 per cent in some tropical countries. Homozygous persons HbA/HbA are likely to die of malaria (if untreated), homozygotes HbA/Hbs have the best chance of survival. When heterozygous have offspring, they keep both the Hbs and HbA gene in circulation. This situation, not rare in population genetics, is known as >heterozygous advantage. The Hbs gene is only advantageous in areas with endemic malaria, and this can be seen in its distribution: Hbs occurs in tropical Africa (with gene frequencies of about 15 per cent in West Africa, 40 per cent on Mount Rumwenzori in Zaire, and up to 20 per cent in Madagascar, to take some examples), in Greece and Sicily, and in the southern extremes of the Arabian peninsula. It also occurs in lack people of African origin, with an estimated 10 per cent frequency in the United States, where there was a notoriously ill-thought-out scheme in the 1970s for the >genetic screening of black people for sickle-cell anaemia.

Significance, the threshold level of >probability at which the experimenter decides to accept the results of statistical analysis. This will vary according to circumstances, but P = 0.05 (1 in 20) would be a minimum.

Silent mutation, any >mutation that does not have a detectable effect. See neutral allele.

Silk, the fibre made by some insects (for cocoons) and spiders (for webs). Silk consists mainly of a protein called fibroin which contains two amino acids: alanine and glycine. The silk worm, **Bombyx mori**, produces the greatest quantities of the fibre, and a curious genetic mechanism enables it to do so. The moth has only a single copy of the gene for fibroin, and it would not be physically possible for this to be transcribed fast enough to make silk at **Bombyx's** speed; instead, in the silk-gland cells the entire genome is replicated about 18 times to give the extra copies of the gene. The more usual mechanism for coping with localized demand for high output from a gene is >gene amplification.

Skin colour. See pigmentation.

Smell, a sense of great importance to the many animals that communicate by means of >pheromones. In humans the sense of smell is much weaker than in other mammals. It is supposed that in human evolution sense has been sacrificed to the anatomical adjustments required for the enlarged brain, but this seems odd, given that humans still have physically quite as much space for the apparatus, external and neurological, as better-smelling animals have. Perhaps humans have a better smelling ability than they think they do, the sense of smell being under-used in human societies; but that would be odd too, given the advantage to every human group to make use of whatever resource it can. Maybe the sense of smell became discused after humans began to walk upright, when it became difficult to get the nose to the ground where most of the smells are. There is genetic variation among humans in the ability to smell particular things, and a curious example of this is that not everyone can smell every colour of that fragrant flower the freesia.

Smith, Hamilton (1931-), American molecular biologist; professor of molecular biology at Johns Hopkins University, Maryland since 1981. Smith made a crucial contribution to the technical repertoire of genetic engineering by isolating and identifying numerous restriction endonucleases in >*Escherichia coli*. These are the enzymes that can cut a >DNA molecule at one exact site. Smith and two of his colleagues shared the Nobel prize for physiology or medicine in 1978.

Snails, a group of molluscs consisting of thousands of aquatic and land species. The latter are hermaphrodites. The land snail *Limnaea* provides the best example of a >maternal effect.

Social Darwinism, the social and political ideas that were developed, in the second half of the nineteenth century, from Darwin's theory of evolution by natural selection, by Herbert >Spencer and others. Spencer believed that human societies evolved as species do, from 'undifferentiated hordes' to military societies evolved as species do, from 'undifferentiated hordes' to military societies and ultimately to industrial societies. It was also believed that the 'struggle for existence' described by Darwin as happening in nature was mirrored by the struggle for survival within human societies, so that social and economic inequalities were viewed as not only justifiable but necessary.

Social insects, a term covering the varios orders of insects that live in organized colonies: ants, wasps (some species), and termites. They are interesting not only because of their astonishingly elaborate 'societies' (including in the case of some ants the keeping of other species of ants as 'slaves') but also for their peculiar mechanism of sex determination.

The sex of an ant, bee or wasp does not depend on its sex chromosomes (they appear not to have any) but on whether the individual is haploid or diploid. The diploid insect is female; if she is fed suitably she becomes a queen and is fertile. She lays eggs; if these eggs have been fertilized they are diploid, and are therefore inherit all his genes from his mother. Males become fertile seasonally, producing gametes by mitosis rather than meiosis (their cells being haploid already). The advantage of this system is that the sex ratio in the colony is not fixed but can fluctuate to meet the environmental demands upon the colony, as the number of eggs fertilized can be raised or lowered. Termites, though also fully social, have the ususal sex-chromosome mechanism for sex determination. There are kings as well as queens in termite colonies, and all termites are diploid.

Sociobiology, 'the systematic study of the biological basis of all social behaviour'. That is E.O. Wilson's definition in his book Sociobiology: The New Synthesis (1975). The essence of sociobiology is the belief that all behaviour patterns are genetically determined, and have evolved through natural selection; sociobiologists are thorough-going Darwinians. Among the questions that sociobiologists are thorough-going Darwinians. Among the questions that sociobiology are >territoriality, >aggression, mate selection, and parental behaviour. Its most controversial topic has been >altruism, for which there does not at first sight seem to be a plausible explanation in terms of natural selection.

Criticism of sociobiology has been strongly voiced, and is of two types. Firstly, geneticists have criticized it because of its reliance on >kin selection, the mathematics of which are only tenable in the case of very closely related kin. Secondly, anthropologists, psychologists, socioligists and others have criticized it because of its oversimplistic view that human behaviour can be thought of as consisting of a series of separate items determined genetically. The sociobiological expanations of homosexuality and incest have come under particularly heavy fire for this. Partly in response to the arguments of social scientists, some sociobiologists (and especially those in Britain) have stated that human social behaviour is a special case, with cultural inheritance as important, as biological inheritance (see exogenetic inheritance). Richard Dawkins, in The Selfish Gene(1976), has proposed the concept of the >meme in order to maintain the idea of selection as the determinant of social behaviour but without being committed to genetic determination. A third type of criticism, that sociobiology is unacceptable because it is intrinsically fascist when applied to humans, is not a scientific argument but a political one. See also game theory.

Somatic cell, any cell in the body of an organism other than its >gametes, the sperm in its testes or the eggs in its ovary.

Somatic mutation, a mutation that occurs in a chromosome within a somatic cell. Such a mutation is not passed on to the individual's offspring, but it will be perpetuated in all cells descended from the one in which the mutation occurred. Some cancers are the result of a somatic mutation (see Philadelphia chromosome,.)

Southern blotting, a technique for separating and identifying fragments of DNA. The DNA sample, perhaps the whole of a genome or a chromosome, is digested with restriction endonuclease and the fragments separated by electrophoresis in a gel. DNA is transferred (or blotted) on to a membrane filter, usually made of nitrocellulose or nylon. Specific sequences of DNA are then found by adding a radioactively labelled probe of RNA or DNA which forms a heteroduplex where the sequence is complementary. After excess probe has been washed away, the bound sequences of DNA can be detected by autoradiography (exposing a photographic plate to radiation from the sample) or scintillation counting (measuring the radiation with a device that shows particle collisions as flashes of light). Southern blotting is called after its inventor, but the parallel technique for RNA - > Northern blotting - is not.

Spacer DNA, a non-coding sequence of DNA between sequences that are gene-coding. An example is found in the gene for >ribosomal RNA, which has spacer sequences to separate the different RNA molecules that will eventually develop.

Spallanzani, Lazzaro (1729-99), Italian priest and physiologist. Spallanzani was the most brilliant experimental biologist of his day. He discovered that animals consume oxygen and excrete caron dioxide (though these gases had not then been identified). He carried out experiments to disprove the idea of spontaneous generation; his sealed jars of previously boiled broth did not go mouldy, so he deduced that the agent causing the mould was something in the air (his experiments were very like those later carried out by Pasteur). By filtering semen, he discovered that spermatazoa are necessary for fertilization, and he carried out the first artificial insemination (on a spaniel bitch). Spallanzani investigated the circulation of the blood, and digestion (using himself as a guinea pig). He also researched the ability of bats to fly in semi-darkness without collisions, and realized that this was not a visual skill (the fact that bats use sonar was not discovered until 1941).

Speciation, the process of formation of a new species. Although it can refer to phyletic evolution, in which one species transforms itself into another over time, it more usually refers to cladogenic evolution, the splitting of a single species into two or more contemporary ones. 'Every species has come into existence coincident in time and space with a pre-existing closely allied species', wrote Wallace. Some mechanism of isolating the incipient species is needed (see isolating mechanism).

Species, a group of related organisms, all of which share the same gene pool and are potentially able to interbreed (producing fertile offspring). The species is the basic unit in classification and in the study of evolution. The test of whether two forms are able to interbreed cannot of course be applied to forms that are only known for their succession of each other in the fossil record; these have to be assigned to species rank using the same sort of criteria that are used for grouping species together into the same genus, etc. Every species has been given a two-part Latin name (see binomial nomenclature). The total number of species at present known to science is just under 1.4 million (about 1 million insects and other invertebrates, 250,000 vascular plants, 74,000 fungi and algae, 37,000 microorganisms, 19,000 fishes, 9000 birds, 9000 reptiles and amphibians, and 4000 mammals). It has been estimated that there may be as many as 30 million species, two-thirds of them tropical and the great majority of them insects.

Spencer, Herbert (1820-1903), British philosopher and sociologist. Spencer published a theory of evolution before the Darwin-Wallace one came out, but his was essentially the same as Lamarck's idea of evolution through the inheritance of acquired characteristics. Following the publication of On the Origin of Species in 1859, he became one of the staunchest champions of the theory of evolution by natural selection, though he did its cause a disservice when he coined the phrase 'the survival of the fittest', which has since been much misunderstood. Spencer expanded the evolutionary idea to include societies, postulating that in the same way as creatures evolve from simple to more complicated life forms so do societies evolve from the simple to the complex. He also drew analogies between the structure of an animal's body and the structure of a society. These ideas, and many more that favoured a rational rather than a religious view of humanity, were set out in a vast work, Synthetic Philosophy, published in ten volumes (1862-93). Spencer's ideas were an important starting point for sociology. See also Social Darwinism.

Sperm, the >gamete produced by male animals, insects, etc. (The equivalent in a plant is a pollen grain.) A spermatozoon (pl. spermatozoa) is the same thing. A human sperm has a head about 2 pm long, a mid-piece about three times longer, and a tail some 50 pm long. About 200 million sperm are present in each ejaculation. After they were first seen by van >Leeuwenhoek, it took a long time for the true role of sperm to be discovered. They were for a long time believed to contain a >homunculus (see pp. 4-5) and as recently as 1833 were thought to be a species of protozoa, with nothing to do with fertilization.

Sperm bank, a collection of sperm frozen and stored in liquid nitrogen, to be used for artificial insemination. Sperm banks for cattle have been routine for many years. The idea of setting up sperm banks for the genetical improvement of humans was strongly recommended by the pioneer American geneticist H.J. >Muller, whose work on >mutations had led him to worry that the quality of the human gene pool was declining because of the effect of medical science in preserving sub-optimal >genotypes; he was also worried about the mutational damage being caused by exposure to X-rays. The difficulty of deciding which men should be chosen to have their sperm in a bank; he later changed his mind completely. There are sperm banks which are used for artificial insemination in the treatment of infertility; the donors are popularly believed to be mostly medical students, and although very brief details on them are known to the doctor carrying out the treatment (notably their colour), their anonymity is strictly preserved and they cannot in law be held responsible for the paternity of the child.

Spermatogenesis, the process by which a male makes gametes, i.e. >sperm. The starting point is in the tubules in the testis, which have a lining of cells called spermatogonia; these divide by >mitosis to form spermatocytes, which are also diploid cells. These do through the >meiosis divisions to produce four haploid spermatids each (all genetically different). The spermatids mature into sperm (properly, spermatozoa). The process of spermatogenesis goes on continuously in humans but is seasonal in most other species. Sperms that are not used in an ejaculation are retained in the testicle and liquefied.

Spina bifida, a congenital condition in which the vertebrae are imperfectly formed, leaving a gap through which the membrances of spinal cord stick out. In many cases, surgery can repair some of the damage, though a child may need many successive operations. About 3 per 1000 babies in the British Isles are born with spina bifida, which is more common in northern and western England and in Eire. There seems to be some genetic predisposition to the condition, probably involving polygenic inheritance, but the mother's vitamin intake may also be a contributing factor.

Spindle, a structure made of minute fibres that appears in the cell's nucleus and is involved in organizing the movement of chromosomes during cell division (see meiosis; mitosis).

Splicing, the technique for joining two molecules of DNA together, for making >recombinant DNA. The basic procedure is to cut both the DNA molecules with a restriction endonuclease that makes a staggered cut in the sequence, leaving a complementary-reading tail (e.g. one that reads AATT; this is complementary to itself, in that A must always pair with T). The two sorts of DNA are then mixed together, and since both have the same complementary tails they stick together, the joins being made by the enzyme DNA ligase. This is the technique used in gene cloning, in which the gene required to be cloned is spliced to the DNA of a plasmid, which can then be taken up by a bacterial cell.

Spontaneous generation, the belief that living matter can arise from non-living matter. This was a universal belief up till its final disproof by Louis >Pasteur in the 1860s.

Spontaneous mutation, a mutation that arises in natural conditions, as opposed to one that has been induced with a mutagen.

Spore, a unicellular reproductive unit that becomes detached from the parent cell and develops into a new individual, either after fertilization by another spore or by >vegetative propagation. In plants, the spore is the haploid product of meiosis (see alternation of generations) and is comparable to a gamete in an animal.

Sporophyte, the diploid phase of a plant's life cycle. In higher plants it consists of the plant itself as visible to the naked eye. See alternation of generations.

Sport, an individual with a visibly different phenotype to the normal, due to a >mutation. The word is from the pre-Mendelian era, having been used in this sense from about 1840, and is now only used in horticulture. A bud sport in botany is a shoot that is different from the plant it grows on, following a >somatic mutation.

Stabilizing selection, the type of natural selection that acts to keep the phenotype of a population stable around a mean. Extremes are selected against, and the amount of underlying genetic variance is reduced. Stabilizing selection only occurs in a population in a constant environment. See also directional selection; disruptive selection.

Standard deviation, a measure of how widely the individual observations in a given sample vary from the mean. It is calculated as the square root of the variance. As a rule of thumb, in a normal distribution 95 per cent of the observations lie within two standard deviations either side of the mean.

Start codon, the codon in messenger RNA that signals the point at which >protein synthesis should start. It is AUG, which codes for the amino acid methionine. See genetic code.

Stasis, the existence of a species in an unchanged form over a long period of time. An unchanging environment is necessary. The coelacanth, a lunged fish ancestral to the amphibians, is an example, being little changed since its first appearance some 350 million years ago.

Statistics, a set of techniques for analysing numerical data. Statistical analysis cannot say whether a hypothesis actually is true or false, merely that it is more or less likely to be true or false. The science of statistics in general, and especially its use in biological problems, was pioneered by >Fisher in the 1920s and 1930s.

Sterility, inability to produce viable gametes. This may be due to absence of defectiveness of gonads, or there may be a chromosomal difficulty that prevents meiosis taking place (e.g. in the case of triploids or interspecies hybrids such as mules). See also infertility.

Sticky end, the end of a DNA molecule where one strand extends beyond the other by a few bases, as opposed to one which ends flush (see blunt end). Two sticky ends can join together if their sequences of bases are complementary.

Stop codon. See nonsense codon.

Structural gene, any gene that codes for a gene product, in contrast to a gene that is a spacing or regulatory sequence (see spacer DNA; regulator gene).

Student's t-test. See t-test.

Sturtevant, Alfred Henry (1891-1970), American geneticist. Sturtevant worked in T.H. >Morgan's 'fly room' at Columbia University, studying the genetics of *Drosophila*. In 1912, while still an undergraduate, he had the idea for >genetic mapping, based on the frequency of crossing-over between genes on the same chromosome. He went on to discover that the effect that genes at one locus have on genes at another locus depends on the distance between the two loci (i.e. position effect). Sturtevant also found that where one chromosome contained an inverted sequence (i.e. an inversion) no crossing-over took place. He did research on various other animals apart from Drosophila: his papers on coat colour in horses (based in part on data from his boyhood on his father's farm) and on left-or rightcoiling snail shells (see maternal effect) are classics.

Substitutional load, the part of the >genetic load of a population which consists of 'old' alleles in the process of being replaced with new better ones.

Substrate, the molecule or compound upon which an >enzyme acts. Each substrate is specific to its own enzyme, and vice versa.

Succession, principle of, the principle that, in the geological record, simpler life forms are replaced in time by more elaborae ones, and that the sequence in which this occurs is the same from one location to another. This was established by the British geologist William Smith in 1799 (see p.14), and was an important of the evidence that was accumulating in favour of the evolution of species. A related finding was that the gaps between present-day life forms can often be filled by intermediate forms that existed in previous ages. See also distribution, principle of.

Supergene, a group of genes in very close linkage, with crossing-over between them suppressed so that they are inherited as a single unit. Supergenes have been found in butterflies, where the several genes involved in mimicry have become closely linked.

Suppressor mutation, a mutation that acts to restore a function that had been lost after an earlier mutation. It may act within the same gene (possibly by changing back a frameshift) or in another gene, by coding for a tRNA that will read the mutated nonsense or missense codon as the normal codon.

Surrogacy, the bearing of a child by one woman to be brought up by another. Let the woman who carries the baby be Mrs A (she is usually married) and the couple wanting to the baby Mr and Mrs B. In paritial surrogacy, Mrs A is made pregnant by Mr B, either by sexual intercourse or artificial insemination. The baby therefore has half its genes from Mr B and half from Mrs A, and is as closely related to Mrs A as any other baby she might have. In full surrogacy, in vitro fertilization is used: an egg from Mrs B is fertilized by a sperm from Mr B, and the embryo is implanted into Mrs A, who has no genetically relationship to it at all (see embryo transfer). The baby is entirely genetically the offspring of Mr and Mrs B who will bring it up. Partial surrogacy has a long history, the earliest recorded instance being in Genesis 16, and as it can be informally arranged without involving the medical profession it may well be more common than official statistics suggest. Full surrogacy became an issue at the beginning of the 1980s, when in vitro fertilization technology became reliable. The British Medical Association's view in 1983 was to doubt whether full surrogacy would ever be acceptale. A succession of sensationalized cases soon followed, including one in South Africa in which a woman gave birth to her own grandchildren - triplets born from her daughter's eggs. With the passing of the Human Fertilization and Embryo Researh Act both partial and full surrogacy are now legal in Britan, though with various safeguards including a prohibition on payment to the surrogate mother other than to cover her expenses.

Switch gene. See oligogene.

Symbiosis, the arrangement in which two species live closely together. It may be for their mutual benefit (mutualism), as with the species of ants that keep aphids as 'cows'; the benefit to the 'cows' is that they are protected against predators. Such a system requires co-ordinated evolution of the two species. In commensalism, two species share a common resource without affecting one another either adversely or favourably; this might be two species that live in the same burrow. Inquilinism is a symbiosis in which are species lives inside the body of the other but without causing the host any harm; this is the relationship between humans and Escherichia coli. The most usual form of symbiosis is parasitism.

Sympatry, the existence in the same area of two or more related species.

Synapsis, the pairing of the homologous chromosomes during the first prophase of >meiosis.

Syndrome, a medical term denoting a number of symptoms that occur together (the word comes from the Greek syn, 'together', dromos, 'a race', 'a running [together]'. The cause may be infective, environmental or genetic. A syndrome caused by a single gene is an example of >pleiotropy, but many genetically caused syndromes result from chromosome abnormalities, not single defective genes.

Systematics, the study of the relationships of the different groups of living things. The term is more or less interchangeable with >taxonomy, but is more old-fashioned. See classification.

T

T Lymphocytes a group of white blood cells involved in the immune system. They have on their surfaces receptors that recognize histocompatibilty antigens on the surfaces of cells, so that they bind on to these antigens on any foreign cell that enters the body. They are therefore the main part of the body's response to transplants (see rejection). T Lymphocytes (so called because they originate in the thymus) recognize the antigens by sequence (unlike >B Lymphocytes).

T4, one of the most intensively studied of the bacteriophages. It only attacks >***Escherichia coli***. The genome is double-stranded DNA, linear, but with matching ends that originate in the thymus) recognize the antigens by sequence (unlike >B lymphocytes).

Tailless, a Mendelia allele occcurring in mice. The inheritance of the tailless gene was, at first, a great puzzle to geneticists, as the trait appeared in succeeding generations as a Mendelian allele would, but with unusual pedigrees. Litters would be either all normal, or all tailless, or one normal and the rest tailless. Never the expected Mendelian rations of 1:1 or 3:1. And was taillessness dominant or recessive? - two normal parents might have tailless offspring (which rules out dominant), but two tailless parents might have normal offspring (which rules out recessive). The penny finally dropped in the mind, not of one the researchers, but of a laboraory technician: the gene is not t = tailless; it is t = tail-eater. What really happens is that the litter can be (a) all normal (no tt pups so no tails eaten), (b) all tailless except one (one tt pup that eats all its siblings' but not its own) or (c) all tailless (two or more tt pups that eat each other's tails as well as those of their siblings). The moral of the story is that what the scientist believes to be the phenotype may be nothing of the kind.

Tandem duplication, a form of duplication in which part of a chromosome duplicates itself, the second copy being inserted in line with the first.

Taste blindness, a condition in which an individual is unable to taste the chemical compound phenylthiocarbamide (PTC). It has a slightly bitter taste, and is found in some members of the cabbage family. The inability to taste PTC is inherited as an autosomal recessive. Non-tasters comprise 25-30 per cent of the population in northern Europe but only 7 per cent of the Japanese and 1 per cent of the South American Indians. There must be some selective advantage to keep the gene in populations at such high frequencies, and it has been suggested that, as PTC is chemically quite close to some substances that affect thyroid function, it may be something to do with the prevention of goitre (enlargement of the thyroid). It is believed that chimpanzees have a similar gene. Taste blindness was discovered by chance in 1931.

TATA box, a sequence found in eukaryote genes. Part of the >promoter region, it occurs about 35 base-pairs upstream from the point at which RNA transcription is to start. Prokaryotes have a >Pribnow box.

Tatum, Edward Lawrie (1909-75), American microbiologist. Tatum provided the microbiology know-how in the joint experiments that he and George >Beadle carried out on the bread mould Neurospora crassa. In 1941 they showed that one gene codes for one enzyme. Tatum next worked with Joshua >Lederberg on genetic recombination in Escherichia coli, and their work was the foundation of the techniques of genetic engineering. Tatum, Beadle and Lederberg shared the 1958 Nobel prize for physiology or medicine.

Taxon (pl. taxa), any group of living things, from the viewpoint of >classification. They key taxon is the species; the higher taxa are:

- For animals: genus, family, order, class, phylum, kingdom.
- For plants: genus, tribe, family, order, class, division, kingdom.

Lower taxonomic divisions into subspecies and varieties can also be made.

Taxonomy, the arrangement of taxa into a system of >classification. The outline of the scheme of classification was laid down in the eighteenth century by >Linnaeus, Cuvier and others (see pp. 13-14), using only anatomical features (and before evolution was accepted, so there was no element of keeping ancestrally related forms together). A modern technique is numerical taxonomy, in which all attributes of the phenotype-metric, biochemical or whatever - are given equal weighting, and groupings are worked out by computer.

Tay-Sachs disease, a human disease in which the baby is born normal but suffers mental retardation after the age of three months, followed by blindness and paralysis; death usually occurs at the age of about two years. It is inherited as an autosomal recessive, the biochemical defect being an inability to produce the enzyme hexosaminidase A. Tay-Sachs disease occurs at a frequency of about 1 in 2500 births among people of Ashkenzi Jewish origin, but only about 1 in 100,000 in other populations (including Sephardic Jews). Carriers of the gene can be detected by a biochemical test. There was a highly successful programme of >genetic screening among the Jewish populations of Washington D.C. and Baltimore, Maryland in the United States to identify couples both of whom were carriers so that they could take appropriate action to avoid having an affected baby; this programme became the model for a number of similar schemes among Jewish people.

Telegony, the belief that the first male that mates with a female has a lasting effect on her subsequent offspring, even when these are fathered by a different male. It is difficult to understand how this was supposed to work: even with blending theories of inheritance one would not expect something in the male's contribution to 'stain' the uterus (or would one?). There was alleged evidence of telegony in the case of an Arab) produced foals that had bristly manes and some stripes on their legs; even Darwin accepted this proof. The experiment was repeated by a professor of veterinary science who bred many such crosses and found no evidence for telegony, but did confirm that quite a few pure-bred horse foals have stiff manes and striped legs anyway.

Many in animal-breeding circles still believe in telegony, though not officially; for instance the Kennel Club in Britain is perfectly happy to registr pure-bread puppies no matter how chequered the reproductive history of the bitch. And even in the time when the zebra story was believed, it was not held to apply to humans: you could marry a rich young widow and take advantage of the money without worrying that her previous mate might reappear in your children. (The converse belief, that the first female he mates with affects all the subsequent offspring of a male, has never been so prevalent - and definitely did not recommened itself in the case of humans.)

Teleology, the belief that there is evidence of purpose in the universe. In the context of biology, it is particularly taken to mean the idea that evolution (especially when seen as the path from the primitive origin of life to the glorious culmination in *Homo sapiens*) has not been purely mechanistic but has been purposively achieved. This is of course not the point of view of most modern biologists, who are quite satisfied with the mechanistic process of evolution by natural selection working on a trial-and-error basis with the raw material provided by random mutations.

The teleologists' 'purpose' has to come from one of two sources. One could be the will of the organisms themselves as they evolve. H.G.Wells and Julian Huxley wrote scathingly of 'the difficulty of ascribing even rudimentary purpose or foreknowledge to a tapeworm or a potato, or collective aspiration to the tapeworm race or the potatoes of the world'. The other possible source of purpose is a Divine Creator, the existence or nonexistence of which is not provable. Biology cannot refute the proposition that natural selection is the tool of a Creator (see creationism). Another point is that a modern biologist would not see evolution as a process ending with the arrival of humans; if God made us in His own image, then that image must include every step of the evolutionary journey from the primeval sludge to modern humanity and beyond.

Telocentric chromosome, one with the >centromere at the extreme end.

Telophase, one of the phases of the two types of cell division, >meiosis and >mitosis.

Teratogen, any environmental factor that causes an embryo or foetus to become deformed. The damage is not genetic but congenital (see congenital defects). Most pharmacological drugs are small molecules that can pass through the placenta into the unborn baby's bloodstream, and are thus always suspected of being teratogenic until proved otherwise. See thalidomide.

Terminal transfrase, an enzyme that synthesizes a 'tail' on to a DNA molecule. It is used in genetic engineering to add a poly-A tail (one that consists only of repeated As) to the end of one DNA molecule and a poly-T tail (one that consists only of repeated Ts) to the end of another so that they join up.

Territoriality, the behaviour by which a group or individual claims rights over a certain space. Almost all vertebrates, and many of the more advanced invertebrates, recognize specific areas as their own. The widest area is the total range, the maximum area (or volume, for all these concepts apply to underwater and aerial habitats as well as purely terrestrial ones) that the individual visits in its lifetime. The home range is the area known thoroughly by the individual and visited regularly by it. This may or may not be the same as the territory, defined as the area or volume that an individual, pair or group actively excludes others from, by marking (see pheromones), display or aggressive behaviour. The key resource that is guarded in the territory may be food supply, shelter, nesting sites or space for sexual display. The territory can change with time, for example during seasonal migration. Sociobiologists hold that territoriality, as other behaviour, is genetically controlled.

Testcross, a cross between an individual with the dominant phenotype but of unknown genotype and an individual homozygous for the corresponding recessive, the cross being made in order to find out the genotype of the first individual. The cross is either AA x aa, and the phenotypes of the offspring will show which it is, by being either all dominant-type or 50 per cent dominant and 50 per cent recessive. A testcross can involve more than one locus. See also backcross.

Testicular feminization syndrome, a condition in which the person has the chromosome complement of a normal male (i.e. 46, XY) but owing to a lack of receptors for the male sex hormone testosterone, does not develop normal testicles. The external genitalia look female, and a boy with this syndrome will be brought up as a girl. The syndrome is inherited as an X-linked recessive, and is the only example of a gene on the X chromosome that affects sexual development (see sex determination).

Tetrad, the four homologous chromatids of a chromosome pair as they lie together during >meiosis. The term is also used for the group of four haploid cells which are the product of meiosis. See tetrad analysis.

Tetrad analysis, a technique for studying linkage and recombination, using the tetrad of meiotic products in a fungus such as *Neurospora crassa*. The tetrad of hour haploid cells lies in a line in the ascus, and a mitosis division takes place so that there are eight cells (>ascospores), lying in four pairs, each pair representing one product of meiosis. Each of the ascospores can be grown separately (as Neurospora is in the haploid state for most of its life cycle) and the presence or absence of particular alleles noted.

Tetraparental mouse, a mouse produced by mixing in vitro two different blastula; also known as an 'allophenic mouse'. The resulting mixed ball of cells is usually capable of developing into an embryo, and can be reimplanted into a mouse for the remainder of its prenatal development. The mouse has four parents instead of two, and these four can all be of different strains. However, each of its cells has only two parents, being derived from one or other of the original 'ingredients'. The mouse is therefore an artificially produced >chimaera.

Tetraploidy, the condition of having four copies of each chromosome. See polyploidy.

Thalassaemia (from the Greek thalassa, 'the sea'), a disease in which the red blood cells are malformed and fragile and, as a result, break down faster than they can be produced; this results in severe anaemia, which often proves fatal. Thalassaemia is probably the commonest single-gene disorder in the world, with high gene frequencies right across the Old World, from Africa to New Guinea, between latitudes 40°N and 15°S.

The biochemical problem is a defect in either the alpha or beta chain of the haemoglobin molecule. In alpha-thalassaemia, the homozygous condition is fatal at birth (hydrops foetalis). In beta-thalassaemia, homozygotes suffer thalassaemia major, a severe anaemia requiring regular blood transfusions (untreated patients die at about two years old), while heterozygotes have thalassaemia minor, a less severe anaemia but one that is still fairly debilitating. There are some 30 different mutations (frameshifts, deletions, insertions or inversions) that can give rise to betathalassaemia. As with >sickle-cell anaemia, the evolutionary justification for populations to have a high frequency of a gene for this disease is that heterozygotes are protected to some extent against malaria, which until recently was one of the major killers in the areas where thalassaemia occurs.

A simple blood test (carried out routnely in many countries) can detect whether a pregnant woman is a carrier of the thalassaemia gene, and couples who are both carriers (the father can also be tested in this way) can be advised that they are at risk of having a thalassaemis major baby. Prenatal diagnosis of the babies of such couples, however, is not so simple, because the babies do not start to make haemoglobin beta-chains until a late stage of gestation; but in some families, it may be possible to use genetic markers to tell with a high degree of probability whether or not the baby has inherited the thalassaemia gene from both parents. In places such as Cyprus and Sardinia where the disease is very common, schemes for the prevention of births of thalassaemia major babies have given interesting examples of the cost-effectiveness of this policy.

Thalidomide, a sedative drug that was put on the market in 1957 after extensive testing on four different species of mammals, including testing for birth defects. It was sold in Europe without prescription, but was withdrawn in 1961 when its association with phocomelia (a deformity in which the hands and/ or feet are joined to the body by very short limbs or none at all) was noticed. Some 8000 babies worldwide were born with (and survived) this defect (congenital, not genetic) after their mothers had taken thalidomide during the first three months of pregnancy; many others with extensive abnormalities died before birth. The reason for mentioning thalidomide in a book on genetics is that the story emphasizes the genetic differences between species: rodents can take thalidomide with impunity, we cannot.

Theophrastus(c. 371-288 BC), Greek philosopher. Theophrastus studied under >Aristotle, and took over the school at the Lyceum from him when he retired in 335 BC. He lectured on as wide a range of subjects as Aristotle had done, but he is chiefly remembered for his book Historia plantarum (On the History of Plants), in which he describes the roles of flower, fruit and seed, and also states that in some plants there are female flowers that must be fertilized by a male (an idea a long way ahead of its time, which was not fully described until Camerarius's work in 1694).

3, ('three prime'), one end of a molecule of RNA or DNA, the name referring to the number of the carbon atom in the sugar that comes at that end (joined to a phosphate). The other end is known as >5', the links between the sugars throughout the molecule being between the 5' and the 3' carbons. 3' is the 'tail' end.

Threshold trait, a trait in which the underlying cause has >continuous variation but in which the phenotype is expressed only at a certain level. An example is cleft palate, in which the disorder is due to a delay in several interacting developmental processes. Up to a certain point this delay is not harmful, but once that critical point is reached one process - fusion of the palate - misses its proper place in the schedule and cannot occur. In threshold traits, the determining factors are both genetic and environmental, so patterns of inheritance are very hard to work out.

Thymine (abbreviation: T), one of the four nitrogenous bases which form the core of DNA (but not RNA, where its place is taken by >uracil). It pairs with adenine (A).

Tissue culture, the maintenance in vitro of a line of cells taken from a tissue of an organism (see cell line). Depending on the type of cell, a culture may be able to go on indefinitely, with cells dividing from time to time. Cancer cells (or normal cells that have undergone transformation in culture) will go on divide after 30-40 divisions; this is known as the >Hayflick phenomenon, and it is not known how it is related to the normal process of ageing.

Tissue typing, the identification of >histocompatibility antigens in donor and recipient prior to transplant surgery.

Tortoiseshell cat, a female cat whose coat colour is made up of random patches of orange and black. The genes for these colours are on the X chromosome, and when random inactivation of one or other X chromosome in each cell takes place (see Lyon hypothesis), either the chromosome with the 'black' gene or the one with the 'orange' gene is left active.

Trans configuration (also known as repulsion), one of the two possible arrangements for the alleles in an individual that is heterozygous for mutations at two linked loci. Trans (from the Latin, 'across') means that one of the mutant alleles is on one chromosome and the other is on the homologous chromosome, i.e. the two chromosomes are a+ and +b. The opposite is >cis.

Transcription, the process by which one strand of DNA is copied into a single strand of RNA; the first step in >protein synthesis. The RNA molecule is synthesized by RNA polymerase, copying the DNA message by following the base-pairing rules (see genetic code). The polymerase first recognizes the promoter region, and actual transcription begins at the start codon a few bases further downstream (going from the 5' towards the 3'end of the DNA strand).

Transfer RNA (abbreviation: tRNA), the type of small RNA molecule that assembles the amino acids for >protein synthesis. Each transfer RNA molecule is shaped roughly like a clover leaf, and consists of an anticodon at one end (which is complementary to the appropriate codon in messenger RNA) and its specific amino acid at the other; there is also a ribosome binding site. In principle, there could be 61 different transfer RNA molecules, one for each 'sense' codon in the >genetic code, sharing the 20 amino acids among them; in practice, each cell contains 40-60 different ones.

Transformation. This has two meanings in genetics. (1) A permanent change in the genetic characteristics of one bacterium by exposure to DNA of a different origin. It was the phenomenon of transformation that led >Avery to the discovery that DNA was the molecule responsible for carrying genetic information. (2) A change in an animal cell in tissue culture so that it grows and divides in the same way as a cancer cell, possibly due to activation of a viral gene. The relationship of this phenomenon to cancer is not understood.

Transgenic organism, any organism that has been transformed (to however slight a degree) by having DNA from another species integrated into its genome. Note that a transgenic mouse - say with a single chimpanzee gene in its genome - is not a mouse/chimpanzee cross; it is still just a mouse. See also genetic engineering.

Transient polymorphism. See polymorphism.

Transition, a mutation in which a purine base (adenie, guanine) is replaced with another purine, or a pyrimidine (cytosine, thymine) with another pyrimidine. See also transversion.

Translation, the process by which the protein coded for in messenger RNA is built up, by transfer RNA and ribosomes. See central dogma; protein synthesis.

Translocation, a piece of one chromosome that has broken off and become attached to a different chromosome. There is usually no adverse effect if this happens ina somatic cell, as descendant cells have all the right genes even if some are in the wrong place (though sometimes the translocation does have a harmful effect; see Philadelphia chromosome). However, if the translocation happens while a gamete is being made, some of the descendant cells will have surplus genetic information and some will have what is in effect a deletion. See also Robertsonian translocation.

Transplants, organs surgically moved from donors to recipients. For the problems of the immunology, see HLA system; immunosuppression. See also graft.

Transposition, the movement of a >transposon or other movable sequence of DNA from one place in the genome to another.

Transposon, a movable genetic element similar to a >jumping gene. They are found in the genome of bacteria, and are involved in the resistance to antibiotics. Unlike jumping genes, transposons leave a copy of themselves in the original position when they move possible for antibiotic resistance to be spread very rapidly, not only within a single strain of bacteria, but from one genus to another. (See conjugation.)

Transversion, a mutation in which a purine base (adenine, guanine) is substituted for a pyrimidine one (cytosine, thymine), or vice versa. See also transition.)

Trihybrid, a cross between individuals that are >heterozygous at three different loci. The genotype of these individuals being Aa Bb Cc, there are eight different gametes that each can produce - ABC, ABc, AbC, abc, abC, aBC, aBc - which leads to 64 (82) possible genotypic combinations in the offspring. The ratio of the phenotypes among the offspring is 27:9:9:9:3:3:3:1 - 27 showing dominant type for all three traits, 9 for each possible pair of dominant phenotypes, 3 for each single occurrence of the dominant, and 1 for the triple recessive. See also hybrid.

Triplet, three bases in sequence in DNA or RNA. It forms a 'word' in the >genetic code. See codon.

Triplets, three babies born together. The frequency with which triplets are born is the square of the frequency of twin births (with quadruplets the cube), though there is no theoretical explanation for this observation (see twinning rate). This rate is no longer the same, given the high frequency of >multiple births following the use of fertility drugs.

Triploidy, the condition of having three copies of every chromosome (a form of >polyploidy). Triploidy does not occur in nature as a permanent feature of a species, as triploid organisms have very low fertility. This is because at meiosis there are three chromosomes trying to find partners for a process that only works with two, so the division is usually a failure. However, triploid organisms are just as capable of normal growth as any others (mitosis is not impeded at all, since the process consists of each chromosome dividing itself in half, irrespective of any partner). Artificially produced triploids can be commercially useful. One example of this is the banana, the trees of which are bred as triploids from stocks of a tetraploid and a diploid strain. The triploid from stocks of a tetraploid and a diploid strain. The triploid hybrids grow well, but their seeds never develop properly, so that the inside of the banana is not full of inedible seeds as it is in a normal banana.

Trisomy, the condition of having three copies of one particular chromosome. This almost always produces some abnormality, as genes are deleterious when present in greater than normal numbers. In humans the best-known example is trisomy 21 (>Down's syndrom), but there are others that occur quite frequently, including trisomy 18 (Edwards syndrome; 1 birth in every 8000 live births, but ten times this number are spontaneously aborted) and trisomey 13 (Patau syndrome; 1 in every 20,000 live births, 20 times more often spontaneously aborted). Trisomy happens as a result of translocation or of non-disjunction of homologous chromosomes during meiosis.

Triticum aestivum, common bread wheat. Primitive (emmer) wheat was first domesticated in about 7000 BC. The present form is hexaploid, i.e. it has six of each chromosome, representing three sets each of different origin (6n = 66). One set is homologous with the wild wheat *T. monococcum* (2n = 22), and another with a related wheat-like grass *Aegilops squarros* (2n = 22), but the third has not been identified and may have come from a species that is now extinct. There are at least 20,000 varieties of *T. aestivum* in cultivation. The wheat used for pasta is a different species, *T. durum*.

tRNA. See transfer RNA.

True breeding, the phenomenon in which a strain of animals or plants produces offspring with the same genotype as its own. The term is most often used in connection with a recessive >allele - for example, Suffolk Punch horses which breed true for chestnut colour - but it could also be used for a strain homozygous for a dominant allele, or for genes at more than one locus.

Tschermak von Seysenegg, Erich (1871-1962), Austrian botanist; one of the three rediscoverers of >Mendel's laws. A member of a distinguished family of scientists, Tschermak specialized in agricultural botany, and his experiments aimed at producing higher-yielding varieties of corn led him to the independent discovery of the principles of heredity first described by Mendel in 1865. Tschermak found Mendel's long-neglected paper in 1900, at the same time as de >Vries and >Correns, who had both discovered the same principles through their own experiments. Tschemak was a very successful breeder of plant varieties: his Aegilotiriticum wheat-rye cross was the first artificially produced hybrid between two different genera, and his rye variety 'Marchfield' is still the highest yielding in the world.

t-test, a statistical technique for finding whether the mean value of one samples is different from that of another sample. It is often used in genetics and ecology; an example would be a comparison of the weights of mice caught in one place with caught in another place. It is also known as Student's-test, from the pseudonym of the mathematician W.R.Gosset who devised it.

Turner's syndrome, the condition in which a female has genotype XO, i.e. only one sex chromosome. A Turner's syndrome girl looks fairly normal, but with short stature and a wide ('webbed') neck; the ovaries are present but undeveloped so the she is not fertile. Although there is seldom any general mental retardation, there is often a defect in spatial perception. The single X chromosome can be inherited either form the mother or the father, but usually it is the maternal X, a normal ovum having been fertilized by a sperm that lacked a sex chromosome. If a Turner's syndrom patient inherits an X-linked recessive gene - e.g. for red-green colour blindness - she will have the abnormality, as there is no corresponding normal X chromosome to 'cover' the recessive (see X linkage). Turner's syndrome occurs in about 1 in 1500 births in Britain.

Twin studies, research using pairs of twins to study the effects of tenotype and environment on human characteristics. Both monozygotic (identical) and dizygotic (non-identical) twins are used. The theoretical basis is that since monozygotic twins have identical genotypes, any differences between them are produced by the environment. Normally the environment is very similar for both twins, particularly as parents of twins often treat their children exactly alike, even to point of dressing them in identical clothes.

Dizygotic twins have (on average) half their genes in common, as do any other pair of siblings. If one assumes that the two kinds of twins experience environments that are to a similar extent identical (probably an oversimplification, though parents are usually carefully to be even-handed with twins), the difference in the amount of similarity or divergence between twin pairs of both sorts can be taken as an indication of the effect of 100 per cent of genes in common versus 50 per cent common. This can be reinforced by looking at siblings reared together, who have the same amount of genes in common as dizygotic twins but a less-than-identical environment (due to difference in age, etc.).

The ideal situation (from the geneticist's point of view, never mind anyone else's) would be monozygotic twins separated at birth and raised entirely separately, preferably (again from the geneticist's point of view) in very different environments. The divergence between such pairs of twins measure the amount of influence that the environment is capable of exerting. Not many such pairs of twins exist: the classic separated-twin studies are those of Newman, Freeman and Holzinger (US, 1937) with 19 pairs and Shields (Britain, 1962) with 44 pairs; >Burt (Britain, 1966) claimed 53 pairs, but was later shown to have fabricated data on many of them (although some were genuine).

Twinning rate, the frequency of twin births, as a percentage of all births. There is considerable variation between populations, with the lowest (0.65 per cent) found in Japan and the highest (4.5 per cent) in Nigeria, the average being between 1 and 1.5 per cent. The proportion of these that are monozygotic (identical) can be calculated from the fact that half the pairs of dizygotic twins (non-identical) are of unlike sex, and that there is a corresponding number of like-sex twins are born per year, and 60 pairs of like-sex, one can deduce that the 'extra' 10 pairs are monozygotic. The birth rates of monozygotic twins is not very different from one population to another, the variation in twinning being due almost entirely to the rate of dizygotic twins.

There is some genetic basis for a tendency of a mother to have twins, but there is also a very strong maternal-age effect, with older women more frequently having twins (again, the difference being in dizygotic twins), and with birth order, later babies being more likely to be (once again, dizygotic) twins. See also triplets; quadruplets.

Twins. See dizygotic twins; monozygotic twins; twinning rate.

U

Ultraviolet light, electromagnetic radiation of a wavelength of about 400 to 4 nm, between the visible spectrum and X-rays. It is present in sunlight, and causes damage to the chromosomes in any cells that it reaches (see DNA repair). It is to protect the skin cells from ultraviolet rays that the brown pigmentation of human skin evolved (see melanin). However, humans need ultraviolet radiation in order to make vitamin D.

Uniformitarianism, the theory that the processes by which the physical structures of the Earth were formed are continuous and are still going on. The idea was first proposed in 1788 by the British geologist James Hutton, who was bitterly attacked because it was counter to religious teaching. Uniformitarianism was later championed by Sir Charles >Lyell and was part of the accumulating evidence in favour of a process of biological evolution.

Unisexual, having only one kind of sex organ; usually used of a plant that has flowers of only one sex (see self-sterility).

Upstream, further back on a DNA molecule, in respect of the direction in which the sequence is being read. See replication.

Uracil (abbreviation: U), one of the four nitrogenous bases which form the core of RNA (but not DNA, where its place is taken by thymine). It pairs with adenine (A).

V

V region, or variable region, part of the heavy and light chains of an >immunoglobulin molecule. It is this region that interacts with the antigen.

Vaccination, the artificial production of active immunity by the injection of a vaccine. The word 'vaccination' is derived form the Latin for 'cow', and was coined by Edward Jenner at the end of the eighteenth century to describe his idea of injecting patients with cowpox virus to protect them against smallpox (a very similar virus). A vaccine now means any suspension of dead or non-virulent viruses or bacteria used in immunization. See also Pasteur, Louis.

Variable number tandem repeats (VNTRs), the repeated sequences of DNA that vary from one individual to another and are the basis for >genetic fingerprintig.

Variable region. See V region.

Variance, a measure of the spred of values of observations around the >mean of a distribution. It is calculated as the sum of the squared of the deviation of each observation from the mean, divided by n - 1 (where n is the number of observation). The total variance is known as the phenotypic variance, part of which is accounted for the genotypic variance. In genetical research, the exact proportion of the variance that can be accounted for by genotypic variance can sometimes be ascertained (but by no means always - see nature versus nurture). See also covariance; standard deviation.

Variation, coefficient of, a measurement of the amount of variation within a population. It equals the standard deviation expressed as a percentage of the value of the mean.

Variegation, patterns, regular or irregular, of different colouring on the leaves of plants. Environmental causes include viral infection and mineral deficiency. genetic causes include alleles controlling regular patterns (often of red or yellow on green, e.g. in *Coleus*), a >chimaera genetical constitution, or local gene-switching by >jumping genes.

Vavilov, Nikolai Ivanovich, (1887-c. 1943), Soviet geneticist. Vavilov learned his genetics from >Bateson in England, and on his return to the Soviet Union, he was soon promoted to director of the Lenin

All-Union Academy of Agricultural Sciences. He researched the origins of cultivated plants, which he believed could be identified from the present-day distribution of wild species. He travelled all over the world collecting live specimens, and eventually had a quarter of a million separate varieties in cultivation. In his book The Origin, Variation, Immunity and Breeding of Domestic Plants (English edition 1951), he stated that the earliest domestication of plants had occured in two places, Ethiopia and Afghanistan. His research also helped to produce improved strains of wheat without which the famines in the Soviet Union would have been even worse. A convinced Mendelian, Vavilov fell of >Lysenko and the official doctrine of the inheritance of acquired characteristics. In 1940 he was sacked and sentenced to death; the sentence was commuted to imprisonment in a labour camp in Siberia, where he is believed to have died of starvation in 1943.

Vector. There are two meanings. First, in >genetic engineering it is the means whereby a gene of DNA sequence is carried into the host cell. Vectors, which must be capable of being replicated by the host cell on arrival, are usually either bacterial >plasmids or viral DNA molecules.

Alternatively, a vector is any organism that is responsible for transferring a pathogen (disease-causing agent) or parasite from one hsot to another. For example, the mosquito that carries the Plasmodium parasite causing malaria is a vector.

Vegetative propagation, reproduction without a sexual process. The simplest form is equal cell division, as in bacteria; after this comes unequal budding, with a small new cell growing after it has budded off the parent cell (as in yeast). In higher plants, vegetative propagation occurs naturally as multiplication of bulbs, corms, and runners, and artificially as grafts. The products are always a >clone of the original individual.

Venoms, the poisons produced by snakes, spiders, etc. In snakes, the venom consists of a mixture of poisonous proteins which are either haemolytic (i.e they break down the blood cells and/or prevent clotting) or neurotoxic (i.e they attack the nerves, particularly those that control breathing). The actual proteins and the balance of haemolytic and neurotoxic effects of smoke venoms vary almost form species to species, so that anti-venoms have to be specific to the snake in question. The genetics of the various venoms is not precisely known.

Vertebrates, all animals with backbones. These include fish, birds, reptiles, amphibians and mammals. See classification.

Vertical transmission, the usual types of genetic inheritance, from one generation to the next. It is contrasted with >horizontal transmission.

Vestigial organ, any organ that persists in a later evolutionary form after its original purpose has disappeared ('organ' here has the meaning of any anatomically distinct structure). Examples are the dew-claws in dogs and the coccyx ('tail') in humans. Vestigial organs were taken by >Buffon, >Lamarck and others as evidence of the evolution of species, as indeed they are.

Virulence, the ability of a pathogen (a virus, bacterium or other micro-organism) to produce a disease. It depends partly on its capacity to invade the host cell and multiply, and partly on its ability to produce toxins. Virulence tends to be favoured by natural selection, as virulent strains are often the most successful at reproducing themselves; but beyond a certain point virulence is a disadvantage to a strain of pathogens, as they kill off their hosts too soon for their own good.

Viruses, the simplest of living things, consisting of a core of either DNA or RNA inside a protein coat. They are not classified as truly living organisms, as they are not capable of self-reproduction but rely on the reproductive capability of their host cell (which may be a bacterium or a cell within a plant or animal). A virus enters a cell by attaching its protein capsule to the outside of the cell wall and injecting just its DNA or RNA into the cell bursts, releasing large numbers of viruses to attack other cells (this is known as >lysis), or the virus integrates its own genes into the host cell's genome and is replicated while remaining harmless itself (this is known as >lysogeny). In this latter state the virus is known as a >provirus. A virus typically has a very small number of genes, perhaps as few as four, two to code for the protein coat and two to help the virus insert its own DNA into the host's gneome.

Those viruses that infect animals include:

- DNA genomes: >adenoviruses, >herpesviruses, >poxviruses, >parvoviruses.
- RNA genomes: reoviruses, retroviruses.

There are also many types of plant virus. The viruses that attack bacteria are called >bacteriophages.

Vitalism, the theory that living matter is intrinsically different from non-living matter. The French philosopher Henri Bergsn proposed (in L'Evolution creatrice, 1907) that the elan vital ('vital spirit') in all living things is the driving force behind >evolution. This idea was taken up by George Bernard Shaw. Similarly the German biologist Hans Driesch taught that the development of the embryo is driven by some vital element in living matter. Vitalism had a tenuous existence in scientific circles in the first half of the twentieth century, because the properties claimed for whatever was the molecule of heredity seemed unlikely to be possessed by any known chemical compound. The discovery of the structure of >DNA must have finally disposed of vitalism.

Vitamins, a group of substances that are necessary for metabolism but only in minute quantities. They are chemically very diverse, but are divided into two groups: (a) water-soluble vitamins - the Bs and C; (b) fatsoluble vitamins - A,D,E and K. Vitamins work as co-enzymes within the >enzyme system, enabling the various essential chemical reactions within cells to take placr, e.g. the production of energy, the synthesis of proteins (for tissue-building, hormones, etc). There are some genetically caused vitamin deficiency diseases, e.g. the form of rickets that results from an inability to utilize vitamin D.

Viviparous, bringing forth live young. This is the latest phase in the evolution of reproductive methods. See also oviparous; ovoviviparous.

Vries, Hugo de (1848-1935), Dutch botanist; one of the three rediscovers of >Mendel's laws. De Vries was professor of botany at Amsterdam for 40 years, and was the first person to study evolution by experimental methods. He introduced the word 'mutations', but in a different sense to its modern one; he believed that evolution can proceed either gradually by natural selection or in sudden jumps ('mutations'). De Vries's experiments led him to discover the same theory to explain it. Searching the literature, he found Mendel's paper published in 1866; he then discovered that >Correns and >Tschermak had come to the same idea and had also found Mendel's paper. In 1900, de Vries arranged for Mendel's paper to be republished and publicized.

W

W chromosome, one of the sex chromosomes in birds and some insects, in which the female is the heterogametic sex (i.e. has two different sex chromosomes). The female has chromosomes WZ, and the presence of the W causes femaleness to develop (see sex determination); the male is ZZ.

Wallace, Alfred Russel (1823-1913), British naturalsit; co-discoverer of the principle of evolution through natural selection. Wallace was selfeducated and was working as a surveyor when he read Darwin's notes of the Beagle voyage, which inspired him to become a naturalist. He went with Henry >Bates to South America, but unfortunately he lost his collection of live specimens when his ship caught fire on the way home. Then he went to the Malay Archipelago, and it was there that he had the idea about natural selection. He had arrived at it by exactly the same mental route as Darwin: during his travels, he had observed the variability of species, and then he had read Malthus. He sent his essy to Darwin in 1858, and Darwin's friends arranged for the two accounts to be presented to the public simultaneously. Wallace was never resentful that the theory became known as Darwinism. He differed from Darwin in doubting whethe natural selection applied to human beings (this was perhaps because Wallace was a spiritualist). Wallace systemstized biogeography (an important area of evidence for evolution) in The Geographical Distribution of Animals (1876) and Island Life (1880), and proposed the 'Wallace line', the imaginary line that divides the faunas of Australasia and South-east Asia.

Waltzing mice, thoe that have an inherited defect in the balancing mechanism in the inner ear, causing them to turn in small circles. There are several mutant genes that give this effect. Waltzing mice were known in China as early as 80 BC, in the Han dynasty.

Warfarin, an anticoagulant drug. Its main use is in treating arterial disease, but it is also well known as a poison used against rodents. In the 1950s rats in Britain evolved resistance to warfarin, and the immune rats spread very rapidly (the rapidity of the spread being related to the severity of the selection pressure, i.e. the amount of poisoning being carried out). They became known as 'Super Rats', an alarmist name for animals that were perfectly normal except for one biochemical trait. There should have been more alarm at the danger of substantial risk to public health that comes when evolution hists back at human attempts to wipe out pests. See resistance.

Watson, James Dewey (1928-), American microbiologist; director of the Cold Spring Harbor Laboratory, Long Island, New York, since 1968 and, more recently, one of the prime movers of the Human Genome Project, from which he resigned as director in 1992; co-discoverer with Francis >Crick of the structure of >DNA. Watson studied at the University of Chicago and then researched bacteriophage genetics at the University of Indiana, but at a very early stage he became obsessed with the idea of solving what he saw as the fundamental problem in biology: what is a gene? He went to Cambridge University in 1951 to do a second doctorate, and joined the Medical Research Council's unit at the Cavendish Laboratory. he and Crick had some clues to follow in their quest for the gene, including >Avery's discovery that DNA is the genetic material, >Chargaffs rules on the ratios of the bases in DNA, and >Franklin's X-ray diffraction photographs, but their breakthrough only came when they tried building a scale model of the DNA molecule out of pieces of cardboard and wire. He and Crick shared the 1962 Nobel prize for Physiology or Medicine with Maurice Wilkins (a colleague of Franklin's). Watson tells the story ot the discovery of the DNA structure in his entertaining book The Double Helix (1968), which is however, ungenerous to several people, including Crick and Franklin. His scientific writings include the important Molecular Biology of the Gene, now in its fourth edition (1986).

Weismann, August (1834-1914), German biologist. Weismann trained as a doctor but became professor of zoology at Freiburg. He was a brilliant embryologist, but had to give up his researches on the development of insects when his eyesight deteriorated. Turning to theoretical biology, he proposed that heredity works by means of 'germ plasm' which is unchangeable and is not formed anew but is passed from one generation to the next. This 'germ plasma', according to Weismann, controls all aspects of the development of the organism (see germ plasm theory); he was consequently a firm opponent of the theory of the inheritance of acquired characteristics. Weismann saw a difficulty in that, if germ plasm were inherited equally from each parent, the amount of it would double in every generation unless there were some mechanism that reduced it. He was vindicated when >Hertwig observed meiosis, which porvides exactly the appropriate reduction mechanism and locates the 'germ plasm' in the chromosomes. Weismann then proposed that one source of biological variation is the different combination of chromosomes at meiosis. His ideas were far ahead of their time, and in perfect accord with Mendel's work (then published but unknown).

Wheat. See *Triticum aestivum*.

Whole-arm fusion. See Robertsonian translocation.

Wild type, the allele that is deemed to be the one that occurs in the species in its original state, as opposed to subsequent mutations, either spontaneous or induced.The wild-type allele may be either dominant or recessive in relation to a mutant allele. Its symbol is +.

Wilson, Edward Osborne (1929- (1929-), American entomologist; has taught at Harvard University since 1953; father of >sociobiology. Wilson was already one of the world's leading authorities on social insects when he published Sociobiology: The New Synthesis (1975). He was unshaken by the acrimonius controvrsy that followed, and has continued to develop the same ideas about the importance of genetics in deter,omomg social behaviour.

Wobble, the fact that a single >anticodon of tRNA can recognize different codons in DNA, provided that the difference is in the third base. For example, the anticodon for cystine, ACG, can recognize and bind itself to either UGC or UGU (see genetic code). Wobble was first recognized by Francis >Crick.

Wright, Sewall (1889-1988), American geneticist. Wright became interested in genetics via agriculture, and his early work was on the best ways to use inbreeding and cross-breeding to improve animal stocks.Turning to the study of populations of animals in the wild, he discovered that if a population is very small, genes can be lost for ever from the gene pool, not because they were inferior but because of mere chance. This is now known as the >Sewall Wright effect.

X

X chromosome, one of the chromosomes. In humans and many other creatures, the female has two X chromosomes and the male an X and a Y (see sex determination). The human X chromosome is the seventh largest in the human karyotype, nearly three times the size of the Y chromosome. It carries a large number of genes, which are said to be X-linked (see X linkage). See also Lyon hypothesis; Ohno's rule.

X linkage, the situation in which a gene is located on the X chromosome. In males, with sex chromosomes XY, any allele on the X chromosome behaves as a dominant, because there is no corresponding allele to mask it on the other chromosome (the Y does not behave as homologous to the X). A male passes on to all his daughters any gene that he carries on the X chromosome, but his sons will not inherit it. A female who carries the same allele on one of her X chromosomes will not be affected, if the gene is recessive as most X-linked genes are. Such a female is known as a carrier, and passes on the gene to half of her offspring, sons and daughters alike: males who receive the gene will inherit the condition, females will be carriers.

It is possible for a female to inherit an X-linked gene simultaneously from her father (he will have the trait in question) and from her carrier mother, but this is rare (how rare depends on the frequency of the X-linked allele in the population). The ginger colour in cats is X-linked, and most ginger cats are male, but the occasional female ginger cat can be found (see also tortoiseshell cat).

Over 100 genes are known to be X-linked in humans, including >colour blindness, >Duchenne muscular dystrophy and >haemophilia. Occasionally an X-linked gene is dominant; its pattern of transmission is similar,

in that a father cannot pass it on to his sons. An X-linked trait is not to be confused with a >sex-limited trait. See also Barr body; Lyon hypothesis; Ohno's rule; sex chromosomes.

Xenopus laevis, the African clawed toad. It has been used for experiments in developmental genetics, because its tadpole (which is in effect the embryo) is robust and easy to keep. It is also useful in experiments which involve the removal of the cell nucleus, as Xenopus's nucleus is large; these experiments have included the earliest successful clones. There are now a number of inbred strains for genetic use. Xenopus has a diploid chromosome number of 36.

Xeroderma pigmentosum, a human disease in which the skin is excessively vulnerable to bright sunlight, resulting in a high incidence of skin cancers. The underlying cause is an inherited inability to repair the damage that ultraviolet rays do to the DNA in the skin cells. The disorder is inherited as an autosomal recessive, and is one of the few clear-cut examples of a genetic basis for cancer.

X-ray crystallography, the technique used for investigating the three- dimensional structure of large molecules such as proteins and nucleic acids. The wavelength of the X-rays used is similar to the inter-atomic distances in these molecules, and the position of atoms can be deduced from the patterns of diffraction seen when X-rays are passed through them. X-ray diffraction pictures taken by Rosalind >Franklin contributed to Crick and Watson's discovery of the structure of >DNA X-rays, electromagnetic radiation at frequencies between ultraviolet light and gamma-rays, about 10-8 to 10-11 m wavelength. X-rays were discovered by the German physicist Wilhelm Roentgen in 1895. The American geneticist H.J. >Muller discovered in 1926 that they were a potent means of inducing >mutations in Drosophila and other organisms, and he became very concerned tht the widespread use of X-rays in medicine would cause serious mutations in humans. He was heeded; no longer were X-rays used on every medical pretext however slight or even as a commercial gimmick (e.g. there used to be X-ray machines in shoe shops for fitting children's shoes). Nowadays X-rays are used as sparingly as possible, particularly in the region of the testicles and ovaries. See also X-ray crystallography.

XYY syndrome, the condition of having an X and two Y chromosomes. Boys born with this develop physiclly as normal males and are fertile, but are taller than average (by about 6 in/15 cm). However, there is evidence that XYY men are five times more likely than XYs to be mentally subnormal. What is more, they are more likely to start displaying criminal behaviour as boys. According to findings in the United States, XYY boys tend to commit crimes of impulsive aggression (especially those leading to criminal damage rather than attacks on people).

When this research was first published, there was much talk about XYY being a 'criminal karyotype', and in the 1960s there were notorious murder trials in four different countries at which the defendants pleaded that they were not criminally responsible because of their XYY karyotype. In the French case, the man was convicted but given only a light sentence owing to that mitigating circumstance; in the English and German cases, the plea was ruled out of order and the men convicted; in the Australian case, the man was found not guilty by reason of his extra Y. It must be emphasized that the American research showed only statistical trends among the whole XYY population in the US; the great majority of XYY people (about 90 per cent) have no behavioural or mental difficulties at all.

Y

Y chromomsome, one of the sex chromosomes. In humans and many other creatures, the male has an X and a Y, while the female has two XYs; it is the presence of the Y that causes maleness to develop (see sex determination). The gene responsible for this acts by accelerating the development of the previously undifferentiated urinogenital tract of the embryo to form first the testes and then the male ducts and penis; it is an example of a >genetic switch. The sex-determining gene is on the short arm of the Y chromosome; individuals who are XY but have lost the short arm of the Y develop as female. The longer arm of Y can apparently be lost with no effect on the Phenotype, so it is believed that there are no general-purpose genes on it. The Y chromosome is the third smallest in the human karyotype.

Yeast. See ***Saccharomyces cerevisiae***.

Z

Z chromosome, one of sex chromosomes in birds and some insects, in which the male is the homogametic sex (i.e. has two identical sex chromosomes). Males are ZZ and females are WZ. See sex determination; W chromo-psome.

Z DNA, DNA in which the double helix is wound left-handed rather than right-handed as normal. The sugars do not fit together so well that way round, so the molecule has a jagged appearance. Z DNA has been studied in *Drosophila* chromosomes, and it has been suggested that it has a role in gene regulation.

Zea mays , the staple crop known in American English as 'corn' and in British English as 'maize' or 'sweetcorn'. It was the subject of early genetical research because of the economic importance of producing improved strains. It also happens to be a convenient plant to study in that large numbers can be grown on small plots of land, and the variants in the kernels are easy to observe and measure. Some important basic genetic discoveries have been made in *Zea*, including jumping genes.

Zygote, the individual as it exists at the moment of fertilization, i.e. the diploid cell that results from the fusion of two gametes. It can also be a theoretical entity: one could calculate the number of possible different zygotes that could result from a given matting (see hybrid).

Zygotene, one of the stages in >meiosis (one of the two processes of cell division).

Part II

(Classical Genetics, Molecular Genetics and Cytogenetics)

A

Abortive transfer - any DNA transfer from bacterial donor to recipient cells that fails to establish the incoming DNA as part of the hereditary material of the recipient. A. t. has been observed following transduction, transformation, and conjugation. In all cases the transferred fragment is diluted out as the culture grows. Failure of integration of transferred DNA into the hereditary material of the recipient cell may be due to: 1. the failure of incoming DNA to form circular molecules; 2. circulrization takes place, but the circular molecule fails to take up its maintenance system. A. t. of extrachromosomal elements (plasmids), as opposed to chromosomal fragments, is relatively uncommon since plasmids are genetic elements of autonomous survival in a bacterial cell. It is only when a mutation in the recipient or a resident plasmid makes the host component of the plasmid maintenance system inactive that a. t. of a plasmid occurs. Genes carried on abortive pieces of DNA may be expressed in the recipient cells.

Acentric - ref. to chromosomes or chromosome segments without a centromere.

Achiasmata - of a meiosis without crossing over and chiasmata. Where a. meiosis occurs, it is usually confined to one of the two sexes. Generally, the other sex has a chiasmata type of meiosis. A common feature of a. meiosis is the absence of any opening-out of paired chromo-somes (bivalents) at diplotene; the four chromatids remain parallel until the beginning of first metaphase. A. meiosis is a distinct and characteristic mechanism which has arisen independently in a large number of organisms. In higher plants it is rare and not yet observed in vertebrates (White 1974). (cryptochiasmate).

Achromatic (Fleming 1879) - those parts of the nucleus which may not be stained with dyes characteristic for chromosomes (chromatic).

A-chromosome (Randolph 1928) - any of the standard chromosomes of the chromosome comple-ments of all eukaryotic organisms. A chromosomes represent a delicately balanced system and each of them usually has to be present to secure normal viability. In many species, the standard complement of A-chromosomes may be supplements by varying members of supernumerary or B-chromosomes. These are derived from the A.C. set and have become functionally subordinate.

Acquired character - a modification impressed on an organism by environmental influences during development.

Acrocentric (White 1945) - of chromosomes where the centromere is very close to one end so that one chromosome arm is small or minute and the other very much longer (metacentric).

Acrosome (Lenhossek 1897) - a cap-like structure which invests the front part of the sperm head. It is limited externally by an outer membrane that is reflected forwards at the posterior boundary of the a. to form an inner membrane contiguous with the nuclear membrane of the spermatozoon (Hancock 1966).

Activator (Huxley 1935) - a gene-dependent substance which stimulates the development of a certain embryonic tissue or organ, i.e., is morophogenetically active.

Activator RNA (Britton & Davidson 1969) - a regulator molecule capable of recognizing specific sites on DNA. A. RNA is assumed to play a role in the regulation of gene expression in eukaryotes. According to the Briton-Davidson model of genetic regulation in eukaryotes, the genetic material is in its normal state repressed. Particular control systems functions specifically to overcome this repression and to switch on appropriate genes by the functioning of a. RNA. A. RNA is assumed to coded for by "integrator genes which are active only when an adjacent segment of DNA (= "sensor" gene) interacts with a specific protein.

Adaptation - any change in an organism's structure or function that allows it to better cope with conditions in the environment.

Adaptive peak (wright 1932) - a symbolic representation (from an evolutionary point of view) of the relations between organism and environment in the form of a topographic map. Groups related genotypes which make their carries able to occupy certain ecological niches are said to occupy "adaptive peaks" situated in different parts of the map and separated from each other by "adaptive valleys" comprising unfit gene combinations. Any such peak represents a particular configuration of gene (allele) frequencies and genotype frequencies characteristic for a population equilibrium and (relatively) stabilized by a complex interaction of opposing forces. If one of these interacting factors becomes modified in its action, three reactions may occur (Dobzhansky 1951, Lerner 1958). The construction of adaptive peaks is as follows: all genotypes are plotted as points on a plane - the more closely they lie the less they differ from each other - and as ordinate the prevailing adaptive value of these genotypes within the appropriate environment is used. In this way a "mountain range" is set up with "peaks" separated by "valleys" and "saddles". Each of the various genotypes is then represented by a point on this "mountain surface" and each population occupies a region on the same.

Adaptor hypothesis (Crick 1958; Hoagland 1959) - an hypothesis (now experimentally proved) which States that the amino acid sequence of a protein is determined, during the course of genetic translation, by the alignment of aminoacyl-transfer RNA's at corresponding nucleotide codons in messenger RNA. The specificity of the translation mechanism depends on base pairing between a nucleotide region of transfer RNA (the anticodon) and nucleotides in the codon.

Adaptor molecules - genetic translation.

Adjacent distribution (McClintock 1945) - the orientation and distribution of chromosomes lying adjacent in the ring or chain configuration of translocation heterozygotes (translocation) to the same pole in the first meiotic division (meiosis) in contrast to alternative distribution, in the course of which alternating chromosomes of the configuration are distributed to the same pole. It is a consequence of a.d. that either all or a part of the products of meiosis (gones or gametes) are genetically unbalanced and contain duplications or deletions. The proportion of unbalanced meiotic products is determined by the position and number of the chiasmata (in the pairing segments or interstitial segments). A. d. is responsible for partial sterility of translocation heterozygotes (semisterility). Two types of a. d. may be distinguished:

Adventitious embryony (Strasburger 1878) - form of apomixis (Agamospermy; the production of seeds without a sexual process.

Agamospermy - a form of apomixis: the formation of seeds without sexual processes through adventitious embryony, diplospory, or apospory. The process can begin autonomously or after pollination (pseudogamy). In pseudo-gamous apomixis the pollen does not make any genetic contribution to embryo formation, but is necessary to initiate the growth of the ovary or to fertilize the endosperm nucleus.

Alien addition line (Leighty & Taylor 1924; O'Mara 1940) - a line with an extra chromosome (monosomic addition) or an extra chromosome pair (disomic addition) from a related species (substitution line; alien substitution line).

Alien substitution line (Kattermann 1938; Unran et al. 1956) - a line in which an alien chromosome or a pair of alien chromosomes from a donor species replace one chromosome or a pair of chromosomes of the recipient species (alien addition line; substitution line). If a pair of alien chromosomes compensates for the missing chromosomes they are considered homoeologous to the pair they replace.

Allele (Johannsen 1909) - one of two or more alternate forms of a gene occupying the same locus on a particular chromosome or linkage structure and differing from other alleles of that locus at one or more mutational sites whose number per gene is between 10^2 and 10^3.

Members of a set of alleles are mutually exclusive genetic markers and arise by gene mutation. Their activity is concerned with the same biochemical and developmental process. A haploid organism or phase of the life cycle has a single representative of each a., a diploid two, a polyploid more than two from the total number existing in the population as a whole.

The individual alleles of a group consisting of more than two are called a series of "multiple alleles" (Morgan 1914). A diploid individual or phase of the life cycle may be homozygous for any one allele or heterozygous for a combination of any two. Segregation in meiosis results in gametes with only a single a. The dominance relations among multiple alleles vary from one group to another. For some groups of alleles every homo- and heterozygous genotype produces a different phenotype. In others, the alleles may be arranged in a descending series in which every a. is dominant over all alleles below it. The number of genotypes in a series of multiple alleles is ½ [n.1(n + 1)] where n is the number of alleles of the group (for four alleles: ½ (4.5) = 10). Multiple alleles of a gene affect similar parts or process in an organism and are indicated by attaching distinguishing letters or numbers as superscripts to the base symbol (genetic nomen-clature).

Allelic (Johannsen 1909) - ref. to the relationships between alleles of the same gene.

Allelic complementation - the production of a non-mutant phenotype when two independent mutations at the same gene locus, but on different homologous chromosomes, are introduced into the same cell. A.c. can be accounted for in terms of the sub unit structure of multimeric protein molecules and occurs in a wide variety of microorganisms and in some higher organisms (genetic complementation).

Allocycly (Darlington 1941) - ref. to differences in chromosome coiling which are shown by chromosome segments, whole chromosomes, or even whole chromosomes complements and may be environmental, genotypic, or cellular in origin.

Allopatric (Poulton 1903; Mayr 1942) - of populations or species which are mutually exclusive, but usually inhabit adjacent geographical regions. Geno-typically distinct allopatric populations of a species are called geographic races, subspecies, or local varieties. Gene exchange between allopatric populations is either limited or completely absent (sympatric).

Allophenic (Hadorn 1955; Mintz 1962, 1967) - of characters (allophenes) which in a given cell system arise by intercellular gene action, i.e., through the action of genes manifest in other cell systems of the organism (Hadorn, 1955). In such cases, a gene has produced an "autophene" in one cell system and the allophene characteristic of another system is the direct or indirect result of the gene which is manifest in the first cell system (gene activation). For allophenes a nonautonomous behavior on transplantation is characteristic. If an autophene secondarily gives rise to one or a series of allophenes, this is called "relational pleiotropy" (Hadorn 1954, 1955) and does not represent a true case of pleiotropy.

Alloploid (Clausen, Keck & Hiesey 1945) - of individuals (called "alloploids" or "amphiploids") which arise after natural or experimental ("synthetic alloploids") which arise after natural or experimental ("synthetic alloploids") crossing of two (or More) species or genera and contain the (structurally and genetically different) chromosome sets of the particular parents, each set being present either once ("allodiploids") or in greater number **("allopolyploids").**

1. Genome allopolyploids: Chromosome pairing in meiosis is exclusively in the form of bivalents. Pairing restricted to exactly homologous chromosomes ("homogenetic pairing") may be the consequence of the striking structural dissimilarity between the chromosome sets of the parents or may result from the activity of specific genes which prevent the pairing of partially homologous chromosomes. The sterility of the diploid hybrid from which the allopolyploid arose is usually complete in this category; fertility is fully restored with the production of the genome allopolyploid derivative. Once formed this derivative is usually completely isolated by sterility barriers from its nearest relatives and may show all the criteria of a new species.

2. Segmental allopolyploids: Chromosome pairing in meiosis is characterized usually by bi- and multivalent formation. The Parental chromosome sets combined in the diploid hybrid are in this case partially homologous (homoeologous and correspond in a considerable number of segments permitting "heterogenetic pairing". The sterility of the diploid hybrid is thus not pronounced. In contrast to category I., segmental allopolyploids are usually unstable. Through segregation they may give rise to stable segregants, which in regard to their chromosome constitution are either autoploid, genome alloploid, or stably segmental alloploid.

3. Autoallopolyploids: Chromosome pairing in meiosis is in form of bi-and multivalents as in 2. because homologous chromosome sets are present in more than double dose. Polyploids of this type are possible from the hexaploid

level upwards and combine characteristics of auto-and allopolyploidy.

Allopolyploid (Kihara & Ono (1926) - alloploid.

Allosome (Montgomery 1906) - a chromosome (=sex chromosome) deviating in size, form, or behavior from the other chromosomes (autosome).

Allosteric (Monod & Jacob 1971) - of an enzyme whose activity is altered when its structure is distorted by an organic compound at a nonsubstrate site. A Enzymes possess at least two distinct non-overlapping receptor sites. one of these, the active (or catalytic) site, binds the substrate and is responsible for the biological activity of the protein. The other, the allosteric site, is complementary to the structure of another low-molecular molecule, the so-called allosteric effector, which binds specifically and reversibly. This bond modifies the catalytic activity of the active site (selective increase or decrease) without the allosteric effect relying on a direct chemical connection between the substrate and the allosteric effector, and operates through introduction of conformational changes in the enzyme molecule which in turn lead to changes in the catalytic site (allosteric transition).

Allosteric effect - the effect of a modifier bound at a site of an allosteric enzyme which is topologically distinct from the active site at which reaction is occurring and which influences the binding or activity at this site by indirect means.

Allosterism - a mechanism which regulates enzyme activity by means of specific activators or inhibitors (allosteric effectors). These combine with the allosteric enzyme at sites other than the substrate catalytic site. Such allosteric effectors are usually compounds of low molecular weight, and are believed to exert their influence by changing the conformation of the protein.

Allosyndesis (Ljungdahl 1924) - in polyploids and aneuploids, meiotic chromosomes which have been introduced into the zygote by different parental gametes at fertilization (allosyndesis).

Amber mutation (Epstein et al. 1963) - a suppressible genetic change which results in the terminator codon UAG in messenger RNA (ochre mutation; opal mutation). The corresponding polypeptide chain is terminated at the site of the amber codon if no phenotypic or genotypic suppression (amber suppressor) occurs.

Amber suppressor - a suppressor mutation which allows amino acid insertion into the polypeptide at the site of an amber codon. (UAG) in messenger RNA. The efficiency with which completion of the peptide chain occurs depends (in E. coli) on the particular a. s. Functional protein may result depending on the amino acid inserted in place of the amber codon. In E. coli, different a. s. have been shown to insert serine, glutamine, tyrosine, lysine, or leucine at the site of the amber codon. A suppressors result from single base substitutions in the DNA sequences coding for transfer RNA s specific for these amino acids. The mutations result in changed anticodons of the tRNA molecules (suppressor tRNA), allowing them to pair with UAG at the ribosome (nonsense suppre-ssion).

Ambiguity - the phenomenon in which a given codori can be translated in more than one way (genetic code; coding ambiguity; translational ambiguity).

Am DNA - abbreviation for antime-ssenger DNA.

Amino acid activation - the first step of protein synthesis (genetic translation), whereby an amino acid reacts with adenosine triphos-phate. The reaction is catalyzed by a specific aminoacyl tRNA synthetase. The first reaction product is an amino acid adenylate, in which the necessary energy is conserved for the second stage - the attachment of the amino acid to a specific transfer RNA molecule. The adenylate remains bonded to the enzyme until transfer RNA molecule. The same enzyme also catalyzes the transfer reaction.

Aminoacyl transferase - two soluble protein factors (TI and T II) which are involved in the transfer of amino acids from tRNA to ribosome-bound polypeptides in mammalian systems. Aminoacyl transferase I catalyzes the interaction between aminoacyl-tRNAs and 80 S ribosomes, placing the aminoacyl-tRNA at the A site of the ribosomes. Aminoacyl transferase II aids in the translocation of the peptidyl-tRNA at the ribosomal A site to the P site (peptidyl transferase).

Aminoacyl-tRNA - the product of the reaction between aminoacyl adenylates and the appropriate transfer RNA. The reaction is catalyzed by specific aminoacyl-tRNA synthetases.

Aminoacyl-tRNA synthetase - any of about 20 specific synthetases which activate the 20 amino acids to from bound aminoacyl-adenylates as the initial step in the synthesis of proteins (genetic translation). The activated amino acid is then transferred to a specific transfer RNA (tRNA) to form the aminoacyl-tRNA. The specific aminoacyl-tRNAs are recognized by a triplet base sequence in the messenger RNA and the activated amino acid is then incorporated into a polypeptide.

Amitosis (Fleming 1882) - nuclear division by a process other than mitosis ("direct nuclear division") which in typical cases involves a dumbbell shaped cleavage of the cell nucleus, in the course of which no chromosomes are recognizable, nor is a spindle formed. A. occurs inter alia in ciliates, certain protists, and in specialized animal tissues. Part of the phenomenon described by a. involves so-called nuclear fragentation.

Amniocentesis - a procedure for diagnosing genetic abnormalities in uterur. By puncturing the abdominal and uterine walls with a needle a little amniotic fluid, containing fetal cells, is taken from a pregnant woman. From these cells it is possible to definitely establish whether or not the developing fetus is affected with certain specific genetic and chromosomal defects. A. has made for significant advance in genetic counselling.

Amphidiploid (Navashin 1927) - of tetraploid (genome allotetraploid) species hybrids (amphidiploids) whose somatic cells contain the diploid chromosome complements of both parent species (syn. allotetraploid, didiploid, double diploid). The diploid hybrid from which the amphidiploids originate is generally sterile owing to the significant nonhomology between the chromosome sets and the consequent difficulty with chromosome pairing during meiosis. With the doubling of the chromosomes number the sterility barrier is removed. Amphidiploids commonly possess a better viability and competitive ability than the original species and the diploid hybrid. These properties frequently result in the establishment of amphidiploids as new species (alloploid).

Amphiplasty (Navashin 1928) - a term proposed to denote morphological changes occurring in chromosomes following interspecific hybridization. Changes affecting individual chromosomes of the complement are called "differential a".

Amplicon (Lima-de-Faria et al. 1973) - a chromosomal (DNA) region which takes part in gene amplification and becomes amplified (repetitious DNA).

Anaphase (Strasburger 1884) - mitosis, meiosis.

Anaphase movement - chromosome movement.

Anaphase separation - during anaphase of mitosis and meiosis II, the disjunction of the chromatids of each chromosome; during anaphase of meiosis I the separation of chromosomes paired as bi- or multivalents. In all cases the disjunction results with causal participation of the spindle.

Anauxotrophic - = prototrophic.

Aneucentric - ref. to chromo-some mutations in the course of which chromosomes with more than one centromere (dicentric, tricentric chromosomes) are formed from monocentric (= eucentric) chromosomes.

Androecious (v. Uexkiill-Syllenband 1901) - of plants having only male flowers (gynoecious).

Androgenesis (Verworn 1891)- male parthenogenesis, i.e., the development of a haploid embryo from a male nucleus (= ephebogenesis; merogony), as opposed to maternal haploidy (gynogenesis) originating from the egg by incomplete fertilization, from other nuclei of the embryo sac in plants, by degeneration of the fertilized or unfertilized egg, or from other nuclei or other embryo sacs in connection with polyem-bryony.

Androgenous - producing only male offspring (gynogenous).

Androgynodioecious (Darwin 1877) - dioecious.

Aneuploid (Tackholm 1922) - 1. of cells, individuals, or populations having one or more addition to that complement. In the case of diploidy this type of chromosomal aneuploidy (Dyer et al. 1970) includes inter alia nullisomics (loss of a chromosome pair, 2 n - 2), monosomics (loss of a single chromosome, 2 n - 1), trisomics (presence of a single extra chromosome of the standard set, 2 n + 1), and tetrasomics (occurrence of an extra chromosome pair, 2 n + 2.) If the loss or addition concerns more than one specific chromosome pair, the situation may be characterized by the terms "doubly monosomic" (2 n - 1 - 1) or "doubly trisomic" (2 n + 1 + 1), etc. (pseudoaneuploid).

Aneuploid reduction - the reduction of genetic variability by decreasing the number of linkage groups. A. r. is supposed to play a role in the evolution of plant species in which outcrossing prevails and which occupy pioneer habitats. Such conditions produce a high degree of heterozygosity on the one hand and require repeated colonization of new areas on the other hand (Stebbins 1966).

Anisogamete - anisogamy; gamete.

Anisogamy - the gametes fusing during fertilization vary in respect to size, shape, and behavior. The distinction may include a) the behavior at fertilization only, b) size only or C) both size and shape (the most extreme case: oogamy). A. can occur in the form of gametogamy, gamet-angiogamy, or somatogamy.

Annealing - the "incubation" of a mixture of DNAs in single-stranded form and the quantitative determination of the helical material formed. The method permits the detection of homologous regions in any two species of DNA, provided the genetic complement is relatively small as in bacteria and viruses.

Anticodon - a sequence of three nucleotides in transfer RNA next in sequence to an invariant uridine in the a. loop of the molecule. The a. serves to recognize the appropriate nucleotide triplet that constitutes the codon in messenger RNA (genetic code). Mutations in the a. occur that change codon-anticodon pairing without affecting amino acid aminto acid being placed in many polypeptide sequences during genetic translation and many, therefore, be expected to be deleterious or lethal in many cases (suppressor tRNA).

Antigen - any agent (substance) which, when introduced into a vertebrate, elicits the production of antibodies which bind specifically to the inducing substance (immunogenetics).

Anti-messenger DNA - a DNA strand (symbol: and DNA) copied from messenger RNA by RNA-dependent DNA polymerase (reverse transcriptase, reverse polymerase, revertase) by a process called reverse transcription (= cDNA).

Antimutagenic - of agents that can reduce the rate, not only of induced, but also of spontaneous mutation. Antimutagens act either as "protective agents" or promote the repair of premutational DNA lesions. Generally, mutagens and antimutagens are counteracting factors; type, extent, and result of the interaction process are dependent on the mutagen and the biological material. Some antimutagens are a. Under some conditions, and mutagenic under other conditions.

Antimutator gene (Drake & Allen 1968) - any mutant gene which lowers the mutation rate. A. alleles have been found which suppress both errors of replication and errors of incorporation (gene mutation). Mutator and a. genes both affecting DNA polymerase demonstrate the profound role of this enzyme in determining the spontaneous mutation rate.

Antirepressor (Oppenheim et al. 1970) - a gene product capable of abolishing repression due to a repressor (in lambda phage). The a. may inactivate the repressor, compete for the repressor binding site on the DNA, or allow the transcription of the operons repressed by the repressor.

Anti-sigma factor - a protein that prevents recognition of initiation sites by the sigma factor of RNA polymerase.

Antisuppressor - a mutation which counteracts suppressor effects (nonsense suppression). A. mutations can occur in any of the following components affecting the genetic translation system (Cox 1973):

Antitermination factor - a protein (specified by an infecting phage DNA) that prevents normal termination of RNA synthesis, perhaps by interfering with the action of the rho factor. The a.f. allows the host RNA polymerase to " read though" sites at which it would otherwise terminate on the phage genome.

Apogamy (de Bary 1877; Winkler 1908). - 1. = apomixis (de Bary 1877).

Apomixis (Winkler 1906) - in plants, the replacement of sexual reproduction (= amphimixis) by various types of asexual reproduction which do not result in fusion of the gametes. Owing to failure or suppression of fertilization and meiosis, no alternation of nuclear phases is coupled with a. provided it occurs regularly in successive generations as obligatory a. Obligate apomicts have a completely closed recombination system and heterozygotic genotypes are preserved at the cost of evolutionary flexibility. In the case of facultative apomicts, apomictic and sexual modes of reproduction coexist. The mode of reproduction in animals equivalent to a. is parthenogenesis.

A., like amphimixis, is a genetically controlled reproduction system and is brought about by mutations which modify the course of sexual reproduction to the point of non-function. This occurs inter alia by transformation of meiosis into an apomeiosis, through hindrance of gamete fusion, the introduction of parthenogenetic egg cell development, or development of synergides. Apomictic reproduction is divided into two main types (Gustafsson 1935): 1. Agamospermy: all forms of a., in which reproduction proceeds via asexually produced embryos and seeds. Fertilization is always bypassed, meiosis usually. The embryo generally has the same chromosome number and the same genotype as the mother individual provided there is no autosegregation.

asynapsis (Beadle 1931) - desynapsis; asynaptic.

Asynaptic (Beadle 1931) - ref. to chromosomes in the first meiotic division in which pairing (chromosome pairing) either fails completely or is incomplete. Asynapsis is recognizable by the presence of a number (Varying from case to case) of univalent and can be used by incomplete homology of the chromosomes (e.g. in species hybrids), by gene mutations (asynaptic genes), or environmental influences. As a rule, asynapsis leads to the formation of hypo-and hyperploid meiotic products and frequently to fetility disturbances (desynaptic).

Autoallopolyploid (Kostoff 1939) - of cells or individuals whose chromosome complement shows the characteristics of auto - and allopolidy (alloploid, autoploid). Generally, autoallo-polyploids are hexaploid or of even higher ploidy and contain chromosome sets of different species: e.g. AAAABB or AAAABBBB, where each capital represents a chromosome set.

Autophene (Hadorn 1955) - a genetically controlled character which is manifested by the cell's own genotype and which shows "autonomous" behavior in transplants and explants (allophene). Gene action leading to an a. is designated "intracellular"; an a. is the direct or indirect result of such action confined to the single cell. The term "mosaic pleiotropy" is used to describe the case of a gene influencing or controlling more than one a. (pleiotropy, gene action).

Autoploid (Clausen, Keck & Hiesey 1945) - of cells or individuals with chromosome sets characteristic of the species in question; the chromosome sets are homologous to each other and thus pair fully in meiosis (alloploid).

The terms autodiploidy, autotriploidy, autotetraploidy, etc. are used when there are two, three, four etc. homologous chromosome sets per cell. The presence of more than two (diploidy) of the monoploid chromosome sets characteristic of the species is generally referred to as autopolyploidy (Kihara & Ono 1926). It is associated with a corresponding increase in the number of genomes and genetic information content.

Owing to the structural identity of the chromosome sets of autopolyploid forms, the homologous chromosomes can pair during meiosis and participate in crossing-over. Since each chromosome has more than one pairing partner, crossing-over produces multivalents (depen-ding on the paired groups consisting of more than two chromosomes, referred to as tri- (three paired chromosomes), quadri- (four paired chromosomes), pentavalents (five paired chromosomes multivalent), etc.

The polyploidization of a diploid species may occur spontaneously or be induced by spindle poisons; it represents a genome mutation and, like allopolyploidy, it is the result of the formation of a restitution nucleus in somatic (somatic polyploidisation) tissue.

Autoregulation (Calhoun & Hatfield 1973) - a regulatory mechanism of gene expression (= autogenous regulation; self-regulation) in prokaryotic and eukaryotic organisms in which the product (protein) of a structural gene modulates its own expression or the expression of the operon to which that structural gene belongs. The regulatory gene product has several functions; it may act as regulatory protein and also as an enzyme, structural protein, or antibody. Often the regulatory protein is the first enzyme in the pathway and participates in control by feedback inhibitions; its regulatory function is not dependent upon catalytic or other function this protein may have and it is not involved in the transport or metabolism of an inducer or corepressor.

Autosome (Montgomery 1904) - any chromosome with the exception of the sex-chromosomes which are referred to as allosomes or heterochromo-somes. The genes localized in the autosomes follow the mode of distribution of these chromosomes to the gametes in the course of meiosis; they show autosomal inheritance, those localized in the sex-chromosomes show "sex-linked" inheritance.

Autosyndesis (Ljungdahl 1924) - in contrast to allosyndesis, the pairing (chromosomes pairing) of complete or partial homologues (homologous chromosomes) contained in the same gametes at fertilization, during the prophase of the first meiotic division (Stebbins 1947). A. can occur as homo - or heterogenetic pairing and is associated with the presence of polyploidy or aneuploidy.

Axoneme - the central thread or bundle of fibrils in a cilium or a flagellum.

B

Backcross - the crossing of a heterozygote (heterozygous with one of its parents. The first, second etc., b. generations are symbolized by B_1, B_2, etc. The second and following backcross are made by crossing of individuals of the B_1, B_2 generations, etc., with the same parent.

Back mutation - a heritable change in a mutant gene (gene mutation) resulting in a revertant which has regained the enzyme or function that was lost with a socalled "forward mutation". A true back mutation restores the original nucleotide sequence which had been changed by forward mutation. The observation of a wildtype phenotype in a mutant line does not necessarily mean that this reversion has occurred by an actual "reversal" of the original mutational event, that is, by true back mutation.

Bacterial sporulation - the conversion of a bacterial cell from the vegetative to the sporulating state as a result of cellular differentiation.

Bacteriocin (Grazia 1925; Jacob et al. 1953) - any substance (protein) produced by bacterial strains which, after adsorption at specific receptors on the surface of sensitive bacterial cells, is capable of killing these cells.

bacteriophage - a virus whose host is a bacterial cell. Bacteriophages (= phages) may have a complex morphology. They consist essentially of a core of genetic material (DNA or RNA) carrying the particle's genetic information which is surrornded by a protein coat. Phages exhibit differences in their life cycles and may be grouped as either virulent or temperate. The introduction by infection of the genetic material of virulent phages into a susceptible host results invariably in the death and dissolution of the cell with the release of (100-10000) new phage particles ("lytic cycle"). Temperate phages can become integrated into the bacterial chromosome at specific points and replicate with it as a unit, the viral functions then being indefinitely unexpressed ("lysogenic cycle"). In the integrated stage the viral material is called a "prophage" and each bacterium of the "lysogenic" clone carries a copy of the Prophage in question. Once integrated into the bacterial chromosome, the viral genes can be considered additional bacterial genes and studied by the usual methods of bacterial genetics. The lysogenic cycle can become converted into a lytic one spontaneously or by induction. In that case the prophage (episome) is released from the bacterial chromosome, multiplies independently and the cycle ends with the liberation of infectious phages by lysis of the host cell.

Bacteriostasis - inhibition of bacterial multiplication without immediate killing.

Balbiani ring (Beermann 1952) - a large RNA puff of giant chromosomes which is present during a greater portion of larval development in the chironomid family of Diptera. This structural modification of specific loci is remarkable for its large size and characteristic ring shape. Similar to RNA puffs observed in the multistranded polytene chromosomes of a large number of dipteran species, the Balbiani rings have a high content of RNA and show a rapid uptake and turnover of RNA precursors. The B. r. structure and its extreme degree of puffing can be understood as the result of unraveling of a thick, folded cable into individual filamentous elements, probably in the form of extended loops.

Baldwin-effect (Simpson 1953) - the exchange of "non-hereditary" by "hereditary" characters via mutation and selection (genetic assimilation).

Band (Painter 1939) - the specific association of a large number of chromomeres in polytene chromosomes of Diptera (the polytene equivalent of "ordinary" chromomeres). Chromosome bands usually have clearcut outlines and characteristic dimen-sions. Each b. is well defined in relation to neighboring bands and separated from these by regions called interbands (banding pattern).

Barr body - sex chromatin.

Basal granule - = basal body.

Basic number - the number of chromosomes in a basic chromosomes set. The b. n. represents the smallest monoploid chromosome number (symbolized by x) of a polyploid series. All chromosome numbers being exact multiples of the b. n. are called euploid. All numbers deviating by single chromosomes from x and from multiples of it are called aneuploid. The specific karyotype comparing the b. n. is the basikaryotype (Sinoto & Sato 1940).

B-chromosome (Randolph 1928) - any chromosome of a heterogeneous category of chromosomes (also called supernumerary, accessory or extra chromosomes) present in many species of both plants and animals which differ in their behavior from normal or A-chromosomes in the following characteristics (some of them may be missing in the special case under observation):

Behavior genetics - a branch of genetics which deals with the inheritance of different types or forms of behavior, such as intelligence, personality traits, etc.

Binemic - of chromosomes that contain or are assumed to contain two DNA helices per metaphase chromatid, as opposed to uninemic chromatids with one DNA helix, or polynemic chromatids with more than two DNA helices.

Biochemical genetics - a branch of genetics dealing with the chemical nature of hereditary determinants and the manner of their action in both development and function (gene, gene action).

Biome - a community of plants and animals (ecosystem). Adjacent biomes usually blend into one another (no sharp demarcation line through a fairly broad transition region called an ecotone.

Biometrical genetics - a branch of genetics concerned with the study of quantitative inheritance.

Biometry - statistics applied to biological problems.

Bipolarity (Margolin 1965) - the phenomenon wherein both strands of a DNA helix (Chromosome) serve in genetic transcription, although at different regions.

Bivalent (Haecker 1892) - a pairing configuration (chromosome pairing) during the first meiotic division which consists of two completely or partially homologous chromosomes (pseudobivalent, quasibivalent). The number of bivalents per cell (meiocyte) normally corresponds to half the somatic chromosome number of diploid and genome-allopolyploid species. In chiasmate meiosis the pairing partners are held together after crossing-over by chiasmata, in achiasmate meiosis by functionally equivalent structures (collochores) until the beginning of anaphase I. The formation of bivalents is prerequisite for the reduction of the somatic bivalent interlocking.

Bouquet stage (Eisen 1900) - a meiotic prophase stage (leptotene to pachytene) in those species where the chromosomes are oriented by one or both ends towards one point of the nuclear envelope. The polarization process responsible for this arrangement originates in the cytoplasm. Where centrosomes occur, there is an orientation relationship between these and the chromosomes.

bridge-breakage-fusion-bridge cycle (McClintock 1951) - a process which may arise from the formation of dicentric chromosomes (chromosome type of the BBFB-cycle) or chromatids (chromatid type of BBFB-cycle), whereby daughter cells are formed (via breakage of the dicentric bridge between the cell poles during the anaphase of meiosis and/or mitosis) differing in their content of genetic information due to duplications and deletions in the chromosomes. A prerequisite for the maintenance of such a cycle is that the daughter cells produced be viable and that the chromosome ends formed by breakage of the bridge should not be stabilized but should fuse (Fig. 3). In the course of a BBFB cycle there may bridge-breakage-fusion-bridge cycle.

C

cAMP - cyclic adenosine-3', 5' monophosphate.

canalizing selection (Waddington 1940) - selection.

CAP (Zubay et. al. 1970) - abbreviation for catabolite activator protein.

Capsid (Lowoff, Anderson & Jacob 1959) - the protein coat of a virus particle or virion which may take different forms (either helical or isometric), depending on the nature of the protein subunits from which the c. is formed. Components (protein subunits) of capsids observed in the electron microscope on axes of symmetry are referred to as capsomeres. Composite capsids may consist of more than one kind of capsomere.

Cascade regulation (Pontecoruo 1963, Scherrer & Marcaud 1968) - 1. in prokaryotes, a form of genetic regulation on the transcriptional level where one several operons codes for the production of an internal inducer which turns on one or more of the other prokaryotic operons.

Catabolite activator protein (Zubay et al, 1970) - a protein (abbreviation CAP) which acts as a transcription-initiation factor necessary for the efficient transcription of bacterial operons which are subject to catabolite repression (bio-degradative operons). CAP is activated by cyclic adenosine monophosphate (cyclic AMP) and the combination of CAMP and CAP is bound to the DNA enabling RNA polymerase to bind to the promoter of these operons is composed of two sites that interact with CAP and RNA polymerase holoenzyme. In the presence of the appropriate repressor, the operon cannot synthesize mRNA, even in the presence of CAP and Cyclic AMP.

Catabolite repression (Magasanik 1961) - the inhibition by glucose and some biochemically related compounds of the synthesis of certain inducible enzymes in bacteria. C. r. is a mechanism of genetic regulation at the DNA level (transcriptional control). The extent of inhibition varies from a few percent to more than 90% and is dependent on the bacterial strain, the source of the catabolite, and the growth conditions. Most genes classified as inducible show c. r.

cDNA - DNA complementary to a purified messenger RNA and produced By RNA-dependent DNA polymerase (reverse transcriptase, revertase).

cDNA (= anti-messenger DNA) can be used as very sensitive and specific probes in molecular hybridization studies. These allow one to distinguish between nucleotide sequences that bear a high degree of similarity and also provide a practical means of assessing the purity of any RNA population containing the sequence reverse-transcribed.

Cell (Hooke 1665) - the basic structural and functional unit of all living matter which represents a multimembrand system, compartmentalized into specific fnctional areas or organelles with discrete but complexly interrelated and coordinated roles.

Cell cycle - the life cycle of the individual cell. In proliferating somatic cells it consists of four phases: the mitotic phase (symbol M) and a three-stage interphase consisting of the pre-DNA synthesis or G1-phase, the DNA-synthesis phase (S) and the post-DNA synthesis or G2-phase. During G1 the chromosomes are unreduplicated; chromosome reduplication occurs in S (Howard & Pelc 1953). A similar subdivision may be applied to the interphase preceding meiosis. In this case the G2-phase is generally either short or completely absent, i.e. meiosis takes place shortly or immediately after the synthesis period (S) of the interphase nucleus. Besides the orderly sequence of replication of the genome, the development of a cell from one division to the next includes the formation of cell organelles and the morphological events of cell division. C. Cycles reprogramming the synthetic repertoire of daughter cells are called "quantal cell cycles", those yielding replicas of the mother cell are termed "proliferative cell cycles" (Ishikawa et al. 1968).

Cell division - the reproduction of cells by karyokinesis and cytokinesis.

Cell hybridization (Barski et al. 1960) - in eukaryotes, the fusion of somatic cells in cell structure and the formation of viable hybrid cells. In animals the hybrid cells produced by co-cultivation of cell lines or inactive virus mediated cell fusion possess a nucleus containing chromosomes of both parents. (Virus-mediated cell fusion is about two orders of magnitude more efficient than cell fusion by co-cultivation). In interspecific somatic cell hybrids, segregation occurs by species-dominant loss of the chromosomes of one parental cell. This makes these cells particularly suitable for genetic analysis (synteny; synkaryon).

Cell line - a population of cells which arises from a primary cell culture at the time of the first subculture (cell strain). A cloned line is a c. l. descended directly from a clone. A diploid c. l. is one in which, arbitrarily, at least 75% of the cells have the same karyotype as the normal cells of the species from which the line was derived. If the karyotype has changed, but the diploid number of chromosomes remains the same, such cells may be referred to as "pseudodiploid". A heteriploid c. l. is one having less than 75% of cells with diploid chromosome constitution (Feodoroff 1967).

Cell lineage (Wilson 1882) - the derivation of a tissue or part of tissue from a definite blastomere of the embryo.

Cell membrane (Nageli & Cramer 1855) - a component of the cell surface (= plasma membrance). The c. m. is a tri-laminate structure (usually 8 to 10 mm thick) with two dense lines spaced 4 to 6 mm apart on each side of an intervening element of lesser density in the electron microscope (unit membrane). Differences in thickness, in symmetry, and in spacing of the two dense lamellae have been found. C. membranes possess pores with diameters greater than 0.4 nm but less than 0.8 nm. Size of pores and charge density within them are the main factors that control the fluxes of material. Hormones, calcium and polybasic substances are assumed to regulate the size of membrane pores.

Cell plate (Strasburger 1882) - in plant cell, the structure formed between daughter nuclei (cytokinesis) after karyokinesis. The c. p. complex or phragmoplast is composed both of fibrillar elements (microtubules), which arise at the plate and are oriented at right angles to its plane, and numerous cytoplasmic droplets which eventually accumulate and fuse in the mid region of the cell. Presumably the vesicles originate from dictyovesicles move before they fuse.

Cell transformation - a heritable change in the biochemical and morphological properties of eukaryotic cells due to infection by (oncogenic) RNA and DNA viruses. Viral transformation of mammalian cells is manifested by the loss of regulatory restraints on their growth potential most frequently accompanied by alteration of cell shape, karyotype and cells surface architecture, by loss of contact inhibition, changes in glycolytic rate, production of increased amounts of organic acids and acid mucopolysaccharides as well as acquisition of malignant potential (production of tumors by transformed cell when inoculated into host animals). These viral induced changes reflect alterations in gene expression. Transformed cells do not contain or produce infectious virus or infectious virus nucleic acid; they do, however, harbor the information of the virus genome incorporated into their chromosomes (synthesis of infectious virus is induced by fusion with premissive cells)

Cell wall - a rigid external coat which surrounds plant cell and some animal cells (one of the variants of the glycocalyx). It is formed outside the cell membrane and contributes to the mechanical strength of plants and to the ability of cells to withstand turgor and osmotic variations.

C. walls normally contain carobohydrates, lipids, proteins, and minerals; usually carbohydrates lipids, proteins, and minerals; usually carbohydrates predominate in the form of polysaccharides (cellulose chitin, xylans, mannans, and lignin). Ultrastructure of the c. w, nearly always comprises long, thin threads (microfibrils 10 to 20 nm wide) embedded in amorphous material. Lipid substances (suberin, cutin, waxes) are deposited either on the inside (suberin) or mainly on the outside (cutin, waxes) of the microfibrillar wall layers (adcrusting layers).

The c. w. of plants is perforated in many places by bordered pits and palsmodesmata which provide direct contact between neighbouring cells of a tissue.

Centric - of chromosome structural change (chromosome mutation) converse to centric fusion or Robertsonian translocation, i. e., the origin of two acrocentric or telocentric chromosomes from one metacentric chromosome (= dissociation). C. f. may be due to centromere misdivision or may involve a donor chromosome providing one centromere and two telomeres. Evolutionary replacement of a metacentric chromosome by two telocentrics has occurred in some animal species.

Centric fusion (Robertson 1916) - the fusion by reciprocal translocation of two acrocentric chromosomes to give a metacentic chromosome and a small fragment which is usually lost. C. fusions are the main way by which chromosome numbers have been changed in (animal) karyotype evolution. In the heterozygous condition, c. fusions show quite regular meiosis with the two acrocentrics invariably passing to one pole and the metacentric to the other. Aneuploid gametes may result, if the three centromeres orientate in a linear manner (centromere orientation) on the meiotic spindle instead of in and alternate configuration.

Centriole (Boveri 1895) - s cylinder some 120 to 150 nm wide and of variable length in the cells of animals, some algae and fungi. The cylinder wall is composed of 27 microtubules arranged in triplets parallel to the long axis of the c'.

Centromere (Waldeyer 1903) - the region (or regions) of each chromosome with which the spindle fibres become associated during mitosis and meiosis. This association is a prerequisite for ordered chromosome movement in the ccurse of karyokinesis. Recent studies support the conclusion that the c. is directly involved in the organization of microtubular subunits into chromosomal spindle fibres.

2. Neo-centromere (Rhoades 1952): chromosome ends (telomere) which, under certain conditions, show movement on the spindle during mitosis and meiosis, just as do the localized centromeres of these chromo-somes. They are secondary centromeres whose activity results in the chromosome ends moving first during anaphase movement.

3. Non-localized centromere: in this case the attachment of the spindle fibers is not confined to a strictly localized segment of the chromosome but takes place along the entire length of the chromosome. As a result all parts of the chromosome have active spindle-mobility and acentric fragments are lacking. Distinction is made between two variants of this type:

a) The multiple or polycentric (polykinetic) condition: each chromosome is furnished with many c., separated by small non-centric segments. The only certain example of this organizational type is that of the germline chromosomes of some ascarid nematodes (nuclear differen-tiation).

b) The diffuse or holocentric (holokinetic) condition (Hghes-Schrader & Ris 1941): every point along the entire chromosome shows active mobility on the spindle. This type of chromosome occurs inter alia in Hemiptera, Homoptera, a few Protista and, among higher plants, in the genus *Luzula*.

4. Semilocalized centromere (Vaarama 1954): the active spindle mobility is couples to a localized primary c. in mitosis but in meiosis is displaced to another localized site (secondary, tertiary c.). Chromosomes with semilocalized c., as are found in Pleurozium, are chromosomes with potentially multiple centromeres and behave functionaly like those with localized c. but the position of spindle activity undergoes a regular shift (centromere shift). Semilocalized centromeres are regarded as intermediate between the multiple and the strongly localized centromere types.

Centromere orientation - The process of orientation of centromeres during prometaphase of mitosis and meiosis. C. o. is achieved as follows: each chromosome (mitosis) or bivalent (meiosis I) exploits two independently acting centromeres, each of which is orientated towards one of the two spindle poles. A stable orientation (congression) of the centromeres of all the chromosomes along the spindle equator ("metaphase plate") results from the interaction between the spindle poles and centromeres via the spindle fibres.

Centromeric fusion - a type of centric fusion.

Cf - = colicinogenic factor.

C-factor (Muller 1926) - any gene or heterozygous structural change (in most cases inversions) reducing the frequency of crossing-over.

CG value - GC value.

Chain-initiating codon - terminator codon.

Chain-initiating mutation - a mutation which gives rise to an initiator codon.

Chain-terminating codon - terminator codon.

Chain termination - termination of polypeptide synthesis during genetic translation which requires both terminator and release factors.

Chiasma (Janssens 1909) - that place in a bivalent multivalent, or in the case of intrachromosomal (internal) pairing within a chromosome (isochromosome), from late prophase of meiosis up to first anaphase where a) pairs of homologous chromatids remain in contact although they are separated in other parts, and b) an exchange of homologous parts between nonsister chromatids of the chromosomes has taken place by crossing-over ("chiasmatype theory"). A site of contact, however, does not in all cases necessarily mean exchange of parts. In achiasmate meiosis so-called collochores may take over function a) of chiasmata.

With normal chiasmate meiosis at least one c. is necessary to hold the pairing partners in contact after repulsion ("chiasma theory of chromosome pairing", Darlington 1929) between them begins in diplotene. Premature separation of the pairing partners is the consequence of failure of chiasma formation bringing univalents into being (asynapsis; desynapsis). In organisms with large chromosomes to the other at each c.

Classes of single chiasmata (Darlington 1929) are:

1. Interstitial chiasma: on each side of the c. is a length of chromatid.

2. Lateral chiasma: the chiasma is terminal as to two chromatids and interstitial as to two others (two kinds, symmetrical and asymmetrical.

3. Terminal chiasma: the chiasma is localized at the ends of the pairing partners following chiasma terminalization.

4. Multiple chiasma: a terminal chiasma in which three or four pairs of chromatids participate (as in multivalents).

5. Imperfect chiasma: a chiasma in which one of the four cross-wise associations is broken prior to anaphase (="incomplete chiasms").

If two chiasmata are present in one bivalent (its 4 chromatids represented by A, A', B, B') there may be three different relationships between them (crossing-over) (Darlington 1937):

1. Reciprocal (comparate) chiasmata: the chromatids A and B or A' and B' participate in both chiasmata (= "regressive chiasmata", "two strand double crossing-over").

2. Complementary (comparate) chiasmata: chromatids A and B participate in one chiasma, chromatids A' and B in the other ("digressive chiasmata", "four-stand double crossing-over').

3. Diagonal (disparate) chiasmata: the chromatids involved in one chiasma are e.g., A and B those in the other A and B' or A' and B (= "progressive chiasmata", "three-strand double crossing-over").

Chiasma localization (Darlington 1931) - chiasma termina-lization.

Chiasma terminalization (Darlington 1929) - a progressive shift between diplotene and metaphase of the first meiotic division in the distribution of chiasmata along the arms of paired chromosomes from their points of origin to more distal positions. The terminalization process may be complete, partial, or entirely lacking. In the last case the chiasmata are called localized (Darlington 1931).

Chiasmatype theory (Janssens 1909, 1924; Darlington 1929) - a theory postulating that true chiasmata are the direct result of crossing over, being formed at precisely the points where the exchange of segments between (non-sister) chromatids took place. In contrast to this "one plane theory" the two-plane theory" postulates (Sharp 1934) that the cross-figures appear because the chromosomes of the pairing configuration split in contiguous regions along perpendicular cleavage planes. Both the "two-plane" and its variant, the "neo-two-plane theory" (Matsuura 1937), are only of historical interest now since all experimental evidence favors the chiasmatype theory.

Chimera (Winkler 1907) - in most cases a plant, more rarely an animal, composed of tissues of two or more idiotypes owing to mutation, somatic segregation, or grafting (= mosaic).

Chimeras form a heterogeneous group and in plants may be classified in three different ways (Tilney-Basset 1963):

1. According to their origin into those arising by:

a) spontaneous mutation,

b) induced mutation,

c) sorting out from variegated seedlings (variegation) or

d) grafting.

2. According to their structure into:

a) sectorial (different tissues grow side by side and occupy distinct sectors of varying size),

b) periclinal (different tissues are disposed one with the other, one occupies the center of the plant structure as a core, the second tissue grows around it as covering layer),

c) mericlinal (actually an interrupted periclinal c. with one kind of tissue of limited extent placed in a peripheral position and covering a central core).

Chloroplast (Schimper 1883) - in plants, a cell organelle (plastid) characterized by a limiting double membrane and an internal, chlorophyll containing lamellar structure embedded in a protein-rich stroma. The chloroplasts are capable of light-dependent photo-phosphorylation and of photo-synthetic CO_2 fixation.

Chloroplast DNA (Chun et al. 1963; Sager & Ishida 1963) - deoxyribonucleic acid (organelle DNA) localized in chloroplasts (contour length 40 - 45 um); its information content is about 4 to 8 X 10^8 daltons (10^{-15} to 10^{-16} g or approximately 2 X 10^5 base pairs) and equals that of a bacterial cell. The c. DNA is circular and one chloroplast may contain as many as 15 to 30 molecules. About 2 to 3 % of the c. DNA codes for chloroplast ribosomal RNA, about 1% codes for c. specific transfer

RNAs. Other c. cistrons code for c. ribosomal protein, for the proper organization of c. membranes, and one or more cistrons code for the synthesis of a functional photosynthetic apparatus. Calçulations suggest that a chloroplast contains more potential genetic information than a mitochondrion.

Chondriome (Meves 1907) - 1. collectively the mitochondria of the cell; 2. all hereditary determinants localized in the mitochondria (cytoplasmon; plasmon; inheritance).

Chromatid (McClung 1900) - one of the two visibly distinct longitudinal subunits of all reduplicated chromosomes (a half-chromosome) which become visible between early prophase and metaphase of mitosis and between diplotene and the second metaphase of meiosis. After these stages the chromatids are known as "daughter-chromosomes". The chromatids are those longitudinal units of the chromosome which become separated during anaphase of mitosis and anaphase II of meiosis. Sister chromatids derived from homologous chromosomes. Normally, chro-matids of homologous chromo-somes carry identical genetic information (identical linkage groups), apart from allelic differences between nonsister chromatids, and are similar morphologically. In all probability, each c. contains a continuous DNA double helix.

Chromatid aberration - a category of chromosome structural changes (chromosome mutation) produced in one and not the other chromatid of a chromosome as opposed to chromosome and subchromatid aberrations in which both chromatids or subunits of one chromatid are involved., chromatid aberrations arise spontaneously or are induced experimentally by mutagens during and after chromosome reduplication in the interphase nuclei.

Chromatid bridge (smith 1935) - a dicentric chromatid with the two centromeres passing to opposite poles during anaphase (bridge-breakage-fusion-bridge cycle). Chromatid bridges arise primarily from spontaneous or experimentally induced chromatid aberrations and secondarily (secondary structural change) through crossing-over between the chromosomes of structural heterozygotes in meiosis.

Chromatid non-disjunction (Sansome 1933) - 1. the distribution of both sister chromatids of a chromosome to one cell pole during anaphase of mitosis or in anaphase II of meiosis;

Chromatid segregation - the term c. s. or "double reduction" as opposed to "chromosome segregation" is used in the case of polysomy (autopolyploidy, various forms of aneuploidy) and heterozygosity for one or more biallelic loci. when two sister-chromatid segments of a chromosome are included in the same meiotic product (gone or gamete), the chromosome are included in the same meiotic product (gone or gamete). the chromatid segments having identical alleles ("equational exceptions").

Chromatin diminution (Boveri 1899) - = chromosome diminution.

Chromatoid body (Morgan & Uzmann 1966) - a cytoplasmic juxtanuclear clump characteristic for spermiogenetic cells in a wide range of organisms. It is either formed by extrusion of nucleolar material or represents an aggregate of cytoplasmic material. The c. b. contains ribonucleoprotein mateial.

Chromatophore (Schmitz 1882) - any particle (veslcular structure; diameter about 15 nm) isolated from photosynthetic bacteria and containing photosynthetic pigments. Such structures whose pattern may vary with the physiological state of the cell are surrounded by a membrane. They are scattered throughout the cytoplasm.

Chromatoplasm - in Cyanophyceae which possess no discrete plastids, a cell region of lamellar structure in which the thylakoids are located. The membranes, to which the assimilation pigments of these cells are bound, arise through the invagination of the plasma membranes which surround the protoplasts.

Chromocenter (Baccarini 1908; Tischler 1920) - any of the tightly coiled (chromosome coiling) chromosomes or chromosome segments within the interphase nuclei. Chromocenters consist of heterochromatin and are positively heteropycnotic (allocycly). Their number per nucleus may vary since heterochromatin has a tendency to fuse (chromosome pairing).

Chromogene (Serra 1955) - any gene localized in chromosomes, as opposed to cytoplasmic genes.

Chromomere (Fol 1891; Wilson 1896) - any of the linearly arranged, bead like and compacted segments in mitotic and meiotic prophase chromosomes and in interchromomeres results in chromosome-specific c. patterns (chromomere size gradient) which may serve to differentiate between the chromosomes of a chromosome set.

Chromonema (Wilson 1896) -The smallest light microscopic strand in chromosomes and chromatids.

Chromosomal aberration - = chromosome aberration.

Chromosome (Waldeyer 1888) - a linkage structure (linkage) consisting of a c. specific linear sequence of genetic information (gene). chromosomes of eukaryotes and prokaryotes are essentially involved in two main activities:

1. Those concerned with the transmission of genetic information from cell to cell and generation to generation.
2. those concerned with he ordered release of this information to control cellular function (gene expression) and development.

The principal synthetic activities related to these functions are the synthesis of DNA and RNA, respectively.

chromosomes are autoreduplicating structures of differential complexity in eukaryotes and prokaryotes whose number (per cell), morphology and organization are organism-specific characteristics.

a) Chemical components of the chromosomes of eukaryotes. The chromosomes contain almost equal amounts of DNA and histones, a variable amount of non-histone chromosomal proteins, and a small quantity of chromosomal RNA. In addition, lipids, polysaccharids and metal ions have-been reported to be components of the c.

Chromosomal DNA forms the backbone of the chromosome fibers. Each chromatid probably contains a single DNA double helix associated with chromosomal proteins and condensed or coiled along its length. Nonrepetitive (unique DNA) and reptitious DNA sequences (serially repeated sequences) are interspersed in a generally ordered manner (g-region).

Chromosomal proteins include two classes, the basic chromosomal proteins or histones, and the acidic or non-histone chromosomal proteins (hertones). Both interact with DNA by forces which include ionic ones. The non-histone proteins vary in quantity and there appears to be a limited heterogeneity between different tissues and species. Histones (which are absent in some fungi and dinoflagellates, and replaced by protamines in the sperm chromosomes of some animals) may be separated into five major fractions which do not vary significantly in type or quantity over a large or organisms.

Chromosome bridge - a double bridge linking the cell poles, formed by dicentric chromosomes (with two centromeres) during anaphase of mitosis and meiosis (bridge - breakage - fusion - bridge cycle). The dicentric chromosomes are produced as a result of primary or secondary chromosome mutations. If only one of the two chromatids or chromatid subunits participate in bridge formation, the terms chromatid bridge and subchromatid bridge are used.

Chromosome coiling - the spiral or helical coiling of the chromonemata of the chromosomes (coil) - one of the most striking phenotypic changes in the chromosomes during mitosis and meiosis. The coiling cycle transforms the functional form of the chromosome (maximum despiralization) into the transport form (allocycly, spiralization coefficient, packing factor). Two types of coils may theoreticall be formed between two or more chromonemata: the "paranemic coil" has freely separable subunits, the "plectonemic coil" has intertwined subunits.

Chromosome complement (Darlington 1932) - the group of chromosomes derived from a particular gametic or zygotic nucleus. It may be composed of one (haploid or monaploid nucleus), two (diploid nucleus), or more (polyploid nucleus chromosome sets.

Chromosome contraction - the thickening and shortening of chromosomes during mitosis and meiosis (= chromosome condensation) by chromosome coiling.

Chromosome diminution - the elimination of whole chromosomes or heterochromatic chromosome segments in mitosis, meiosis or interphase (i.a. in Ascaridae, in certain species of the genus *Cyclops* and various Diptera). The process leads to differences in the chromosome complements of the germ line and the soma and, in a few cases, within the gonads themselves, the to differences between the gametes and the remaining cells. There are various mechanisms leading to c. d. (nuclear differentiation).

Chromosome disjunction - the assortment of chromosomes during the first meiotic division.

Chromosome elimination (Seiler & Haniel 1921) - the loss of chromosomes from nuclei during certain early cleavage divisions in Dipteran and Hemipteran insects. Only the germ-line cells retain the full chromosome complement; all somatic cells have a 'reduced' number. C. e. is a mechanism of somatic versus germ-line differentiation (chromosome diminution).

Chromosome fusion (Seiler & Haniel 1921) - the union of two (or more) chromosomes by means of chromosome structural changes to form a single chromosome (chromosome mutation). among chromosomes with localized centromeres, stable unions are possible only if the fusion takes place in the centromeres region (centric fusion).

Chromosome gradient (Lima-de-Faria 1954) - the gradual decrease (or increase) in chromomere size, stainability, and thickness of the chromosomal fibrillae observed in specific regions of the chromosome. The gradients exhibit a variation of defined form which bears definite relations to particular regions of the chromosomes:

Chromosome map - the graphic representation (in linear form) of a chromosome in which the genes (genetic markers) belonging to a particular linkage group are plotted according to their relative distances. A distinction is made between genetic maps and cytological maps.

Chromosome mottling (Tijio & Levan 1948) - = chromosome erosion.

Chromosome movement - the movements of chromosomes during mitosis and meiosis which are a prerequisite for the anaphase separation of chromatids (mitosis and meiosis II) or chromosomes (meiosis I). There is generally little movement of chromosome during interphase. The random behavior of the chromosomes at anaphase which gives rise to chromatid or chromosome distribution to the cell poles depends upon the movements takes place on the spindle prior to mitotic and meiotic metaphase.

During mitosis, the chromosomes undergo orientation (centromere orientation) relative to the poles (auto-orientation) leading to the "congression" of single chromosomes on to the metaphase plate. After that the polar movement of the chromatids takes place.

Chromosome mutation - any structural change (= intergenic mutation or chromosome aberration) involving the gain, loss or relocation of chromosome segments. Chromosome mutations arise spontaneously or are induced experimentally by chemical or physical mutagens (mutation) and can be detected cytologically during mitosis, meiosis or in giant chromosomes of Diptera. The ability to detect different kinds of chromosome mutations will depend on the size, structure and number of chromosomes concerned and the ease with which they can be handled.

chromosome mutations are either intrachromosomal (= homosomal) and then call "interchanges", or intrachromosomal and then called "interchanges" (allelosomal when in homologous, or heterosomal, when in nonhomologous chromosomes). Intrachanges are either in one arm (= paracentric or homobrachial) or involve both chromosome arms (= pericentric or heterobrachial).

Descriptively, all such structural changes include deletions or deficiencies (the actual loss of terminal or intercalary portions of the chromosome and the included genes), duplications (the presence of a group of genes more than once in the genome), inversions (a chromosome segment becomes inverted and reinserted into its original chromosomal position), and translocations (relocations of chromosome segments within one or between chromosomes). Deletion and inversions are intrachanges, duplications and translocations may be intra- or interchanges.

Chromosome polymorphism - the occurrence of one or several chromosomes in two or more alternative structural forms within a population.

Chromosome puffing - puff; polytene chromosome.

Chromosome set - in eukaryotes, a minimum viable complement of indispensable (B-chromosome) chromosomes, each represented once. The number of chromosomes in a basic c. s. is called the basic number.

Chromosome sterility hybrid sterility.

Chromosome stickiness (Beadle 1932) - stickiness.

Chromosome substitution (Karpechenko 1935) - the exchange of single chromosomes or chromosome pairs by other chromosomes of the same chromosome complement ("autosubstitution" or by chromosomes of the complements of other species of genera ("allosubstitution" or "alien substitution").

Chromosome theory - the chromosome theory of Mendelian heredity states that the chromosomes, as carriers of genetic information, represent the material basis of "nuclear inheritance". The regularity of the gene distribution from cell to cell (mitosis) and from generation to generation (meiosis) and the concomitant genetic recombination depend on the characteristic behavior of the chromosomes in the course of nuclear division.

Cis-configuration (Haldane 1942) - the alleles of two linked genes or the heteroalleles of a cistron are said to be in the cis-configuration (coupling phase) if in doubly heterozygous or heterogenotic cells the two (dominant) wild - type alleles are located in one, the two mutant alleles in the other two homologous chromosomes (AB/ab or + +/ab).

Cis-trans test (Lewis 1951; Benzer 1957) - a genetic test (= complementation test) to decide whether two gene mutations (m^1 and m^2) occur in the same functional gene ((mutations (m^1 and m^2) occur in the same functional gene (cistron) and to establish the limits of this genetically active region. To this end, the heterozygotes (or heterogenotes) in which the concerned mutations are located in the same chromosome (cis-configuration: + +/m^1m^2) and those in which they are present in separate chromosomes (tran-configuration + m^1/m^2 +) are compared. Two recessive mutations are assigned to the same cistron if their trans-configurations exhibit the mutant phenotype and their cis-configuration the normal phenotype (wild-type). If one or both mutations are dominant they are considered to belong to the same cistron if the cis - and trans-configurations differ in phenotype. on the other hand, should the cis - and trans configurations be phenotypically indentical this is evidence that the mutations occurred in different cistrons. The fact that among mutations of a cistron "intra-cistronic complementation" (genetic complementation) can occur whereby independent mutations, in the trans-configuration, interact to produce the normal phenotype demands that the c. t. be used critically. However, in "intra-cistronic complementation" the normal phenotype is often more fully expressed with cis-configuration than with trans. Since cis - and trans-configurations are not equivalent in all cases complementation, it is reasonable under these conditions to assign the independent mutations to the same cistron.

Cistron (Benzer 1957) - segment of DNA or RNA which codes for a specific gene product, either a protein (messenger RNA) or an RNA involved in genetic translation (ribosomal RNA; transfer RNA) within cistrons, pairs of mutations in trans-configuration are either deficient for a particular protein (enzyme), or produce that protein in a structurally abnormal form

(Fincham 1959). The c. may be equated to the gene as a functional unit and is defined operationally by the results of unequivocal cis-trans tests (complementation tests). Any c. consists of a cistron-specific number and sequence of nucleotide pairs. If a protein consists of a single polypeptide chain or a series of identical chains then it is determined by a single c.; if a protein is composed of two or more nonidentical polypeptide concept'). Within each c. genetic recombination (recon) and mutation (muton may take place at many sites (transcription; operon).

Cladogenesis (Rensch 1947) - mode of evolution which consists in the splitting of an evolutionary (phylogenetic) line, such as a species. The two or more descendent species may then follow independent evolutionary courses. C. not only involves species splitting and species divergence but also the origination of genera, families, and other higher taxonomic categories. C. and anagensis (phyletic evolution) are combined in varying proportions in most lines of descent.

Cleavage map - a chromosome map of viruses based on the fragments produced by restriction endonucleases of bacteria (DNA modification restriction system).

Cline (Huxley 1939) - gradient (decrease or increase) within a continuous population in the frequencies of different genotypes in different localities ("genocline") or phenotypes ("phenocline") (Darlington & Mather 1949).

Clone (Webber 1903) - a population of cells or organisms derived from a single cell or common ancestor by mitosis. The mode of reproduction giving rise to a c. is asexual. A c. is not necessarily homogeneous and therefore the terms c. or cloned should not be used to indicate homogeneity in a population.

Cloning - clone; molecular cloning.

C-meiosis (Levan 1939) - a modified form of meiosis which occurs following the action of colchicine ("C-meiosis" = "colchicine meiosis") and other spindle poisons. The spindle mechanism which normally directs chromosome distribution. At the same time, chromosome coiling and chiasma formation are inter alia functions which are indirectly affected. The complete suppression of the spindle mechanism, which may involve only the first meiotic division or only the second, or both divisions, gives rise (via the formation of restitution nuclei) to meiotic products with more than one chromosome set.

C-mitosis (Levan 1938) - a modified form of mitosis, caused by partial or complete inactivation of the spindle mechanism and concomitant disturbance in anaphase. The modifications is produced by the action of colchicine ("C-mitosis" = "colchicine mitosis") and other spindle poisons. Parallel with the effect on chromosome movement is an increased chromosome contraction. The centromeres divide late and the chromatids associated at the centromere mutually repel one another. In this way cross-shaped, so-called "C-pairs" are formed.

Coadaptation (Dobzhansky 1951) - the selection process (= integration) by which harmoniously collaborating genes become accumlated in the gene pool of a population.

Code - genetic code.

Coding ambiguity - the phenomenon whereby one nucelotide triplet (codon) codes for more than one amino acid. Ambiguity in the genetic code may result from: 1. A codon may be misread owing to environmental conditions existing around the tRNA mRNA ribosome complex which interfere with the codon recognition process (mistranslation). 2. A codon may be misread because of a specific change in the primary, secondary, or tertiary structure of either the transfer RNA or the ribosome such that it affects the codon anticodon interaction. 3. An aminoacyl-tRNA synthetase may misrecognize either an amino acid or a tRNA and form a wrong aminoacyl tRNA.

Codominant - of alleles not mutually related in terms of dominance recessivity; instead, their gene products occur independently, and both are manifest phenotypically. If, for example, the heterozygote of the alleles A^1 and A^2 belonging to a multiple allelic series and dominant to a, shows no phenotype intermediate to that of A^1A^1 and A^2A^2 then the alleles are termed codominant.

Codon (Crick 1963) - any triplet of nucleotides (coding unit) in DNA or RNA (if RNA is the carrier of primary genetic information as in some viruses) that codes for a particular amino acid or signals the beginning or end of the message (anticodon). The term c. is also used for the corresponding (and complementary) sequences of three nucleotides in the messenger RNA (nodoc) into which the original DNA sequence is transcribed (genetic transcription).

Codon recognition - the relationship between the anticodons of transfer RNA and the codons in messenger RNA. A site on the ribosome m RNA complex (initiation complex) at which codon-dependent binding of aminoacyl-tRNA occurs during genetic translation is called the codon recognition site (Watson 1964).

Coefficient of coincidence (Muller 1916) - coincidence.

Coenocyte - a multinucleate cell produced by repeated nuclear division without cytokinesis (= multikaryon).

Coenogamodeme (Gilmour & Heslop-Harrison 1954) - = coenospecies.

Coenospecies (Turesson 1922) - a group of individuals of common evolutionary origin comprising more than one taxonomic species. The c. is made up of ecospecies which following hybridization may participate to a limited extent in mutual exchange of genes. Gene exchange between different c. does not occur. Sterile hybrids are produced by hybridization.

Coil - one complete revolution of the internal spiral of a chromosome (chromosome coiling). Darlington & Mather 1949).

Coincidence (Muller 1916) - the coefficient of c. denotes the proportion of observed double crossovers to the number expected from the random combination of single crossovers among three or four linked genes (Darlington & Mather 1949).

Colchiploidy (Dermen 1953) - polyploidy induced following the application of the spindle poison colchicine (polyploid).

Colicin (Frederiq & Beltz-Barean 1953) - any of a class of bactericidal protein antibiotics which are specified by extrachromosomal plasmids (colicinogenic factor) in certain bacterial strains, can absorb to the surface of susceptible bacteria and, in the majority without penetrating the surface, kill these cells.

Colicinogenic (Fredericq 1953) - of bacterial strains able to produce colicins.

Colicinogenic factor (Fredericq 1953) - any of the extrachromosomal genetic elements (plasmids, symbolized by Cf) responsible for the production of colicins in colicinogenic strains of Enterobacteriaceae (= colicin factors; col-factors). Col-factors are transferable from colicinogenic (col+) to noncolicinogenic (col-) strains by means of cell contact or by transduction. Each col-factor represents an independent replicon and consists of double-stranded (circular) DNA.

Colicinogeny colicin.

Colicin tolerant (Hill & Holland 1967; Nomura & Witten 1967) - of bacterial mutants that are tolerant to the killing action of a number of colicins. Such mutants still possess normal colicin receptor sites but the sequence of biochemical changes that normally follow colicin adsorption does not occur. The mutation results in widespread changes in the structure of the cell surface.

Colinearity (Crick 1963) - the exact point-by-point relationship between the order of amino acids along the polypeptide chain and the order of the corresponding codons along the polynucleotide chain of the nucleic acid (sequence hypothesis; genetic transcription; genetic translation).

Complementation - 1. all cases in which interaction between viral gene products or gene functions in multiply or mixedly infected cells results in an increased yield of infective virus of one (or both) parental types (helper virus)

Complementation analysis - determination of whether cells with the same phenotype have the same genotype (genetic complementation).

Complementation test - = cis-trans test.

Complex heterozygous (Renner 1917) - of a special type of genetic system in plants (called complex heterozygotes) based on heterozygosity for multiple reciprocal translocations (as exemplified by *Oenothera*). C. h. species are heterozygous for numerous allele pairs linked to different chromosomes but segregating as a unit during meiosis.

Compton effect - in an atom, the interaction of a high energy photon with a single electron resulting in the ejection of the electron and secondary emission of a lower energy (longer wave length) photon.

Conditional mutation - any of a class of mutations whose viability is dependent on a set of (permissive) conditions.The perish under non-permissive restrictive conditions. A special class of c. mutations is the temperature-sensitive mutation (lethal factor). At the restrictive temperature, the mutant dies or further development is arrested; at permissive temperatures the mutant survives and reproduces. In micro-organisms, the majority of temperature-sensitive mutations are single basepair substitutions of the missense type of mutation. At the permissive temperature, the mutant protein is functional; at the restrictive temperature it is not.

Conduction (Clark & Adelberg 1962) - in bacterial conjugation, the transfer of the bacterial genetic material and of the conjugon (= "promoter") as linked elements (donation).

Conjugation - the term c. denotes several events that connect, couple, or unite chromosomes, unclei, cells or individuals.

C) in bacteria (bacterial conjugation or bacterial mating), a parasexual mode of unidirectional (polarized) transfer of genetic information involving direct cellular contact between a donor ("male") and a recipient ("female") cell. The contact is followed by the formation of a cellular bridge physically connecting both cells. Subsequently a segment (sometimes all) of the male's chromosome may be transferred into the female and may undergo genetic recombination with a homologous chromosomal segment of the recipient cell. The conjugation process invariably occurs between cells of different mating type (sex factor) with or without subsequent recombination. With chromosome transfer almost always a temporary diploid, partial zygote ("a merozygote") is produced. Only those portions of the donor segment which following recombination become part of the recipient's genome will participate in the subsequent replication process.

Conjugational DNA synthesis - the synthesis of plasmid DNA during bacterial (*E. coli*) conjugation mediated by F-like or I-like conjugative plasmids. A specific single strand of DNA is transferred to the recipient bacterium where the complementary strand is synthesized. DNA homologous to the transferred material is synthesized in the donor upon the template of the remaining strand. The bacterial functions necessary for this c. DNA s. differ from the requirements for bactrial DNA synthesis.

Conjugative plasmid - any F-like or I-like plasmid which mediates bacterial conjugation. A specific single strand of DNA is transferred during conjugation to the recipient bacterial cell where the complementary strand is synthesized. DNA homologous to the transferred material is synthesized in the donor upon the template of the remaining strand.

Conjugon (Luria 1963) - any of a number of different genetic elements (plasmids) in bacteria that promote bacterial conjugation and gene transfer. These include the F palsmid, colicinogenic factors (Col), and resistance transfer factors (R). Conjugation due to the presence of F and certain R-factors is brought about by the synthesis of a special kind of pilus (sex pilus) at the bacterial surface which also acts as receptor for F-specific (male pilus) bacteriophages.

D

Double telotrisomic (Kimber & Sears 1968) - an individual deficient in one chromosome but which has two telocentric chromosomes, one for each arm of the missing chromosome (= pseudotrisomic).

Drug resistance factor - = resistance factor.

Dubinin effect (Stern & Kodani 1944) - a form of variegated-type position effect which leads to the reduction or inactivation of the action of a normal allele and is conditioned by translocated heterochmatin in the immediate vicinity of the allele. This type of position effect was first found for the cubitus interruptus gene in Drosophila by Dubinin & Sidorov (1935).

Duplicate genes (Shull 1914) - two identical allele pairs showing the same phenotypic action but localized in different chromosomes. D. g. are of relatively frequent occurrence and are probably due to secondary polyploidy (polyploid). In gene formulae, the d. g. are generally given the same symbol but characterized by different suffixes or subscripts.

Duplication (Bridges 1919) - a chromosomal structural change (chromosome mutation) resulting in the doubling of a section of the genome of pro- and eukaryotes. The size of the duplicated segment may vary considerably.

Duplications are considered to play a role in origin of new genes through functional diversification of duplicate members.

Cytologically, the existence of relatively small duplications in the chromosome complements of eukaryotes may be proved by the appearance or regional alternations of pairing (chromosome pairing) in first meiotic prophase and (somatically paired) polytene chromosomes of Diptera in that member of a pair of homologues which is heterozygous for the duplicated segment.

The smallest duplications that can be investigated cytogically are those of single bands of polytene chromosomes. However, whole chromosomes arms may be duplicated, thus giving rise to isochromosomes. Where whole chromosomes (linkage groups) are doubled in the haploid chromosome set, this constitutes a genome mutation (aneuploid). Duplications may occur, inter alia, as primary structural changes of chromosomes, as a result of disturbances in the crossing-over process (unequal crossing-over), or as secondary chromosome mutations due to crossing over in inversion-or translocation hetero-zygotes.

E

Elongation factor (Lucas-Lenard & Lippmann 1966) - soluble protein factors (= transfer factors involved in the polymerization of peptides (elongation of peptide chains) during genetic translation. Unlike termination and initiation factors which catalyze single steps in protein synthesis, e. factors are required for the formation of every peptide bond.

Endomitosis (geitler 1939) - a form of somatic polyploidization (-polyploid) which is fairly common in differentiated or differentiating tissues. It takes place within an intact nuclear envelope and gives rise to endopolyploidy, i.e., duplication of the chromosomes without cell or nuclear division in the usual sense. E. is characterized by an increase in DNA content.

In some organisms, e. has been found to resemble a normal mitotic cycle. Endoprophase is followed by endometaphase. In endoanaphase the chromatids separate within the nucleus, and in endotelophase an interphase state with double the original chromosome number is attained. The formation of the mitotic apparatus and of a metaphase plate is omitted. The autonomously separated chromatids normally remain in a typical parallel arrangement.

An e. is termed "masked e." (Resch 1952) or "endoreduplication" (Levan & Hauschka 1953), where no stages formally comparable with the normal mitotic cycle are observed. An endoreduplication may be established with certainty if it is followed by a normal mitosis in which the products of endomitotic chromosome reduplication reveal themselves as diplo- or quadruplochromosomes which separate before anaphase producing a doubled or quadrupled number of individual chromosomes, as the case may be.

Endopolyploidization associated with e. may lead to very high degrees of ploidy. In general, fixed degrees of ploidy are characteristic of certain tissues. In most cases, endopolyploidy represents an ultimate state of cytodifferentiation and is not a factor of cytodifferentiation.

If separation of the reduplicated chromosomes products formed by e. fails to occur, so-called polytenic or polynemic chromosomes result (polytene chromosomes).

endoplasmic reticulum (Porter et al. 1945) - a complex membraneous (tubular and vesicular) system (symbol ER) in the cytoplasm of enkaryotic cells which is a component of the cell's endomembrane system, delimits cell compartments, and conveys metabolites, macromolecules, and perhaps electrolytes into, out of. and within the cytoplasm. The ER consists largely of lipid (30 to 50%) and protein (40 to 60%) with some RNA. The lipids are mostly phospholipids (50 to 60%) among which lecithin predominates. The protein is partly structural and partly enzymatic.

The lipoprotein membranes (unit membrane) differ in configuration and constitution from the cell membrane and bring the nuclear envelope into close association, if not in continuity, with the Golgi apparatus and the cell membrane, Many of the ER membranes are studded with ribosomes (rough ER); others are devoid of ribosomes (smooth ER). Membranes of the rough ER measure approximately 5 to 6 nm in width, whereas the cell membrane is 7.5 to 10 nm wide. The membranes of the ER are symmetrical, but show dense cross bridges probably representing protein bands separating lipid micelles

The ER is most variable cell organelle and forms a dynamic, rapidly changing structure. Its formation is cell-type specific. It develops moderately in very young meristematic cells, but can develop strongly in growing cells and during mitosis. ER elements are sparse in mature, metabolically relatively inactive cells, but highly developed in cells actively engaged in protein synthesis.

Enzyme - a protein which even in low concentration speeds up, enables or controls chemical reactions in living organisms without being used up in the reactions. Enzymes are typically composed of a protein part (the apoenzyme) which confers specificity and a nonprotein part (the coenzyme) necessary for activity. Enzymes bind to substrates thereby affecting their breakdown or union with other substances. They are molecules of complex chemical constitution produced in living cells under genetic control (genetic transcription; genetic translation).

Most enzymes have molecular weights substatially higher than 17000 and contain more than one peptide chain. Normally, these polypeptide subunits show no catalytic activity by themselves. It is essential that the subunits be fitted together exactly in a specific, three dimensional conformation to preserve enzyme activity.

Epistasis (Bateson 1907) - a form of gene interaction whereby one gene interferes with the phenotypic expression of another nonallelic gene (or genes), so that the phenotype is determined effectively by the former and not by the latter when both genes occur together in the genotype. Genes whose expression is altered by nonallelic genes are said to be "hypostatic", or to exhibit "hypostatic". Epistatic action may depend upon the presence of a dominant allele A, the recessive allele a, or the heterozygous genotype Aa. The hypostatic reaction of nonalleles may depend upon the presence of B, b or Bb.

1. E. of dominant genes: if gene A is dominant over a and gene B is dominant over b, then A is recognized as the epistatic gene if the expression of B and b is suppressed and the genotypes A.B. and A.bb appear the same. In this case, the normal F2 dihybrid - segregation ratio (9:3"3:1) is modified to 12 (9A.B + 3A.bb): 3aaB.:1aabb.

2. E. of recessive genes : in this case recessive gene a suppresses the manifestation of B and b. Under this condition the F2 segregation ration of 9A.B.:3A.bb 4(3aaB.+ 1aabb_ results.

In population genetics and quantitative genetics, the term epistasis is sometimes used to refer to all nonallelic gene intgeractions)-interaction deviation.

Euchromatin (Heitz 1928) - those chromosomes or regions of chromosomes are said to consist of e. and called "enchromatic" which show the normal cycle of chromosome coiling, normal staining properties and do not become heteropycnotic, as opposed to heterochromatin. The euchromatic portions of the chromosome appear to be almost free of rapidly-reassociating, repetitious DNA. They show the same DNA-histone ratio as heterochromatin but three to four times more acidic chromosome proteins (hertone).

E. is the genetically active, dispersed part of chromatin in eukaryotic - interphase nuclei.

Eugenics (Galton 1883) - the study of agencies under social control that may improve or impair the hereditary qualifies of future generations of man, either physically or mentally (Stern 1960). E. may be subdivided then into negative (preventive) and positive (progressive) eugenics. The aim of negative e. is to prevent the increase or the presence of alleles which produce undesirable phenotypes. Positive e. is concerned with furthering the increase of alleles which cause desirable phenotypes or, at least, guarding against the decrease of such alleles (euphenics).

Euploid (Tackholm 1922) - of cells, tissues and individuals with one complete chromosome set ("monoploidy") or with whole multiples ("diploidy", "polyploidy") of the basic, monoploid number of chromosomes characteristic of the species. Each chromosome is represented once in each chromosome set (aneuploid).

Evolution - the transformation of the form and mode of existence of an organism in such a way that the descendants differ from their predecessors (Zimmermann 1953).

Evolutionary divergence - the degree of divergence, at the intra- and interspecific levels, of two (or more) populations, which presumably have evolved from a common ancestor. A direct test of the degree of evolutionary divergence of two compatible populations is the relative variability of their F2 generation.

Excision repair (Setlow & Carrier 1964; Boyce & Howard Flanders 1964) - the repair of DNA lesions (single stand breaks, damaged bases, and other structural defects) by a dark repair mechanism which has been found in many biological systems, including human cells. E. R.. consists in the excision of defective regions (oligonucleotides) in one of the two DNA strands and the subsequent resynthesis of the excised nucleotide sequence (repair replication), utilizing the complementary base pairing information in the intact strand. In most cell types, the amount of DNA replaced by E. R.. involves less than 1% of the genome. Unscheduled DNA synthesis in eukaryotes is an expression of e. r.

Exogamy (McLennan 1865) - includes all those systems of sexual reproduction (mating system) in which there is a greater frequency of mating between unrelated or only distantly related partners than would be the case for purely random mating (panmixia) within the population (cross-fertilization or outbreeding). The genetic consequence of e. is an i9ncreased incidence of heterozygosity in the population (endogamy).

Extrachromosomal - of structures or processes that are not part of the chromosomes or proceed outside the chromosomes, respectively (inheritance' plasmid).

Extrachromosomal inheritance - inheritance.

Extranuclear - of structures or processes which are located, proceed or are active outside the nucleus in contrast to intranuclear structures or processes.

F

Feedback inhibition (Umbarger 1956; Yates & Pardee 1956) - the inhibition of an early enzyme in a biosynthetic pathway by the end product of that pathway. In f.i., substrate and inhibitor may bind at two different sites on the enzyme (allosteric) and the inhibitor binding appears to produce a change in the tertiary or quaternary structure of the protein. The inhibitor specific site and the substrate binding site may be on different polypeptide chains of a multichain enzyme. The obligatory requirement for excesses of two or more end metabolites to achieve inhibition is called "concerned" (Datta & Gest 1964) or "multivalent feedback inhibition" (Paulus & Gray 1964).

F episome - = F plasmid.

F incompatibility - incompatibility which operates after an F-prime factor (F plasmid) has been successfully transferred to a recipient bacterial cell already carrying an F plasmid. If the resident F plasmid is in the autonomous state, either this or the incoming F-prime element is inherited unilaterally at cell division, ultimately giving a population of cells carrying one F plasmid or the other. If the resident F plasmid is integrated into the bacterial chromosome, it is always the incoming F-prime element that is inherited unilaterally. Both F.i. and F surface exclusion are determined, at least in part, by the F plasmid itself.

Fi+ plasmid (Watanabe et al. 1964) - any of a class of resistance factors (R plasmids) able to repress the bacterial F plasmid or sex factor. Those plasmids which do not repress the fertility of F cells are called fi- (fertility inhibition). fi+ plasmids produce a cytoplasmic repressor which switches off synthesis of sex pili.

Flexibility (Thoday 1953) - 1. Genetic f. (Thoday 1953): the ability of a genotype to vary and adapt itself to changing conditions and thereby survive. This faculty stems from the cryptic or potential genetic variability (flow of variability).

forward mutation - a gene mutation from wild-type to the mutant condition, as opposed to back mutation. The terms f.m. and back mutation are purely descriptive.

Founder effect (Mayr 1942) - the establishment of a new population by a single or a few individuals. The limited number of founders of a new group can give rise to a large drift effect (genetic drift). The f.e. is an important event in determining the genetic make-up or gene pool of natural populations. The f.e. and subsequent isolation of a new colony may lead to drastic reorganization of the gene pool. It may precede speciation; especially in insular situations.

F plasmid - a genetically distinct unit (= F episome; F factor; sex factor) of circular, duplex DNA in E. coli with a molecular length of approximately 94,500 base pairs (plasmid). Its presence in the cell, either extrachromosomal or integrated into the bacterial chromosome, determines the "sex" of the bacterium, the host chromosome mobilization and transfer via conjugation (conjugon) of bacterial marker genes, and the formation of sex pili.

Based on their constitution with respect to the Fp. bacterial cells may be subdivided into four types (Demerec et al. 1966):

1. F-cells: Bacteria lacking the sex factor F. This confers on the cells activity as a genetic recipient ("female") in conjugation, lack of activity as a genetic donor ("male"), failure to transmit F, ability to be infected with F and resistance to male-specific phage.

2. F+ cells: Bacteria harboring an autonomous sex factor F which does not carry any genetically recognizable chromosome fragments. Such cells transfer F with high frequency to F- cells which become converted to the F+ state.

3. F' (F prime) cells: Bacteria harboring an autonomous sex factor F which carries a genetically recognizable segment of the bacterial chromosome. F' factors are also called "substituted sex factors" (Hayes 1964), "F-merogenotes" (Clark & Adelberg 1962), or "F-genotes" (Ramarkrishnan & Adelberg 1965).

F' plasmids are formed in two steps from a F+ bacterium;

Frameshift mutation (Brenner et al. 1961; Crick et al. 1961) - a class of mutations (= reading frame mutations, sign mutations) that arise from the insertion (+) or deletion (-) of a nucleotide, or any number of nucleotides other than three or multiples of three into or from DNA (frameshift mutagen). F. mutations displace the starting mRNA is "misread" by the translation process (out-of-phase translation) from the point of nucleotide addition or deletion. Thus once a f.m. is introduced into a gene the reading frame is shifted so that all codons distal to the mutation are read out of phase and reading frame, or unless the particular genotype also carries a f. suppressor mutation (frameshift suppression).

A closely positioned f.m. of opposite sign results in a double f.m.(+-) and restores the normal readng frame in mRNA except for the segment between the two frameshifts.

Frameshift suppressor - any of a class of external suppressor mutations specific for frameshift mutations, i.e., capable of suppressing such mutations by changing (correcting) the phase of translation. Almost all suppressible frameshift mutations are of the addition type, most likely including an extra GC pair in the DNA, containing repeated GC pairs. F. suppressors are genetic alterations outside the mutated cistron (external suppressor) and may be dominant or recessive to their corresponding wild-type allele. Dominant f. suppressors give rise to altered transfer RNA molecules that recognize quadruplets of bases in mRNA. Their anticodon contains a base **quadruplet** instead of a base triplet found in the wildtype tRNA. This allows the suppressor tRNA to read non-triplet codons and to correct for single base pair additions in the message. Base deletions are probably not suppressible. The suppressors are specific for frameshift mutations, exhibit allele specificity and can be classified according to their patterns of cross-suppression. The efficiency of suppression is between 1 and 10%.

G

Gene (Johannsen 1909) - a particular sequence (cistron) of nucleotides along a molecule of DNA (or on occasion RNA: certain viruses) which represents a functional unit of inheritance (defined operationaly by the cis-trans test). Genes can be divided into those which code for polypeptides (structural genes), those which are translated (genetic transcription, genetic translation) into proteins, and possibly those whose functional significance does not demand that they are transcribed at all (DNA sequences which may govern the punctuation or regulation of genetic transcription).

A g. consists of a linear array of potentially mutable units (mutational sites or mutons) between which intragenic genetic recombination (recon) can occur and permits the genetic mapping of mutational sites.

Gene activation - the differential expression of gene action at the chromosome level brought about via mechanisms of repression and derepression. Specific products of the cell determine which genes of the genome will function. The cell, in turn, is modified by. the action of these genes with the result that particular genes are then derepressed in succession (operon). In higher organisms, differential g.a. is a fundamental component of the processes of differentiation and development and follows a definite pattern in time and space.

Gene amplification (Brown & David 1968) - the selective production of multiple gene copies without proportional increase in others. G. a. is known to occur in the DNA puffs of Rhynchosciara and a few others species of flies, in the ovaries of several insects (crickets and water beetles), and during oogenesis of many organisms such as fish, mollusks, worms, insects, and amphibians.

Gene conversion (Winkler 1930; Lindegren 1949) - unequal recovery of genetic markers (alleles) in the region of the exchange during genetic recombination in those cases in which all products of meiosis can be recovered (as in fungi). In principle, unequal recoveries of alleles (departures from the 2A: 2a or, if the haploid products of meiosis undergo a post-meiotic division, from the 4A: 4a pattern) could result from asymmetry in the mechanism of genetic exchange or from revision of a heteroduplex joint.

At the molecular level, g.c. inolves the accurate transfer of DNA regions of 100 to 200 nucleotides from one chromatid to another. This results in homozygosis of small chromosome regions which originally were heteroallelic. Meiotic g.c is assumed to be a two-step process: 1. the formation of short hybrid DNA segments in corresponding parts of two homologous chromatids in meiotic prophase; 2. correction of mismatched base pairs in these segments by excision and repair (mismatch repair) of regions of heteroduplex DNA either during or immediately after intergenic recombination by crossing over during which the damage requiring such repair may arise.

Gene duplication - the duplication of DNA sequences representing genes which seems to have played a dicicive role in the evolution of higher organisms. G.d. may occur by: 1. unequal crossing over between two homologous chromosomes during meiosis; 2, unequal exchange between two sister chromatids of the same chromosome during mitosis; 3. chromosome structural changes of the same chromosome during mitosis; 3. chromosome structural changes of the intrachange type; 4. redundant duplication of DNA in certain regions; 5. duplication of whole genomes resulting in polyploid cells and individuals; 6. duplication of whole individual chromosomes resulting in aneuploid cells and individuals.

Gene mutation - any heritable change within the limits of a single gene ("intragenic mutation" or "point mutation"), as opposed to changes of chromosome structure ("genome mutation"). Discrimination between gene mutation and chromosomal structural changes (especially minute ones) is, in practice, frequently impossible.

"Transition" (Freese 1959) means the replacement of one pyrimidine base by another pyrimidine, or of one purine by another purine at some site of the DNA polynucleotide:

"Transversion" (Freese 1959) is the replacement of a pyrimidine by a purine, or a purine by a pyrimidine, in the DNA polynucleotide:

Adenine	— Thymine	⇄ Thymine	— Adenine
	⇅		⇅
Cytosine	— Guanine	⇄ Guanine	— Cytosine

Both kinds of substitutions should be possible at both the nucleotide and subnucleotide (base) level.

The consequences of base pair substitution mutations are same sense, missense and nonsense codons in the messenger RNA transcribed from a structural gene.

Gene pool (Dobzhansky 1951) - the total genetic information encoded in the sum total of the genes in a breedihg population existing at a given time. The gametes of all breeding individuals in a population furnish a pool of genes from which the genes of the next generation are chosen.

A g.p. which preserves a constant composition throughout a series of generations is deoted as existing in equilibrium. Such a g.p. possesses a balanced structure and its components are "coadapted". i.e. changes of the allele frequencies of other loci until either the original balance is restored or a new one established.

Genealogy - a family pedigree or lineage.

Genetic code - all those regularities (coordination principles) according to which the genetic information is encoded in DNA (or RNA in some viruses), transcribed to messenger RNA in the course of genetic transcription and translated into proteins with specific sequences of 20 amino acids via genetic translation. The g.c places the amino acid adaptor molecules (the transfer RNAs) in the correct positions during protein synthesis.

Genetic complementation (Fincham 1966) - the complementary action (cooperation) of homologous sets of genetic material (genomes) involving the interaction of mutant genes, or their products, in double mutants. Those combinations that result in market improvement in the function under study or in the development of a character which cannot be realized by the individual action of single mutants are said to complement each other.

Genetic death (Muller 1950) - preferential elimination (by selection) of genotypes which are carriers of mutationally arisen alleles (genetic load) which reduce their adaptive value or fitness (=genetic extinction). G.d. removes mutant genes from the gene pool of the population and may be the consequence of a reduction in physical vigor, sexual drive and/or fertility, or a high mortality rate in embryonic or juvenile stages of individuals possessing the critical genotype. G.d. is described as "overlapping", if 1. an individual is eliminated as a consequence of the presence of one particular gene, but would also have suffered g.d. by one or more other independent genes in its genotype ("independent overlapping"), or if 2. g.d. results from the synergistic effect of two or more mutant genes ("dependent overlapping").

Genetic drift (Wight 1921) - 1. any change, eihter directed ("steady drift") or undirected ("random drift") in gene frequency in a population (Wright 1955)

Genetic engineering - genetic manipulations (by-passing the sexual cycle) by which an individual having a new combination of inherited properties is established.

G.e. presently follows two major approaches: 1. The cellular approach involves the in vitro culturing of (haploid) cells and the hybridization of somatic cells (cell hybridization); 2. the molecular approach involves the direct manipulation of the DNA (Heynetal. 1974). (molecular cloning i plasmid e.).

Genetic load (Muller 1950) - for any genetic locus of a genotype (individual) or gene pool (population), the proportion (symbol L) by which the fitess of the average or optimum genotype at the locus under observation is decreased owing to the presence of deleterious (lethal, semilethal, and other subvital) genes.

G.I. thus denotes the genetic disability sustained by an individual or population and is expressed as the average number of potential genetic deaths per individual. The total of all contributions to the g.l. of human populations is about four lethal equivalents on the average and, for example, may be represented by four genes which are lethal as homozygotes or forty genes with a coefficient of selection of 0.1. Mixing of populations, intermarriage, change of environmental factors which favour a heterozygote advantage, and change of mutation rate are among those factors which may influence the magnitude of genetic load.

The principal components of g.l. are the following:

1. Mutational loads due to deleterious mutations at loci which are homozygous for nondeleterious alleles (load maintained by recurrent mutation).

2. Segregational loads due to genes segregating from favored heterozygotes and giving rise to unfit homozygotes (load maintained by segregation from advantageous heterozygotes).

Minor components of g.l. include "input loads", i.e. the load of inferior alleles in a gene poolcaused by mutation and immigration (gene flow); "substitutional loads", i.e. the cost to a population of replacing a particular allele by another in the course of evolutionary change; "selection loads" due to selection for an intermediate optimum of a quantitative character, and "interaction loads" due to the interaction of deleterious genes. A g. 1. is termed "balanced load" where decerases in the overall fitness of a population due to the segregation of inferior genotypes occur and whose component genes are maintained in the population because they improve fitness in different combinations (e.g., as heterozygotes).

Genetic map - the representaion of the genetic (relative) distance separating nonallelic gene loci in a linkage structure (the genes of one linkage group or chromosome), or the arrangement of the mutational sites (alleles) of a particular gene by utilizing the frequency of genetic recombination via crossing-over between genes (chromosome map) and within genes (gene map; fine structure map).

Genetic maps are linear, unbranched structures reflecting the sensitivity of genetic recombination tests in most genetic systems. In some bacteriophages and bacteria, genetic maps take the form of a circle (genetic circularity). The units of distance on the map are expressed as probabilities of recombination, or percentages. In many cases, genetic maps and "complementation maps" are parallel (genetic complementation). This is assumed to be the result of co-linearity of the gene and the polypeptide chain whose amino acid sequence it determines.

Genetic polarity (Jacob & Monod 1961; Franklin & Luria 1961) - the phenomenon wherein certain gene mutations (polar mutation) in bacterial or viral operons, in addition to inactivating the product (protein) of the gene in which they are located, also reduce the expression of all the operator-distal genes of the operon. G.p.seems to be a general characteristic of all operons transcribed into polycistronic mRNA. Mutational changes which produce polar effects (in most cases nonsense mutations, in some cases frameshift, insertion, and deletion mutations) appear to have as a common alteration the introduction, either directly or indirectly, of a polypeptide chain terminating codon (terminator codon) within the gene.

Genetic regulation - the overall regulation of the type (specificity) and rate of cellular processes by regulation of the activity of specific genes controlling individual biochemical reactions, or of a block of functionally related gens which control a particular reaction pathway (gene action).

Since the major steps in the production by "structural genes" of enzymes (proteins) that control the biosynthetic, metabolic and catabolic processes are genetic transcription (DNA RNA), genetic translation (RNA polypeptide) and assembly (polypeptide protein), gene activity as well as the resultant enzyme activity may be controlled at any of these stages. G. r. as described from experimental work with bacteria may involve:

1. Control of enzyme synthesis by stimulation (positive control) or repression (negative control) at the transcription of operons, i.e. messenger RNA formation from DNA. In microorganisms, this is interpreted according to the "operon concept". Any of the four steps of RNA synthesis (initiation, continuation, termination, and release) may be controlled.

Genetics (Bateson 1905) - the science of heredity and variation. It uses viruses ("virus genetics"), microorganisms ("microbial genetics"), plants ("plant genetics"), animal ("animal genetics"), and man ("human genetics") as objects of study. The subject matter is the phenomenology and physiology of heredity ("classical genetics") as well as the nature of the genetic material and the storage of genetic information, its replication, mutation, transmission, recombination, and translation into systems by which the genetic material mediates its control over metabolism and development and determines the reappearance of parental characters among progeny ("molecular genetics"). "Population genetics", as distinguished from studies of inheritance at the familial level, describes in mathematical terms the consequences of inheritance on the population level attempts to predict the behavior of future generations. It deals with the frequencies and interactions of genes in interbreeding populations and studies the agencies (e.g., mutation, natural and artificial selection, gene flow, migration, and chance factors) which tend to alter gene frequencies and thus to cause evolutionary changes.

Genetic transformation (Griffith 1928; Avery et al. 1944) - the intraspecific and interspecific (between members of different but related species) transfer of genetic information by means of "naked" extracellular DNA in bacteria. With g.t., in contrast to transduction, the DNA not integrated into the recipient cell's genome cannot function and manifest its genetic information phenotypically. Uptake of transforming DNA requires that its minimum molecular weight be of the order of about 105. Single-stranded (denatured) DNA is much less readily incorporated into the recipient cell than undenatured, double-stranded DNA.

Genetic variance - that portion of the phenotypic variance which is caused, by differences in the genetic constitution of the individuals in a population (variation). G.v. can be divided into two components: 1.g.v. resulting from differences between homozygotes (additive g.v.); 2. g.v. resulting from specific efects of various alleles in heterozygotes (dominance variance). Dominance here refers to deviations of the heterozygote from the two homozygotes.

Genome (Winkler 1920) - in eukaryotes, the basic (monoploid) chromosome set of an organism, consisting of a species-specific number of linkage groups; hence the sum total of its genes In prokaryotes (bacteria, viruses) all genetic information is contained in one linkage group and g., therefore, denotes the complement of all hereditary factors included therein.

Genophore (Ris 1961) - the physical equivalent of a linkage group (the linkage structure or chromosome) in prokaryotes, bacterial plasmids, mitochondria and chloroplasts (=genoneme) which is represented by a "naked" nucleic acid molecule. The term g. was proposed to emphasize the differences in structure of chromosomes in prokaryotes and eukaryotes.

Genotype (Johannsen 1909) - 1. the sum total of the genetic information (genes) contained in the linkage structures (chromosomes) of the pro- and eukaryotes, as distinguished from their phenotype (idiotype). The g. determines not a unique phenotype, but a range of phenotypic capacities referred to as an individual's "norm of reaction" to the environment. The share of one particular g. in the gene pool of the next generation is determined by the viability and reproductive success of its carriers.

Golgi apparatus (Golgi 1898; Ramon & Cajal 1908) - in animal and plant cells, a series of concentrically bent double membranes enclosing submicroscopic spaces (flattened, membrance-bound sacs also referred to as "cisternae", "saculles", or"lamellar units") which can be individualized and may be called dictyosomes or golgiosomes. Size and number of these "cisternae" are subject to considerable variation; their relative length may differ from species to species, with stage of ontogeny, and in relation to activity.The extent of the G. a. is determined by the extent to which its component dictyosomes are interassociated.The G. a. may be localized in the vicinity of the cell nucleus or the cell periphery (cortex), or the dictyosomes may be distributed throughout the cell.

G period (Howard & Pelc 1953) - periods of the interphase nucleus in eukaryotes.

G1 is the period during which the chromosomes are released from their condensed mitotic state and protein and RNA synthesis resume at a rapid rate. Particular enzymes have been shown to be specifically synthesized in G1, which ends with the beginning of the S period of interphase.

G2 is the interphase period which bridges the intrnal between the end of DNA synthesis (S period) and the initiation of mitosis. Condensation of chromosomes presumably begins during G2 by supercoiling involving complexing histones with DNA and subsequent folding up of te fibers.

Gynandromorph (Goldschmidt 1915) - an individual of a dioecious (unisexual) species which is a sexual mosaic (a chimera), typically male in certain portions (sectors) of the body and typically female in others. Gyandromorphs result from the presence of chromosomes of both sexes in different parts of the body.

Gynandromorphs have been observed in animals (not yet in plants) where sex differentiation is apparently intracellular and not hormonal and may arise in several ways:

1. by loss of an X-chromosome during early cleavage in cases of an XX-XY- mechanism of sex-determination, provided XO-cells are male in character (as in Drosophila);

2. by suppression of polar body formation giving rise to a binucleate egg cell. In cases of female heterogamety (as in silkworms), one of these nuclei contains an X-chromosome, the other a Y-chromosome. Fertilization of each

of the nuclei by a sperm containing one X-chromosome may result in a g.;

3. by fertilization of only one of the two nuclei present in a binucleate egg cell in cases where the diploid nucleus gives rise to female and the haploid nucleus to male tissue (as in honey bees);

4. by loss of a whole chromosome set during the first cleavage division. Both the size and position of the male and female sectors are dependent on the time and position of the onset of these processes. In animals with sex hormones, the production of clear-cut gynandromorphs is prevented since these hormones give rise to an intersexual development irrespective of the different constitution of the tissues.

H

Hairpin loop - any double-strandet region in sigle-strandet RNA and DNA caused by base pairing between complementary base sequences. Such regions might play a role in the stability of the molecule and their formation is due to the ability of the nucleotide chain to turn back on itself and form a particular secondary structure. (palindrome)

Haldane's rule - when among the offspring of crosses between species and varieties, one sex is absent, rare, or sterilé, it is the heterogametic sex.

Haploid (Strasburger 1905) - of cells or individuals with a single genome or chromosome set provided more than one linkage structure (as in eukaryotes), distinct in homology and function, is present. Viruses nd bacteria whose genome consists of only one linkage structure are haploid, and so are parts of the life cycle (referred to as "haplophase") of eukaryotes. In diploid species of eukaryotes, individuals may occasionally occur whose somatic cells are haploid. In most cases they result from spontaneous or induced parthenogenetic development of unfertilized eggs (parthenogenesis). Such individuals are referred to as "monohaploids". They are to be distinguished from so-called "polyhaploids", i.e. haploid individulas arising from polyploid species.

Hardy-Weinberg law (Hardy 1908; Weinberg 1908) - a statement (= "H.-W. formula" or "H.-W. equilibrium principle") in mathematical terms concerning the theoretical relationship between gene frequencies and genotype frequencies within populations. It describes the genotypic equilibrium produced by a static gene pool. In a large random mating population, both the gene frequencies and the genotype frequencies are constant from generation to generation in the absence of mutation, migration, and selection (the population is in H.-W. equilibrium). If the frequencies of allele A and a (of equal selective value and occurring together in a population) are q and 1 - q and mating is at random, then male gametes qA + (1-q) a fertilize eggs of qA + (1-q) a producing offspring (genotypes) according to the following binomial law:

$$[qA + (1-q)\,a] \times [qA + (1-q)a] = q2AA + 2q(1-q)\,Aa + (1-q)2\,aa.$$

heritability (Lush 1949) - The

proportion of the total phenotypic variance for which genetic differences are responsible (broad-sense h.), or the proportion of the total phenotypic variance that is due to additive genetic variability (narrow-sense h.). H. is a measure of the amount of genetic variability, excluding that expressed by heterozygotes, and decreases with an increasing environmental component of variance for the character under observation.

Heterochromatic (Heitz 1928, 1929) - originally of chromosomal segments or whole chromosomes which possess a dense, compact structure in telophase, interphase and early prophase, as distinct from euchromatic segments. Today, this phenomenon is referred to as "positive heteropycnosis" and h. chromosomal regions or chromosomes are defined as those which become heteropycnotic at some stage of the cell cycle, i.e. show "allocyclic" (allocycly) behavior.

Heterochromatin is generally recognized by the following features:

1. A condensed state throughout interphase (positive heteropycnois giving rise to chromocenters); 2. out of phase replication of DNA (late replication in polytene chromosomes; 5. susceptibility to chromosome structural changes. Heterochromatin is subclassified into two types (Brown 1966):

1. Constitutive heterochromatin is present in all cells at identical positions on both the homologous chromosomes and forms a permanent structural characteristic of a given chromosome pair. Generally, the DNA of constitutive heterochromatin consists of repetitious DNA, non-repetitive, AT-rich mainband DNA, and non-repetitious DNA (in some species). Satellite DNA tends to be localized to centromeric heterochromatin.

2. Facultative heterochromatin varies in its state in differnt cell types, at different developmental stages, or even from one homologous chromosome to another, i.e., is heteropycnotic in special cell types or at special stages. It represents inactivated and condensed segments of euchromatin (dosage compesation).

Heteroduplex - a double-stranded DNA molecule with a region of noncomplementarity due to chemical or physical influences which alter bases of one of the two strands of the molecules or to occasional faults in DNA replication, or to genetic recombination between molecules within a region of genetic difference (heteroduplex mapping).

Heterogametic (Wilson 1910) - that one of the two sexes which during meiosis gives rise to two types of gametes (male- and female-determining), in contrast to the "homogametic" sex. In the case of the XX-XY or XX-XO-types of sex determination, the h. sex is that with both an X- and a Y-chromosome or with only one X-chromosome (=digametic). In most cases, the male sex is h., the female homogametic with respect to the sex-chromosome constitution.

Heterogenetic (Waddington 1939) - ref. to meiotic chromosome pairing in hybrids (alloploids) when the pairing partners are derived from differnt original ancestors, as distinct from "homogenetic" pairing where the association is between chromosomes derived from one of the original ncestors. The frequency of h. pairing as compared with homogenetic, depends on the relative degree of affinity between the completely homologous, as compared to the partly homologous ("homoeo-logous"), chromosomes, i.e. on their differential affinity.

Heterokaryon - a bi-or multinucleate cell, or a tissue composed of such cells, whose nuclei are of more than one genetic type, as distinct from a homokaryon in which the nuclei are genetically identical. The property of being hetero- or homokaryotic is referred to as heterokaryosis and homokaryosis, respectively. In certain fungi (Asco- or Basidiomycetes), heterokaryosis constitutes the principal vegetative phase of the life cycle.

Heteropycnotic (Gutherz 1907) - of chromosomes or chromosome regions which, in comparison with "isopycnotic" ones (Ostergren 1950), are out of phase (allocycly) in respect to their coiling cycle (chromsome coiling) and staining properties. (heteroch- romatic).

1. Positively h.: chromosomes or chromosome segments tightly coiled by prophase and remain so as others (isopcnotic) are undergoing decondensation at anaphase. During interphase they show up as the chromocenters. Positive heteropycnosis is characteristic of heterochromatin.

2. Negatively h.: chromosomes or chromosome segments condense more slowly or to a lesser extent during prophase than do the positively h. and undergo decondensation more rapidly at anaphase.

Chromosomes and chromosome regions which are positively h. at one stage of the cell cycle, or in one kind of cell or tissue, or in one of the sexes, may by negatively h. at another stage, in another cell, or in the other sex.

Heterostyly (Darwin 1877) - a combination of morphological, physiological, and genetic mechanisms which promote outbreeding in certain angiosperms. H. means the presence in a species of two (distyly) or three (tristyly) different types of individuals, distinguished by the relative positions of stigma and anthers in the flowers, as distinct from homostyly. In the case of distyly, one type of individual has short styles and long stamens the other long styles and short stamens. With tristyly, the alternatives are "short-styled", "medium-styled', "long-styled" individuals. Usually, pollination and fertilization is between flowers differing in the lengths of their styles ("heteromorphic incompatibility").

Heterozygous (Bateson & Saunders 1902) - 1. in higher organisms (eukaryotes), of diploid of polyploid individuals ("heterozygotes") having different alleles at one or more genetic loci (genes) in homologous chromosome segments, in contrast to homozygotes with identical alleles at these loci.

Heterozygosity provides a store of genetic reserves and potential plasticity and permits a large proportion of the individuals to exhibit combinations of phenotypic properties near the optimum. Underlying both is the greater homeostasis of heterozygotes.

Hir cell - a cell harboring the F plasmid integrated into the bacterial chromosome which may or may not (in case of a defective F plasmid) confer on the cell the phenotype of high frequency of recombination. Each Hfr strain transfers the bacterial chromosome with a definite polarity to F- recipient cells during bacterial conjugation. Hfr strains originate by opening up the circular chromosome of the bacterial (F+) cell at one of many possible sites and insertion of the F plasmid at this place.

Histocompatibility gene - any gene regulating the immunological properties of histocompatibility antigens. Such cellular antigens, which represent the transcription products of h. genes, differentiate individual and cause the rejection of tissue transplants. Histocompatibility and histo-incomatibility (acceptance and rejection of transplants, respectively) depend upon the genetic relation of donor and host. Transplants exchanged between members of the same inbred strains (isografts) which have h. alleles in common survive permanently, while grafts exchanged between members of different strains (allografts) which have no h. alleles in common are promptly rejected.

Histone (Kossel 1884) - any of the basic proteins complexed with DNA (= nucleohistone) in the chromosome of eukaryotes (nonhistone chromosomal protein). The close correlation between h. synthesis and DNA replication is presumably related to the fact that within cells there exists a strict requirement for newly synthesized DNA to form DNA-H, complexes (chromatin; nucleosome).

The association by electrostatic bonds of h. with DNA in eukaryotic chromosomes has implicated histones in two possible functions:

1. a structural role in maintaining and controlling the conformation of the chromosome through the cell cycle (coiling and condensation of chromosomes);

2. involvement in mechanisms of genetic regulation (coarse control resulting in tissue-specific gene repression patterns during development) by forming the supercoiled structure of DNA which in turn sterically inhibits transcription by RNA polymerase.

There are five main types of h. in the chromatin of eukaryotes.

These are F1 (=H1) and F 2b (=H2B) (lysine-rich h.), F3 (=H3) and F2 a 1 (=H4) (arginine-rich h.), and F2 a2 (= H2 A) (an intermediate type). Usually, there are nearly equimolar amounts of all the histones except for F1, of which there is about half as much. The five main types are all fully characterized by amino acid analyses and are not - in a given organism - tissue-specific. The sequence studies of histones have delineated well defined regions of the chain rich either in basic and helix-stabilizing residues or rich in apolar acidic and aromatic residues. The latter regions are the sites of secondary structure and h. -h. interaction, while the former are the sites of DNA interaction. The amino acid sequences of these proteins have, during the course of evolution, remained extraordinarily stable (F1 shows greater variability than the other h. fractions).

Homeostasis (Cannon 1929) - the tendency of a system to be specified by context (individual, population, process, etc.) to maintain a dynamic equilibrium and in case of disturbance to restore the equilibrium by its own regulatory mechanisms.

Human genetics - the study of the genetically controlled similarities and difference (physical and mental, normal and abnormal) in human beings, in respect to their causes and modes of transmission (inheritance) from generation to generation (medical genetics, pharmacogenetics).

Hybrid - any offspring of a hybridization or cross between two genetically unlike individuals.

Hybridization - 1. in the broadest sense, any cross-mating (crossing) of two genetically different individuals which leads to hybrid progeny (cell hybridization).

2. The crossing of individuals belonging to two unlike natural populations that have come into contact secondarily. Five kinds of h. may be distinguished when h. is considered as a population phenomenon (Mayr 1963):

I

Idiogram (Navashin 1922) - a diagrammatic representation of chromosome morphology used diagnostically for comparison of the karyotypes of different species and varieties. The construction of idiograms is based on measurements (in several or many cells) of total chromosome length and arm length ratio as well as on centromere position, nucleolus organizer position, position of heteropycnotic segments, etc. for any one chromosome during metaphase of mitosis. Centromerė position is usually calculated from the total length, c, and the lengths of the long and short arms, l and s, respectively, of the chromosome in question and expressed either as a difference d = l = s, as a ratio r = l/s or as a centromere index i = 100s/c.

Immune response - a complex series of events (initiated by the interaction of immuno-competent cells and antigen. The i.r. in multicellular organisms provides the capacity to distinguish and elim̃inate or suppress both foreign (extrinsic) material and, in addition, intrinsic alien patterns arising by somatic mutation or an equivalent process. The immune system therefore plays a key role in resistance to pathogenic micro-organisms, "surveillance" against malignant cells, and control of autoaggressive processes. The cells taking part in i. responses are 1. lymphocytes and 2. various accessor cells (e.g. macrophages, cells of the reticuloendothelial system, various types of blood leucocytes).

Inborn errors of metabolism (Garrod 1902) - a genetically determined biochemical disorder in which a specific enzyme defect produces a metabolic block. The block may have pathological consequences (=molecular disease). Nearly all the i.e. o.m. studied in human genetics are inherited as autosomal recessives, a few as sex-linked traits.

Inbreeding - a mating system (= endogamy) which, in contrast to out- or crossbreeding (= exogamy), involves the breeding together of individuals more closely related than mates chosen at random from a population. The degree of i. in any particular population or mating is a function of the relationship between the mating partners (inbreeding coefficient). The closest form of i. is self-fertilization; other forms are sib matings and half-sib matings.

inbreeding depression - a decrease in fitness and vigor, resulting from inbreeding imposed on individuals that are normally outbreeding (= inbreeding degeneration). The extent of i.d. is not the same in all lines produced by inbreeding. That point after several generations of inbreding at which no further decrease of fitness and vigor occurs is referred to as the "inbreeding minimum". I.d. is the consequence of increased homozygosity for deleterious recessive genes (genetic load) and the break up of balanced polygenic systems (Mather 1941). The crossing of inbred lines which have reached their inbreeding minimum frequently results in heterosis.

Incompatibility (Stout 1918) - the selective restriciton of mating competence by extrasexual "incompatibility genes" which limit or prevent zygote formation (fertilization) by certain classes of gametes and thus in an otherwise freely interbreeding group limits either self-fertilization ("self-incompatibility") or cross-fertilization ("cross-incompatibility") between certain self-incompatibility. Members of the same i. class are typically strictly self-sterile and cross-fertile and the various i. systems which occur in the plant kingdom usually promote outbreeding. Mating between individuals of compatible classes is usually reciprocal.

Initiator (Jacob et al. 1964) - 1. according to the replicon model of bacterial DNA replication, a gene product that interacts specifically with a particualr site of the replicon, the replicator. The result of this interaction is the initiation at the replicator site of a new round of DNA replication.

2. = a promoter, i.e., the starting point for the transcription and/or translation process (Margolin & Bauerle 1966).

Initiator codon - any codon (AUG, GUG, and GUA) in messenger RNA that directs initiation of polypeptide synthesis (genetic translation; initiation complex) by stimulating the binding of initator tRNA and by phasing the translation of messenger RNA. The isolation and sequence analysis of ribosomal binding sites of natural mRNA molecules demonstrated the exclusive use of AUG as the i.c. in all currently known initiation sequences. In in vitro systems, all three codons may initiate translation when their position is at or near the 5' terminus of the message. When the codons are in internal position of the mRNA and preceded by an i.c., they code (genetic code) for methionine (AUG) and valine (GUG and GUA). (translational reinitiation).

Instability (Mather 1953) - a type of variation which appears to be random in direction. It is not genetic in origin and has no identifiable environmental cause. I. is sometimes explained as being due to developmental errors ("developmental noise") arising from sudden changes during development, unconnected with particular environmental influences. "Stability" refers to the opposite condition, i.e. non-occurrence of variations of the type described above, or more generally, to any condition characterized by the absence of plasticity (Bradshaw 1965).

Interference (Muller 1916) - 1. Chromosome interferecne (Muller 1916); the effect of one crossing-over event in either decreasing (positive chromosome i.) or increasing (negative chromosome i.) the probability that a second crossover will take place in its vicinity. In both positive and negative chromosome i., the frequency of multiple (double) crossovers is not in accordance with a random distribution of subsequent crossovers. Chromosome i. is also called "chiasma i." (Mather 1931). "chiasma position i." (Carter & Robertson 1952), "crossover position i.", or "position i." (Whitehouse 1965).

Strengths of chromosome i. are commonly measured as coefficients of coincidence (Muller 1916),i.e. as the quotient of the number of observed double crossovers and the number expected from the Poisson distribution:

K= actual frequencies of doubles

expected frequencies of doubles

Coincidence varies inversely with interference, and coincidence values usually range from o (complete positive i) to 1 (no. i. whatsoever). In those cases where the occurrence of one crossover increases the probability of a second nearby (localized negative i.), the number of double crossovers is in excess of the random expectation and the coincidence values are greater than 1. It is improbable that this kind of i. has anything to do with chiasms formation (as is assumed in positive i., hence "chiasma i."). One explanation for localized negative i. is the assumption that pairing of linkage structures, which actually gives rise to recombination (socalled "effective pairing"), is usually restricted to short regions which pair discontinuously and within which the probability of recombination is high. According to this hypothesis, recombination in two intervals of the linkage structures is greatr than expected on the basis of random probabilities, provided these intervals are so close together that they frequently belong to the same effectively paired region.

Intergenic (Muller 1941) - of mutations (chromosomal structural changes or chromosome mutations) involving more than one gene, in contrast to intragenic mutations (gene mutation).

Interlocking (Mather 1932) - 1. True i. of meiotic chromosomes: The intertwisting of different (nonhomologous) chromosomes which ensues during meiotic pairing (chromosome pairing) if a foreign chromosome passes through a loop between associations of two homologous chromosomes. True i. takes place during zygotene-pachytene of the first meiotic division and chiasma formation is a prerequisite for its maintenance up to metaphase 1.

Intersex (Goldschmidt 1915) - any individual of a unisexual (dioecious) species whose reproductive organs and/or secondary sex characters are partly of one and partly of the other sex without showing genetically different parts (gynandromorph), i.e., sex-intergrades who are neither clearly male nor clearly female but represent a mixture of male, female and intermediate characters. Intersexes are generally infertile or at most produce mature gametes of one sex only. Hence it is always possibel to distinguish them from hermaphroditic individuals which produce mature gametes of both sexes. Intersexes may start their development in conformity with the genetically determined sex and then switch over to the other sex. The degree of intersexuality is determined by the timing of the switch-over in differentiation.

Introgression (Anderson & Hubricht 1938) - the incorporation of genes of one species into the gene pool of another species by hybridization ("introgressive hybridization") and backcrossing. According to Anderson (1953), sympatric i. (in which old and new genotypes produced by i. coexist in the same populations) and allopatric i. (in which new genotypes produced by i. exist separately from the old genotypes) may be distinguished.

Inversion (Sturtevant 1926) - a chromosomal structural change of the "intrachange" type (chromosome mutation), characterized by the reversal of a chromosome (or chromatid) segment, and the gene sequence contained therein, relative to the standard arrangement of the linkage group in question.There is no experimental evidence for the existence of terminal inversions. According to the "breakage-reunion-model", intercalary inversions are the result of two breaks (referred to as inversion points) in one chromosome, 180° rotation of the segment localized between them and reunion at the sites of breakage.

The following i.-types may be distinguished, according to the number of inverted segments within a chromosome and the location of the i. points with respect to each other.

in vitro - of stucies under artificially controlled conditions outside the organism as opposed to in vivo studies carried out in the intact organism.

Isochromosome (Darlington 1940) - a monocentric or dicentric (isodicentric) chromosome with homologous (equal and genetically identical) arms which are mirror images of one another. Stable monocentric isochromosomes may arise by centromere misdivision (transverse breakage and union in the inner zone of the centromere misdivision (transverse breakage and union in the inner zone of the centromere region) and reduplication of the telocentric fragments in the following interphase, provided the chromatids of these fragments do not become separated during mitotic anaphase but are distributed together (united at the terminally localized half of the original centromere) to one of the two cell poles.

Isolation - the restriction, or limitation, of the extent of gene flow (exchange of genetic information) between discrete populations owing to barriers to interbreeding. I. is a factor in evolution and the various mechanisms of i. may be classified as follows (Grant 1963):

1. spatial i.: Gene flow between populations is reduced or prevented by the distance between them;

a) geographical i.: The populations concerned occupy different territories, i.e. are allopatric;

2. environmental i.: Gene flow between populations is limited by the availability of habitats suitable for the growth and subsistence of their hybrid progeny (selection against the hybrids and their derivatives);

a) ecological or habitat i.: The population occupy different habitats in the same territory, i.e. are sympatric;

3. reproductive i.: Gene flow is prevented or restricted by genotypically controlled difference in the reproductive behaviour and fertility relationships of the individuals which are members of different populations;

Isozyme (Markert & Moller 1959) - any of a class of multiple, separable forms of enzymes (= isoenzymes) occuring within the same organism and having similar or identical catalytic activities. Isozymes may have different kinetic properties. The relative concentrations of i. often differ from one tissue to another and may change during the course of development. The molecular multiplicity plays a role in the differentiation of cells during embryonic development and also increases the evolutionary adaptibility of organisms. Sets of i. can be generated by both genetic and epigenetic mechanisms.

K

Karyotype (Eevitsky 1924) - the particular chromosome complement of an individual or of a related group of individuals, as defined both by the number and morphology of the chromosomes usully in mitotic metaphase. A diagrammatic representation of a k. based on the morphological features of the chromosomes (the "endophenotype") is called an "idiogram". The absence, duplication or positional rearrangement of certain morphological features of the chromosomes, as compared with their typical topography (with the standard k.), is visible evidence of chromosome mutation.

Diploid individuals possessing a k. in which two chromosomes of a pair have different gene arrangements as a result of chromosomal structural changes, are referred to as "heterokaryotypic" (or are "heterokaryotypes"), whereas those in which the two chromosomes of a pair are similar in gene arrangement are "homokaryotypic" ("homokaryotypes").

Karyotypes may differ in respect to their basic chromosome number, the shape and relative size of the chromosomes, the number and size of secondary constrictions, the absolute size of the chromosomes and the distribution and size of hetero- and euchromatic chromosome segments.

Karyotypes which consist of chromosomes essentially similar in size (all with median or submedian centromeres), and those with chromosomes differing greatly in size and/or (partly) with subterminal or terminal centromeres, may be called "symmetric" and "asymmetric" karyotypes, respectively (Stebbins 1950). In comparison with plants, the karyotypes, respectively (Stebbins 1950). In comparison with plants, the karyotypes of animals are usually far more asymmetric. Karyotype evolution as a consequence of chromosomal structural changes (inversions, translocations, centric fusion, etc.) is usually associated with a progressive reduction of the basic chromosome number and an increased asymmetry.

Killer (Sonneborn 1938) - any of a number of strains of Paramecium containing symbionts (kappa) in their cytoplasm. These are relatively large, complex particles composed of DNA, RNA, protein and other substances. This genetically controlled symbiosis is the basis for the production of a toxin by killer strains. When released into the culture medium this substance is able to kill other strains referred to as "sensitives".

The killer character is dependent upon two genetic components, as demonstrated, e.g., for the symbiont kappa. One component is the genotype of the Paramecium cell, the other the kappa particles. Kappa particles are reproduced only in cells containing the genes K, s1 and s2, In cells of genotype Kk, only half the number of kappa particles is present, in comparison with homozygous dominant genotype, KK. Kappa is particulate and usually several hundred particles per killer individual are present. Two types of kappa are known: so-called "brights", containing a bright refractile body known as R or R body, and "nonbrights" referred to as N. Nonbrights are self-reproducing and give rise to brights. The killing or P particle is identical with the bright or B particle. Nonbrights are capable of infecting sensitive strains and transforming them into killers.

Kinetoplast (Alexeiff 1917; Lwoff 1949) - a small, spherical or rod-shaped structure (= blepharoplast or parabasal body) localized posterior to the basal body of the flagellum of all parasitic flagellates. The k. is a DNA containing, self-replicating or self-dependent cell organelle which ordinarily divides before the nucleus. According to electron microscopy, the k. is a sac, bounded by a double membrane, containing an array of dense DNA containing fibrils and typical mitochondrial cristae formed from the inner membrane. The k. is a differentiated part of the large, single mitochondrion in Trypanosomatidae. Kinetoplast DNA can be separated from nuclear DNA by CsCl gradient centrifugation and appears to be quite distinct from the nuclear DNA of the same species. The details of structure, as well as the relative proportions of DNA containing fibrils and of mitochondrial material vary from species to species and with the developmental stage of the life cycle. Flagellates with a k. also have a basal body from which the flagellum arises.

Kinetome - kinetid.

Knob - a relatively large, heavily stainable chromomere localized terminally, or (less frequently) interstitially, in the chromosomes of certain organisms (e.g. maize). It may be used as a "landmark" for the identification of particular chromosomes (idiogram).

L

Lamarckism - the erroneous concept, advocated by Lamarck, that organic evolution is brought about by environmental induction of adaptive changes in living organisms. Such changes were thought to be inherited by the progeny ("inheritance of acquired characters") (Darwinism; Neo - Darwinism).

Lampbrush chromosome (Ruckert 1892) - a special type of chromosomes in primary oocyte nuclei, both vertebrate and invertebrate, during the prolonged diplotene stage of the first meiotic division, and in spermatocyte nuclei of *Drosophila*. The lampbrush state of chromosomes is phase - specific and reversible. Lampbrush chromosomes may even be larger than the (multistranded) polytene chromosomes of Diptera; the largest with a length of about 1 mm have been found in urodele amphibia and are characterized by loops projecting in pairs from the majority of the chromomers which are connected by interchromomeric fibers. Frequently these loops exhibit a thin axis (probably consisting of one DNA double helix) from which fibers project which are covered with a loop matrix consisting of RNA and protein. These loops may vary in size and shape from one chromomere to another and may be used to construct tentative cytological maps of the lampbrush chromosomes. Generally, chromomere size and loop size are inversely correlated.

Leaky - of 1. a protein coded for by a mutant gene which shows some residual activity ("leaky protein"),

Lewis effect (Pontecorvo 1955) - position effect.

Ligand - a small molecule (including activators, substrates, and inhibitors of enzyme activity) bound to a protein by noncovalent forces. Interactions between like and unlike ligands may be called homotropic and heterotropic, respectively.

Linkage (Morgan 1910) - the association in inheritance of certain genes (and, in turn, of the phenotypic characters they control) due to their being localized in the same chromosome or nucleic acid molecule (linkage structure). Genes borne on homologous chromosomes belong to the same linkage group. The number of linkage groups is identical with the number of chromosomes of one chromosome set.

Linkage of genes is only rarely complete. It is broken (genetic recombination) when homologous linkage structures exchange corresponding parts by a process referred to as crossing-over ("partial 1."). Crossing over in eukaryotes may occur during meiosis (meiotic crossing-over) or mitosis (mitotic crossing-over). In prokaryotes (such as bacteria and viruses) it can occur without meiosis or mitosis by functionally analogous processes at the level of nucleic acid molecules.

Linkage (rather than independent genetic segregation) of two genetic markers (e.g., allelic pairs A/a and B/b) used in a crossing experiment show a tendency to remain together, i.e., the proportion of so-called recombinants (Ab and aB) is less than the proportion of parental combinations (AB and ab). If two marker genes are borne on nonhomologous linkage structures, independent segregation occurs and the parental and recombination types occur in equal frequency.

Sex-linkage (Morgan 1914) is the special type of 1. of those genes localized in the sex-chromosomes of eukaryotes. According to their position in either nonhomologous (differential) or homologous segments of the X and Y chromosomes, such genes may either be completely (localization in "differential segments" in which no crossing-over takes place) or partially sex-linked. The latter genes, localized in homologous segments of X- and Y-chromosomes, may be exchanged between the sex-chromosomes by crossing-over.

Linkage map - a chromosome map containing in linear order the genes belonging to the linkage group associated with a particular chromosome. Crossover and recombination values are utilized to demonstrate a definite serial order of the genes in a chromosome and provide an index of the relative distance between any two genes (genetic map).

Locus (Morgan, Sturtevant, Muller & Bridges 1915) - the position of a gene on the genetic map. Allelic genes are situated at identical loci in homologous linkage structures (chromosomes). The l. of a particular gene is identifiable as a site of an allelic difference arising by gene mutation. Frequently the term l. is used more or less synonymously with gene, i.e. for a closely linked array of mutational sites concerned with a single function.

Lomasome (Moore & Mc Alear 1961) - any of the cell's boundary formations (border bodies) which protrude into the interior of the cell (evaginations delimited by the cell membrane) and enclose cytoplasmic and multivesicular structures somewhat variable in form. Lomasomes are located between the cell membrane and the cell wall and have been found in fungi, algae and higher plants.

They may play a role in secretion, in increasing the surface area for the diffusion of substances involved in cell wall formation or wall breakdown, in membrane proliferation and in endocytosis.

Lysogenization (Ball 1925; Bordet 1925) - the establishment of lysogeny by infection of a nonlysogenic bacterial cell with a temperate bacteriophage and the attachment of the phage to the bacterial chromosome at a specific region, called the attachment region or attachment site. The infecting phage DNA cyclizes and is then linearly inserted into the host chromsome by reciprocal recombination between what is called the bacterial chromosite and the homologous segment of the phage genome, the episite. For the site-specifc recombination involved in integration a function specified by a phage gene is needed. Phage mutants unable to perform this function are called integration-defective. Integration of the phage genome into the bacterial chromosome and its passive replication with it is called stable lysogenization. If the temperate phage fails to attach to the bacterial chromosome and is inherited unilaterally among the progeny of the infected cell (and eventually diluted out of the population), this is termed abortive lysogenization.

Lysogeny - the potentiality of bacterial cells for production and release of bacteriophages as a stable, heritable character. It is widespread among strains of different bacterial species. Such strains are referred to as lysogenic or lysogens.

Lysosome (de Duve et al. 1955) - any of a class of morphologically heterogeneous cytoplasmic particles (ranging in size from 0.25 to 0.8 pm) in animal and plant tissues characterized by their content of hydrolytic enzymes (e.g. acid phosphorylase, acid ribo- and deoxyribonuclease, cathepsin, xxx-glucoronase), and the structure-linked latency of these enzymes. The intracellular functions of lysosomes depend on their lytic potential. The single unit membrane of the 1. acts as a barrier between the enzymes enclosed in the 1. and the external substrate. The activity of the enzymes contained in lysosomes is limited or nil unless the vesicle in which they are enclosed is ruptured. Such rupture is supposed to be under metabolic (hormonal) control.

M

Malignancy - the ability of tumor cells to grow progressively and to kill their host.

Map distance (Bridges 1932) - the distance, in terms of percentages of crossing-over obtained from genetic experiments, between particular linked genes (genetic map). Since m. d. is based on the total frequency of crossing-over in a chromosome region it is directly translatable into units of m. d. or map units. One unit of m. d. between linked genes is that interval within which 1 per cent crossing-over takes place.

The ralationship between "recombination frequency" and m. d. is not one-to-one, since recombination frequency measures only those crossovers that lead to reciprocal exchanges of genetic markers. The farther apart the genes, the more striking the discrepancies become because double crossovers are scored as nonrecombinants. Expressions that relate m. d. to recombination frequencies and correct for the probable double crossovers which have occurred in the interval under consideration are referred to as "mapping functions".

Marker gene - a gene of known function and known location on the chromosome (genetic marker).

Marker effect - any case in which two mutant (genetic markers) that are at or near the same site give very different frequencies of genetic recombination when both are crossed to the same third mutant elsewhere in the gene. An unambiguous test for the occurrence of m. effects can only be made with a system in which the physical positions of the genetic markers are determined by a method other than recombination mapping.

Meiotic drive (Sandler & Novitski 1957) - any mechanism (generally associated with various kinds of chromosomal structural changes in heterozygous form) which may affect the genetic composition of a population via inequalities in chromosomal transmission during meiosis (non-random assortment of chromosomes). Regardless of their individual selective advantages or disadvantages, the genes on a chromosome favored by preferential segregation make a disproportionate contribution to the gene pool of the next generation.

Mendelian inheritance (Castle 1906) - the mode of inheritance of chromosomal genes, in contrast to "non-Mendelian" or "extrachromosomal" inheritance which is controlled by cytoplasmic hereditary determinants.

Mendelian population (Dobzhansky 1935) - an interbree-ding group of individuals sharing a common gene pool (population). The m. p. is the basic unit of study in population genetics. Usually, there is no clearcut boundary between two adjacent Mendelian populations owing to gene flow between them. If gene flow is limited, such populations may show considerable differences with respect of their gene pools. Gene frequency is the most important single index characterizing m. p.

Meromixis (Wollman, Jacob & Hayes 1956) - a process in bacteria during which only a fraction of the genome of a donor cell is transferred to a recipient cell which contributes both its cytoplasm and its entire genome. The result of m. is a merozygote.

Merozygote (Wollman, Jacob & Hayes 1956) - in bacteria, an incomplete zygote which is diploid only for part of its genetic material and haploid for the remainder. The process of m. formation is called meromixis A m. contains the whole genome of the recipient cell and a merogenote derived from a donor cell. Merozygotes are usually very unstable and may be hetero- or homogenotic with respect to their genetic constitution. The potential for genetic recombination in am. is limited to the diploid part of its genome (genetic transformation; tranduction; bacterial conjugation).

Mesosome (Fitz- James 1960) - any of the intracytoplasmic membraneous structures of bacteria (= plasmalemmasome), which appear to originate from the cell membrane by an invagination (infolding) and "pinching off" process. In sections, mesosomes appear as clusters of vesicles and tubules, as membraneous whorls, or as a combination of these types of ultrastructure. The list of proposed m. functions includes their participation in energy production, DNA replication and nucleoid separation (Fig. 65),

Messenger DNA - single stranded DNA (e.g., DNA isolated from certain bacteriophages) that can act as a messenger in an in vitro system of protein synthesis. It binds initiator tRNA (fMet-tRNA) to ribosomes in the presence of initiation factors and GTP. The ability to translate (genetic translation) a DNA strand depends critically on the Mg2+ concentration.

Metacentric - of chromosomes whose centromere is localized in a roughly median position (lengthwise). This gives them a V or J shaped appearance at metaphase (idiogram).

Metafemale (Stern 1959) - an individual which has according to the balance theory of sex-determination a greater dose of female determiners than normal females (e.g., Drosophila individuals with 3X 2A individuals. Metafemale Drosophilae show hypoplastic ovaries and both malformed wings and eyes; the majority die as larvae or pupae. Human beings with two sets of autosomes and three x-chromosomes are underdeveloped with respect to their primary and secondary sex characters.

Metaphase pairing index (Tobgy 1943) - that fraction of the sum total of all cells observed in which two particular chromosomes are found in the paired condition (chromosome pairing) at metaphase I of meiosis.

Metaphase plate - the equilibrium position in the equatorial region of the spindle (= equatorial plate or nuclear plate) reached by the chromosomes via their congressional movement (chromosome congression; centromere orientation) at metaphase of mitosis and meiosis.

Metaxenia - xenia.

M factor (Davison et al. 1969) - a protein factor (in *E. coli*) which stimulates in vitro transcription by purified RNA polymerase (20 to 30 fold) from several templates. The M. f. is distinct from the rho factor and the sigma factor but has several properties in common with sigma. Both are thermolabile proteins which sediment at approximately 4 to 5 S and both stimulate RNA synthesis from a variety of templates. M might be a continuation factor preventing abortive termination of initiated RNA chains or might confer additional specificity on sigma-directed intiation (psi factor). Stimulation is probably due to interaction of the M. f. with polymerase.

Microtubule (Slautterback 1963) - any of a class of subcellular filamentous components found in a wide variety of eukaryotic cells. Microtubules are straight cylinders (24 ± 2 nm in diameter) with a hollow 15 nm core. Their wall is 5 nm thick and consists of distinct longitudinal profilaments (12 or 13) composed of globular protein submits. Any of these subunits represents a dimer (called tubulin) having a molecular weight of 120,000 (6 S). It assumes a globular shape and the subunit peptide chain is about 600 amino acid residues long. The universal characteristics of its protein (containing 150 to 180 glutamic and aspartic acids), regardless of its origin, are: 1. it binds specifically with colchicine, and 2. it binds with guanosine triphosphat (GTP).

Migration - in genetics, the transfer of genetic information among populations by the movement of individuals or groups of individuals from one population ("emigration") into another ("immigration"). M. may result in changes of gene frequencies in the populations concerned (gene flow) and is thus a factor in evolution.

Missense mutation (Brenner et al. 1961) - a base substitution (gene mutation) that results, during genetic position in the polypeptide chain (samesense mutation; nonsense mutation). Whether a mutant phenotype is readily detectable depends on the particular amino acid substitution. Certain types of amino acid substitutions affect the assigned functions of a cistron less drastically than others (conservative substitution). Some change the net electric charge of the polypeptide chain, so that a mutant chain becomes distinguishable from the original wildtype chain by electrophoresis. A m. m. may escape detection in those cases where the insertion of the incorrect amino acid does not lead to an appreciable change in the conformation and subsequent function of the protein. A m.m. can be suppressed (missense suppressor) by secondary mutations affecting transfer RNA structure and function (suppressor tRNA).

Missense suppression (Yanofsky et al. 1961) - a type of suppression due to the presence of a missense suppressor.

Missense suppressor (Yanofsky et al. 1961) - a suppressor that acts on a missense mutation and inserts the correct amino acid at the site of the missense codon. Missense suppressors represent mutationally altered tRNAs (suppressor tRNA) which accept the relevant amino acid and insert it into a polypeptide chain in response to an altered (missense) codon. Since a codon is missense at some mutated position and sense everywhere else, suppressor tRNA species cause generalized mistranslation. They are, therefore, generally very ineffecient and cells which carry them grow more slowly than normal (nonsense suppressor)

Mitochondrial DNA (Nass & Nass 1962; Luck & Reich 1964) - organelle DNA contained in mitochondria and replicated and expressed by a mitochondrion-specific genetic apparatus. M.DNA (abbreviation mt-DNA; mit-DNA) is at least ten times smaller than the genome of the smallest free-living cell (like mycoplasm) and contains information for mitochrondrial rRNA, mitochondrial tRNA, for at least a few mitochondrial ribosomal proteins and for at least a few proteins necessary for the structural and functional integrity of the ribosomal unit membranes.

Mitochondrion (Benda 1897) - a semiautonomous cell organelle (about 0.5 to 1.0 m in diameter, and 5 to 10 m in length), largely made up of protein (60 to 65%) and lipids (35 to 40%), and present in all cells of eukaryotes (chondrioid). The genesis of the m. requires the function of two genetic systems: the nuclear system and a system localized within the m. The m. contains distinctive ribosomes, ribosomal RNA (21 S and 12 S), tRNA species, aminoacy] - tRNA synthetases, initiation, elongation, and termination factors, i.e. the components of a minor protein synthesizing system coded for, in part at least, by mitochondrial DNA (the mitochondrial genome).

The number of mitochondria per cell is kept rather constant, i.e., there is a tight coordination of cell division and rate of formation of mitochondria which is most likely governed by a strict regulation of mitochondrial DNA replication and transcription.

The primary function of mitochondria in the cell is to synthesize ATP from ADP and Pi and all mitochondria contain the enzymes, coenzymes and other cofactors required by the individual enzyme complexes which catalyze respiration and phosphorylation.

Morphologically, the m. consists of a limiting or outer membrane and an inner membrane separated by an intermembrane space. Outer membrane and an inner membrane separated by an intermembrane space. Outer membrane and intermembrane space make up the mitochondrial outer compartment. The inner membrane encloses an inner space (called the matrix) and is comprised of highly invaginated and structurally distinct cristae. The inner membrane and the matrix space constitute the mitochondrial inner compartment.

Mitosis (Flemming 1882) - a mode of nuclear division (=indirect nuclear division or karyokinesis) providing for the production of daughter nuclei which contain identical chromosome numbers and are genetically identical to one another and to the parent nucleus from which they arose (amitosis; endomitosis; meiosis). M. is a genetically controlled process which is usually followed by cytokinesis and gives rise to new, genetically equivalent cells in the growing somatic regions of eukaryotic organisms. It maintains the continuity of chromosome number and type and is a continuous sequence of events which may be divided into five main stages as follows

a) mid prophase; b) late prophase; c) prometaphase; d) metaphase; e) anaphase; f) telophase.

1. Prophase: The chromosomes which reduplicated during interphase of the cell cycle make their appearance in species-specific number. At the beginning of this stage they are long threads consisting of two chromatids each. During prophase which ends with the disruption of the nuclear envelope, the chromosomes become progressively shorter and more compact owing to chromosome coiling which transforms them from the functional interphase form into the condensed "transport form".

2. Prometaphae: The spindle is formed to which the centromeres of each chromosome become attached. Subsequently the chromosomes undergo congression movement (chromosome movement) and finally come to lie at the spindle equator.

3. Metaphase : The chromosomes reach their metaphase arrangement on the metaphase plate, an equilibrium position with all centromeres aligned at the spindle equator (centromere orientation).

4. Anaphase: The two sister chromatids of each chromosome separate by their movement toward opposite poles of the spindle.

5. Telophase: The chromatids ("daughter chromosomes") of each chromosome arrive at the poles and reconstitution of a nuclear envelope around each of the two daughter nuclei takes place. Each of the two nuclei is identical in its chromosome complement to the other, to the original complement, and similarly in the content of genetic information coded in the chromosomal DNA.

Mitotic apparatus (Mazia & Dan 1952) - the ensemble of structures constituting the so-called "achromatic figure" in the classical description of mitosis. The m.a. includes the asters, which surround the centriole (if present), and the mitotic spindle plus a matrix in which the spindle fibers are embedded and can be isolated as an intact body form the cell. It is a transient structure present only during the division period of the cell cycle.

Mitotic poison (Dustin 1934) - operationally defined as any of many different kinds of noxious agents which act to impair mitosis (and by extension meiosis too), irrespective of whether the impairment takes place during or following the treatment of actively dividing cells. The mitotic disturbances may result in morphological changes of the chromosoems and/or inhibition of mitosis as well as cell death.

Mixoploid (Nemec 1910) - of cell populations whose component cells differ in their chromosome numbers, irrespective of whether these numbers are euploid or aneuploid. Thus, the term mixoploidy covers all types of mosaics or chimeras in which the heterogeneity between different elements involves differences in chromosome number. Mixoploidy may stem from a variety of mitotic irregularities, from cellular and nuclear fusions, or even from amitotic irregularities, from cellular and nuclear fusions, or even from amitotic processes, if these give rise to viable products. A special case of mixoploidy is endopoly-ploidy arising by endomitosis or endoreduplication of chromosomes (Levan & Muntzing 1963).

Molecular biology (Astbury 1950) - a branch of modern biology in which biological phenomena and processes are studied not only from a phenomenological point of view but by physical-chemical and biochemical investigations at the molecular level. M. b. is based on developments in genetics, bioche-mistry, physical chemistry of macromolecules, and chemical physics and represents an effort to account for biological events in terms of the established principles of physics and chemistry.

Molecular cloning (Cohen et al. 1973) - the cloning of DNA molecules from prokaryotic and eukaryotic sources as part of a bacterial plasmid replicon. The procedure employed for m.c.

requires a plasmid vehicle suitable for replication and selection of foreign DNA (in E. coli), a technique for covalently joining DNA fragments from different biological sources, and a method for introducing the resultant hybrid molecules into recipient bacteria without substantially altering bacterial viability (Cohen & Chang 1974). (nuclear cloning).

Molecular genetics - a branch of genetics which strives to account for genetical events in terms of the established principles of physics and chemistry. In the broadest sense, m.g. is the study of those aspects of genetic systems, or components thereof, which can be described at the molecular level. In a more restrictive sense, the study of the molecular aspects of genetic mechanisms and the control of metabolic processes by genetic material (Steiner 1965).

Monoecious (Darwin 1877) - of plants with male and female sex organs on the same individual and producing male and female gametes (dioecious). Plants with male and hermaphroditic flowers or female and hermaphroditic flowers on the same individual are called andromonoecious and gynomonoecious, respectively. M. fungi include (Esser 1971): 1. all species which produce male and female sex organs on the same mycelium; 2. all species without sex organs in which each mycelium can serve both as nuclear donor and nuclear recipient.

Monoisosomic (Kimber & Sears 1968) - of a cell or individual which lacks one chromosome pair but has an isochromosome for one arm of the missing pair (diisosomic).

Monokaryon (Boveri 1907) - 1. a uninucleate cell, a spore, or a tissue consisting of such cells, as opposed to a dikaryo (or multikaryon);

Monosome (Haselkorn & Fried 1964) - 1. a single ribosome bound to messenger RNA (polysome); 2. a chromosome in the chromosome complement lacking an homologous partner (monosomic).

Morgan unit (Haldane 1919) - a unit of genetic map length (= 100 centimorgans), i.e., that length of a chromatid that has experienced, on the average, one crossover event per meiosis. The map distance between two gene loci, when it is expressed in Morgans, is therefore the number of crossover events between the two loci. For short intervals, the map length in M. units is equal to the recombination fraction expressed as a decimal, since double-crossovers will have a neglegible frequency.

Multiple genes - two or more nonallelic genes with similar or complementary cumulative effects on a single character.

Multisite mutation - a mutation of one genetic locus involving changes of two or more adjacent mutational sites. Multisite mutations (gene mutation) occur spontaneously and are inducible experimentally by the application of mutagens. They do not revert to wild-type spontaneously or after mutagen treatment (reversion) and some in all probability are true deletions (deletion mapping).

Multivalent - any association of more than two completely or partially homologous chromosomes whose homologous regions are held together by meiotic chromosome pairing and chiasma formation from prophase to metaphase of the first division of meiosis (as distinguished from univalents and bivalents). Depending on whether three, four, five, six, seven, or eight chromosomes are meiotically paired, the multivalents are called trivalents, quadrivalents, pentavalents, hexavalents, heptavalents, and octovalents, respectively.

Mutagen - any physical or chemical agent ("mutagenic agent") significantly increasing mutational events and thus mutation rates above the spontaneous background level (antimutagen). Mutations resulting from treatment with such agents are called "induced mutatios", as opposed to the normally occurring "spontaneous mutations". The most widely applied mutagens are ionizing radiations, ultraviolet rays, base analogues, and alkylating agents which can induce both gene mutations and chromosomal structural changes (chromosome mutation).

Mutation (de Vries 1901) - any detectable and heritable change in the genetic material not caused by genetic segregation or genetic recombination, which is transmitted to daughter cells and even to succeeding generations giving rise to mutant cells or mutant cells or mutant individuals provided it does not act as a dominant lethal factor. If the descendants of a mutant cell give rise only to somatic cells in multicellular organisms, a mutant spot or area arises. Mutations in the germ line of sexually reproducing organisms may be transmitted by the gametes to the next generation resulting in an individual with the new mutant condition in both its somatic and germ cells. Since the primary carrier of genetic information is usually DNA, a mutation may be any (or a combination of) detectable, unnatural change affecting the chemical or physical constitution, mutability, replication, phenotypic function, or recombination of one or more deoxyribonucleotides; nucleotides may be added, deleted, substituted for, inverted, or transposed to new positions with and without inversion (Herskowitz 1965).

Mutation pressure (Wright 1921) - the continued recurrent production of an allele by mutation tending to increse its frequency in the gene pool of a population (gene frequency). If no other forces intervene, the mutation of allele A to a leads to changed proportions of these alleles which might ultimately result in the replacement of A by a (the entire population would then be aa in respect to the A locus). However, since back mutations from a to A also occur (generally at different rates), the population never becomes homozygous for one type of allele by mutation alone. Instead of "fixation" of one allele, an equilibrium will be established at that point where the number of mutations from A to a equals the back mutation rate from a to A.

Mutation spectrum - the frequency distribution of induced or spontaneous mutations in different genes of a genome or different sites within a gene. M. spectra are usually mutagen-specific (mutagen specificity). Selectivity of mutation induction may be dur to the following mechanisms (Kaplan 1974): the mutagen may react selectively with certain bases or base sequences of DNA because of special affinities to them (mutational hot spot); 2. the attack of the mutagen on DNA may depend on the physiological state of a genome region or its environment, e.g., the state of replication, transcription, repression, host-controlled modification, or certain neighboring cell components; 3. repair systems may act selectively at certain genome regions, or on certain types of premutational lesions distributed non-randomly within these regions; 4. mutational changes at different sites of a gene or in different genes may be phenotypically expressed to differently observable degrees (mutagen specificity).

Mutational load (Muller 1950) - genetic load.

Muton (Benzer 1957) - the basic unit of gene mutation, represented by a mutational site (a nucleotide pair) within a gene or cistron. The muton is the smallest alterable element within the one-dimensional array of mutational sites in the genetic material which when altered may give rise to a mutant organism (recon).

N

Neocentric (Rhoades & Kerr 1949) - centromere.

Neocentric activity - a diffused or terminalized action of the centromere (usually in first and second meta-anaphase of meiosis, not in mitosis) in individuals with an abnormal genotype. The chromosome arms showing n.a. are pointed like telocentric ends and coorientate with their own centromeres and with other neocentric ends of the same pairing configuration (meiosis I) or daughter element (meiosis II) (Darlington 1965).

Niche (Grinnell 1904) - all the relationships of an organism to its physical and biotic environment, i.e., the genetically determined capacity (range of tolerance) and pattern of biological response of an individual, a species population, or the whole species to environmental conditions (ecological niche). The variety of resources a population exploits is called the niche width and may be narrow or wide.

Nick - any single strand break in a nucleic acid molecule. An enzyme able to produce such breaks may be called a "nickase", while an enzyme able to heal these breaks may be called a "sealase" (Hurwitz et al. 1967).

Nonsense mutation (Brenner et al. 1961) - any mutation (base substitution or frameshift mutation) that converts a sense (amino acid specifying) codon into a chain terminating codon (UAG, UAA, or UGA in mRNA). N. mutations result in premature termination of polypeptide chain synthesis at the site where the nonsense codon occurs in the messenger RNA.

N. mutations can lead to a pleiotropic polar effect (genetic polarity; polar mutation), reducing the level of expression of genes operator-distal to be n. m. in the bacterial operon. To select for n. mutations in viral functions it is necessary to isolate the mutations in a genotype carrying a nonsense suppressor (nonsense suppression).

Nonsense suppression - suppression of the polypeptide chain termination effect of a nonsense codon (nonsense mutation) and partial or complete restoration of the normal phenotype. In the case of n. s., nonsense codons can act as sense (amino acid specifying) codons.

This may be due to the presence of a nonsense suppressor gene or a chemical suppressor (such as 5-fluoro-uracil and streptomycin). If, for instance, the nonsense triplet is UAA, then the substitution of 5-fluoro-uracil (5-FU) for U in the mRNA may result in the reading of FUAA as CAA, a sense codon. This will allow translation to continue since the tRNA will mistake FU for C with a higher probability than if U were present.

Nuclear transplantation - the transfer and insertion of the nucleus of a specialized cell into an unfertilized egg whose nucleus has been removed. If the donor-cell nuclei are not taken from the cells of an embryo or larva reared from a fertilized egg but from a young embryo that is itself a result of a nuclear-transplant experiment, this is called serial nuclear t. The effect of n. t. is the production of a clone, i.e., a population of individuals all having identical genomes in their nuclei.

Nucleohistone - a complex of deoxyribonucleic acid and histone with the histone bound to the large groove of DNA. Two models of n. structure have been proposed; 1. Non-basic residues of the histone are looped away from the DNA, enabling the basic residues to neutralize successive phosphate groups of DNA. 2. One histone is stretched along the length of about 12 turns of the DNA helix, a segment in which about 75% of the DNA phosphate group would not be neutralized by histone basic residues and are, therefore, free for interactions with other cations.

Nucleoid (Piekarsky 1937) - the region in the bacterial cell to which the DNA (the bacterial genome) is confined (= karyoid, bacterial nucleus, nuclear body, nucleoplasm, DNA plasm). Folding and supercoiling of DNA makes the n. a more or less compact structure (without membrane) and does not interfere with DNA replication and genetic transcription. Often there are two or more nucleoids per cell (depending on nutrition). Isolated nucleoids may be membrane-associated (mesosome) or membrane-free.

Nucleolus (Bowman 1840) - an intranuclear organelle of eukaryotes which represents the chromosomal site for the genes coding for ribosomal RNA (ribosomal DNA). The primary gene products (ribosomal precursor RNA), derivatives of these products (ribosomal RNA), their associated proteins, and some, perhaps all, of the enzymatic equipment (RNA polymerase, RNA methylase, RNA cleavage enzymes) required for synthesis, conversion, and assembly of ribosomes involve the nucleolus. Subsequently ribosomes are transported to the cytoplasm.

Nucleolus organizer (McClintock 1934) - the chromosome region which is active in nucleolus formation and contains ribosomal DNA, the genes from which ribosomal RNA is transcribed (= nucleolus organizing region; nucleolar organizer). These genes are highly redundant (gene reiteration), being present in multiples of 100 to 500 and more per haploid chromosome set. During metaphase of mitosis, the n. o. of the "nucleolus organizing chromosome (s)" is usually a secondary construction. The number of n. organizers in a haploid chromosome set is characteristic of the organism. As a rule, at least one n. o. is present per chromosome set and if this is removed, abnormal nuclear behavior (frequently leading to the death of the cell) results.

Nucleotide - any of the monomeric units which are the building blocks of the polynucleotids referred to as nucleic acids. A n. is a phosphate ester of the N-glycoside of a nitrogenous base and consists of a purine or pyrimidine base, a pentose (D-ribose in ribonucleic acid, 2'-deoxy-D-ribose in deoxyribonucleic acid) and a phosphate (PO4) group. The combination of nitrogenous base and sugar is called a ribonucleoside (when the pentose is D-ribose) and a deoxyribonucleoside (when the pentose is 2'-deoxy-D-ribose). The nucleotides are formed by the addition of a phosphate group to the nucleotides and are either ribonucleotides (in RNA) or deoxyribonucleotides (in DNA):

Nucleus (Brown 1831) - a membrane-enclosed cell organelle (nuclear envelope) which represents one of the two main compartments of the eucell and contains the bulk of its genetic information (nuclear DNA) in the form of chromatin. Replication of chromosomal DNA, genetic transcription of DNA and processing of DNA transcripts (heterogeneous nuclear RNA; messenger RNA; ribosomal RNA; transfer RNA) occurs in the n.

Nulli-haploid (Riley & Chapman 1958) - haploid.

Nulliplex (Belling, Blakeslee & Farnham 1923) - of a polyploid in which all chromosomes of one homologous type carry the same recessive allele of a particular gene (e.g., aaa in case of triploidy, aaaa in case of tetraploidy, etc.), in contrast to simplex, duplex, triplex, quadruplex, etc., in which the dominant allele is represented one, two, three, four, etc., times (Darlington & Mather 1949).

Nullisomic (Blakeslee 1921; Sears 1941) - of an aneuploid cell, tissue (in the case of chimeras), or individual with both members of one (or more than one: double nullisomic, etc.) particular pair of homologous chromosomes missing from the chromosome complement. Normally, nullisomics (or nullisomes) cannot survive in diploids, but they are recoverable in analogous forms, e.g., in allopolyploids (which are functionally diploid) and autopolyploids. Nullisomy may arise from selfing or intercrossing of monosomics by fusion of two gametes lacking the same chromosome ("n -1") of a particular chrornosome pair.

O

Okazaki piece - any of the short transient nucleotide sequences of DNA formed during discontinuous replication of deoxyribonucleic acid. They have a short stretch of RNA at their 5' phosphate ends which is used as a primer for the initiation of new DNA molecules (terminal deoxynucleotidyl transferase). Okazaki pieces are subsequently joined by DNA ligase to form a complete daughter strand. The RNA primer is probably removed by nucleases after synthesis of new DNA molecules is initiated but before the completion of the daughter molecule by DNA polymerases and DNA ligase.

Oncogene hypothesis (Huebner & Todaro 1969) - an hypothesis which proposes that all cells contain in their DNA the information necessary to specify the complete genome of a tumor virus (the virogene). This virusspecific DNA is transmitted vertically from generation to generation, and certain virogene functions may be normally expressed, e.g., in development. The virogene includes oncogenic information (the oncogene) which is responsible for most tumors. This information is normally repressed. A breakdown in the cellular regulation of these viral genes may, however, result in expression of the virogene and oncogene and, therefore, in neoplasia (protovirus hypothesis).

One gene - one enzyme hypothesis (Beadle & Tatum 1941; Beadle 1945) - according to this hypothesis (based on studies in biochemical genetics) each gene controls the synthesis or the activity of but a single protein (enzyme) with catalytic activity. Since its formulation this concept has been verified in principle.

One operon - one messenger hypothesis - operon.

Opal mutant (Brenner et al. 1967) - a nonsense mutation giving rise to the chain terminating codon UGA in messenger RNA (amber mutant; ochre mutant).

Operator (Jacob & Monod 1959) - according to the operon model, the recognition site at which negative control of genetic transcription is mediated through binding of a repressor. The o. is a stretch of bases (21 bases in the case of the lac o. of E. coli) along a DNA molecule, located at the proximal end of an operon. When the repressor is bound to the o., it prevents the RNA polymerase from initiating transcription of the operon. Repressor interaction is with the native double-strand of DNA in the o. Cases are known where a single repressor can repress structural genes physically separated but sharing one o.

Operon (Jacob et al. 1960) - a group of contiguous structural genes showing coordinate expression and closely associated controlling sites, i.e., elements which determine the genetic transcription of only those genes to which they are attached (cisdominant effects) (Epstein & Beckwith 1968). This definition does not presume any particular mechanism of regulation, negative or positive. In the case of negative control of o. expression the controlling sites are so coded that transcription of the o. occurs in the absence of any ctoplasmic regulatory component (repressor) specific for the particular system. In a positive control system the coding at the initiator controlling site prevents expression of the o. in the absence of any specific regulatory product. The o. represents a unit of polarized coordinated genetic transcription in prokaryotes that is initiated and regulated at the operator site. It consists of the promoter, operator, one or more structural genes coding for enzymes performing related metabolic functions, and the terminator. Transcription of the o. (eventually resulting in a polycistronic messenger RNA) starts at the promoter where the RNA polymerase binds to DNA. Synthesis of mRNA initiates if the operator is free of repressor protein (negative control). Transcription terminates at the end of the o. defined by the terminator site where the mRNA is released. The rate of transcription in a given o. is a function of two parameters: 1. the frequency of transcription initiation, and 2 the rate of transcription propagation.

Some operons seem to have extra regulator loci: in addition to the usual promoter site adjacent to the operator, an extra promoter between structural genes has been found. Since this promoter is not linked to the operator, some transcription of the last few genes always occurs even when the system is switched off.

Outbreeding - in contrast to inbreeding, a mating system (= cross-breeding) in which mating is between individuals less closely relatec than average pairs chosen at random from the population.

overdcminant (Hull 1946) - of two alleles if the heterozygote (Aa) is superior in comparison with the two homozygotes (AA and aa), i.e., lies outsice the range of the homozygotes as to the genotypic value of the character under observation (= superdominant).

P

Pachytene (v. Winiwater 1900) - meiosis.

Pairing block (Darlington & Mather 1932) a chromosome segment of definite length (consisting of a definite number of chromomeres) which acts as a functional unit in meiotic chromosome pairing. The pairing affinities of the chromosomal pairing blocks and their subunits are satisfied by associations in twos. If there are more than two homologous chromosomes in the meiocyte, there may be changes of partners which may lead to multivalent formation (zygomere).

Palindrome (Wilson & Thomas 1974) - any of the hairpin-like structures (folding back of a single linear DNA chain) which result from inverted repetitions of base sequences (e.g., ABCtC'B'A') located at intervals along the chromatid of eukaryote chromosomes. Since no unpaired single-chain regions in the turn-around region (t) have been found they are called doublechain palindromes which read the same both backwards and forwards. Palindromes may serve as specific recognition sites at the DNA level, possibly for binding of proteins. Besides in DNA, hairpin loops also exist in transfer RNA, ribosomal RNA, heterogeneous nuclear RNA and messenger RNA and may be responsible for some of the specificity of interaction of these molecules with the appropriate proteins.

Panmictic - of a randomly inter-breeding population of individuals which mate at random, i.e., each individual is equally likely to mate with any individual of the opposite sex. The population as a whole forms a deme.

Paracentric - of intrachromosomal structural changes (inversion) that do not include the centromere region of the chromosome involved in the "intrachange", as opposed to pericentric ones which include the centromere.

Parasexual (Pontecorvo 1954) - of genetic systems that achieve genetic recombination by means other than the regular alternation of meiosis and fertilization; also of the mechanisms providing for nonmeiotic recombination of the genetic material.

Pedigree - a list of ancestry or genealogical register. Pedigree diagrams are frequently used in human genetics in the analysis of Mendelian inheritance. Females in such a p. are symbolized by either circles or by the symbol ♀, males by squares or the symbol♂. Parent symbols are joined by a horizontal "marriage line", and symbols for their offspring are placed in a horizontal row below this line to which they are connected by verticals. The children of a parental pair represent a sibship, and the individual children are referred to as sibs or siblings, irrespective of their sex. They are listed and numbered in a p. from left to right in order of birth. If an individual possesses the character whose inheritance forms the subject of the p. it is designated by a black symbol. An open symbol denotes absence of the character under observation. Twins are represented by two symbols connected by lines to the same point of the sibship line; a number enclosed in a large symbol shows the number of sibs who are not listed separately.

Pentaploid - of auto - or alloploid cells, tissues, and organisms, respectively, with five chromosome sets in their nuclei (symbol 5n).

Periclinal (Baur 1909) - of a chimera in which the distinct, idiotypically different tissues are arranged in concentric layers.

Peroxisome (de Duve 1965) - a special type of cytoplasmic organelle (= microbody) characterized by the association of one or more hydrogen peroxide-generating oxidases with large amounts of catalase. Peroxisomes occur in certain cell types actively involved in the formation of carbohydrate from non-carbohydrate precursors. They are bound by a single membrane, are smaller than lysosomes (0.3 to 1.5 nm in diameter) and originate from dilated regions of the smooth endoplasmic reticulum. Frequently, they contain in a finely granular matrix a dense core or nucleoid that may have a regular crystalloid structure. Small peroxisomes (micro-p.) are extensively interconnected with the endoplasmic reticulum.

Phage conversion - the acquisition of new characters (e.g., changes in certain antigenic properties, in the production of toxins, and in the sensitivity to infection by unrelated phages) by bacterial cells containing a prophage ("prophage-mediated conversion" or "lysogenic conversion"), a temperate phage not yet integrated in the bacterial genome, or virulent mutants of temperate phages. Converted characters disappear and the cells regain their original characters when the converting characters disappear and the cells regain their original characters when the converting phage is lost. The phenotypic changes associated with the presence of phages are, therefore, due to the addition of new genetic information rather than substitution of new information for old (mutation, transformation, transduction, and conjugation).

Phagosome (Strauss 1958) - any phagocytic or pinocytic vacuole or vesicle before the fusion with lysosomes (= endosome). Phagolysosome and lysophagosome are the terms used for the combined granules.

Phenotype (Johannsen 1909) - the observable properties (structural and functional) of an organism, produced by the interaction between the organism's genetic potential (its genotype) and the environment in which it finds itself. The genotype sets the "norm of reaction" on the environmental influences. The term p. can be applied either to the totality of expressions of the genotype or to only a part, i.e., to particular characters, traits, or phenes.

Photoreactivation (Kelner 1949) - the partial elimination of the lethal and mutagenic effects of ultraviolet light (290 - 310 nm) by subsequent exposure of irradiated cells to other nonionizing radiation (330 -450 nm) leading to a decrease in the effective UV-dose (symbol PR). PR has been found to proceed by at least three distinct mechanisms (Rupert & Harm 1966):

1. direct photoenzymatic repair of UV - damage to DNA by splitting the pyrimidine dimers; apparently this is the major mechanism of PR in bacteria and bacteriophage (Rupert 1960, 1961);

2. indirect processes (related to photoprotection) by which light induced growth delay favors a light independent repair system (Jagger & Stafford 1965);

3. direct nonenzymatic photochemical repair as observed in transforming (genetic transformation) DNA but not in any organism (Setlow & Setlow 1963).

Phragmoplast (Errera 1888) - a defined plasma body representing a modified is identifiable after chromosome movement to the poles. The p. is involved in the formation of the cell plate (phragmosome). Initially, the p. is in lateral view a barrel-shaped plasma mass which then flattens out and assumes the form of a biconvex lens. The spindle fibers traversing the p. are simultaneously shortened as it flattens and the two daughter nuclei are drawn to the middle of the dividing cell. They finally come to rest on both sides of the p. The cell plate is formed in the region of the p. which deposits wall lamellae (primary walls) against both sides of the so-called middle lamella, i.e., the cell plate impregnated with pectin.

Plasmid (Lederberg 1952) - originally, an extrachromosomal genetic element; today, any genetic element which is additional to the normal genome of the bacterial cell and, in particular, may be propagated in one or the other of two alternative states: either independently in the cytoplasm or, following insertion, as an integral part of the host's chromosome. Integrated plasmids were formerly called episomes.

So far all bacterial plasmids have proved to consist of circular DNA (genetic circularity) and to represent (in the extrachromosomal state) cytoplasmic replicons, i.e., genetic elements whose replication and segregation to daughter cells at division in independent of the bacterial chromosome.

Plasmodesma (Strasburger 1882) - any of the fibrillar plasma connections (in all terrestrial plants) which extend through the cell wall and bridge adjacent cells. Plasmodesmata are unbranched cylinders of cytoplasmic origin which penetrate the cell wall perpendicular to its surface and consist (in cross-section) of two unit membranes separated from each other by a thin layer of groundplasm and surrounding a thin lumen. Plasmodesmata, collected together in groups within depressions, or narrow regions of the wall, are referred to as pits. Elements of the endoplasmic reticulum are assumed to play a role in the origin of plasmodesmata.

Plastid - in plant cells, any of a class of cell organelles which may or may not contain pigments.

Depending on the presence or absence of pigment, p. include chloroplasts (containing chlorophyll and active in photosynthesis), chromoplasts (containing carotenoids and common in fruits and flowers), and leucoplasts (containing no visible pigments and like amyloplasts able to synthesize starch). Leucoplasts can be transformed via chloroplasts into chromoplast which may be regarded as degenerative forms of the plastids.

Chloroplasts of higher plants are lens-shaped, membranous structures (about 2 μm thick and 3 to 5 μm in diameter) surrounded by two smooth membranes separated from each other by a distinct space. Closed, flattened sacs (called "thylakoids") are the structural units of internal organization. The larger thylakoids make up the so-called stroma lamellae of the internal membrane system which varies considerably from plant to plant and within one plant depending on its physiological state. The smaller thylakoids are stacked together to form the grana lamellae of the membrane system which contain the light absorbing pigments. Each granum represents a column of flattened appressed discs piled one on the other and each of the discs is composed of a pair of membranes.

Pleiotropy (Plate 1910) - the production, by one particular mutant gene, of apparently unrelated multiple (or manifold) effects at the phenotypic level (= polypheny). Since there is no evidence at the level of primary gene action that one gene (one cistron) has more than one primary function (one gene - one enzyme hypothesis), pleiotropic gene action should be regarded as a reflection of the highly integrated state of cellular and developmental metabolism. With the remoteness of the phenotype from the primary action of the gene and the number of intervening steps which can be influenced by other genes and by environmental factors; secondary effects may give rise to a "genetic syndrome" consisting of a variety of phenotypic effects.

Point mutation (Bridges 1923) - a gene mutation (intragenic mutation) which can recombine with any of a large number of mutants, as opposed to larger alterations where the corresponding mutants, as opposed to larger alterations where the corresponding mutant does not show genetic recombination with at least two other mutants each able to recombine with the other. The precise definition of a "point" is dependent on the number and kind of mutants crossed and on the smallest distance in the genetic map for which recombinants are experimentally, or possibly molecularly, detectable (Freese 1961).

Polar mutation (Franklin & Luria 1961; Jacob & Monod 1961) - any mutation of the nonsense, frameshift, deletion, or insertion mutation type which results in genetic polarity. In the case of a p.m. (= dual effect mutation of Englesberg 1961) in an operon 1. the production of the particular enzyme for which the nonmutated gene codes is prevented (ribosomes carrying nascent peptide chains are released as they encounter the mutant site in the mRNA of the mutated gene), and 2. the functioning of the operator-distal genes of the operon is relatively decreased (mRNA distal to the mutation remains free of ribosomes and is highly susceptible to endonucleolytic attack) (polarity gradient).

Polaron (Lissouba & Rizet 1960) - a segment of the genetic material (a nucleotide sequence in DNA) within which genetic recombination is polarized and due primarily to gene conversion, i.e., mechanisms other than crossing-over. Polarity of recombination within a p. is evident from the observation that in those cases where one of a group of linearly located alleles undergoes conversion more frequently than another placed to its left, then the right-hand member (Whitehouse 1963).

The polarity (gradient) of conversion is interpreted as the formation of a DNA hybrid (heteroduplex) by an initial breaking point outside of the implicated region (polaron hybrid DNA model). The often converted site is the one nearest to the breakage point (linkage point) where crossover and noncrossover recombinations are assumed to initiate.

Polaron hybrid DNA model (Whitehouse 1963) - a model based on the correction of heterozygosity in DNA as a source of intrachromosomal genetic recombination which can explain both reciprocal (crossing-over) as well as nonreciprocal recombination (conversion).

The entire sequence of events, termed a "dissociation cycle" is assumed to proceed as follows (Whitehouse & Hastings 1965):

1. Enzymatic breakage (b in Fig. 75) of complementary nucleotide

chains, one from each of two DNA molecules, takes place at fixed points (outside or at the ends of two DNA molecules, takes place at fixed points (outside or at the ends of the genes), homologous in position at the "linkage points" defining the ends of each chromosomal polaron (a region between neighboring genes in which recombination is by conversion only but may give rise to reciprocal as well as non-reciprocal recombination). Breakage is followed by dissociation and uncoiling of these chains on one or both sides of the break.

2. DNA synthesis as a correction process begins from the points of breakage in one or both directions.

3. The newly synthesized polynucleotide chains uncoil.

4. Pairing and coiling of complementary strands, one from each molecule, lead to the reformation of double helices and, thus, the rejoining of breaks.

5. Breakdown of any unpaired chains.

Polycentric (Darlington 1937) - of chromosomes or chromatids (= polykinetic) with more than one localized centromere: dicentric (with two centromeres), tricentric (with three centromeres), etc. Polycentrics are in most cases the result of chromosome mutations. In special cases (nuclear differentiation), compound polycentric chromosomes represent a normal feature of chromosome organization. (operon).

Polygenic (Plate 1913; Mather 1941) - of traits or characters (quantitative characters), whose expression is controlled by many genes (polygenes) which individually exert a slight effect on the phenotype. The mode of inheritance of p. characters is called p. inheritance and its main properties may be summarized as follows (Lerner 1968):

Polymeric (Nilsson-Ehle 1908) - of genes (gene interaction) whose individual effects are equivalent, but whose actions intensify each other (cumulative genes). Interactions among p. genes which give rise to so-called "polymery" (the production of a particular character by co-operation among several p. genes) are of two types: dominance interactions between alleles, and nonallelic interactions. These interactions may be expressed statistically as components of genetic variance (variation). For a quantitative character controlled by p. genes, there are thus six components of variance:

polynemic - 1. of primarily multistranded chromosomes in eukaryotes, as opposed to binemic chromosomes (polytenic);

Polyploid (Strasburger 1910; Winkler 1916) - of somatic cells and tissues, as well as of individuals (polyploids) having three (triploid), four (tetraploid), five (pentaploid), or more complete chromosome sets instead of two as in diploids. The state of being p. is referred to as "polyploidy" and may arise spontaneously or be induced experimentally by mitotic poisons (C-mitosis; C-meiosis). A polyploid that originated from a zygote containing unlike chromosome sets (usually due to hybridization of two species) is called an allopolyploid, as opposed to a polyploid containing more than two homologous chromosome sets in its somatic cell which is called an autopolyploid.

Polyploidization - the conversion (spontaneous or experimentally induced) of cells of individuals from the haploid or diploid state to a polyploid state (C-mitosis; C-meiosis; endomitosis). Agents capable of inducing polyploidy may be "polyploidogenic" (mitotic poison).

Polysome (Warner, Rich & Hall 1962) - a multiribosomal structure (= ergosome, polyribosome) representing a linear array of ribosomes held together by messenger RNA (the single ribosome bound to mRNA is referred to as a monosome by Haselkorn & Fried 1964). Probably the number of ribosomes is related to the size of the mRNA chain. Polysomes represent the "active complexes" in cellular protein synthesis and are able to incorporate amino acids into polypeptides both in vivo and in vitro. Such ribosomes aggregates are usually 10 to 20 times more active than isolated ribosomes (monosomes) in protein synthesis.

Polysomic (Blakeslee 1921) - of an otherwise diploid cell or individual with one or more chromosomes represented three (trisomic; doubly trisomic) or four (tetrasomic) times instead of twice ($2n + 1$, $2n + 1 + 1$, $2n + 2$, etc.). Polysomic inheritance (Blakeslee, Belling & Farnham 1923) is the mode of inheritance which arises when any chromosome (in the case of polysomy or polyploidy) finds more than one partner for chromosome pairing at meiosis.

Polytene chromosome - a special type of chromosome representing a bundle of numerous fibrils (chromatids) which arise, according to the "polyteny hypothesis" (Koltzoff 1934, Bauer 1935), by repeated cycles of endoreduplication of single chromatids. This type of chromosome in its typical form is characteristic for various tissues of Diptera and may be up to 200 (and even more) times the size of corresponding chromosomes at meiosis or in the nuclei of ordinary somatic cells. In Diptera, homologous chromosomes are somatically paired (chromosome pairing), and, therefore, the number of p. chromosomes per cell corresponds to half the (diploid) chromosome number of somatic cells. The number of fibrils per p. chromosome is species-specific and may ultimately reach about 2000 in extreme cases. The length and diameter of the fibrils varies between 100 - 250 µm and 15 - 25 µm, respectively.

Population (Johannsen 1903) - a community of potentially interbreeding individuals (at a given locality) which share a common gene pool. Such a "Mendelian population" (Dobzhansky 1935) represents a (dynamic) panmictic unit (breeding size). The sum of all factors governing the pattern by which gametes from various individuals unite with each other during fertilization makes up the population structure.

Population genetics - a branch of genetics that describes (in mathematical level. It deals with the frequencies and interactions of genes in so-called "Mendelian populations" and studies the influences of agencies such as mutation, natural and artificial selection, migration, mixing of races and chance factors which tend to alter gene frequencies and thus to cause evolutionary changes (evolution).

Position effects (Sturtevant 1925) - a change in the phenotypic effect of one or more genes due to a change in their position with respect to other genes in the genotype; this may be the result of chromosomal structural changes (chromosome mutation) or of crossing-over. Differences in the phenotypic expression of a gene caused by a change in its spatial relation to other genes on the chromosomes may be brought about by:

a) changes in the production of a given messenger RNA (genetic transcription),

b) changes in the genetic translation of the messenger RNA in question, or

c) changes in the interactions between the polypeptides resulting from the translation of messenger RNA.

According to Lewis (1950), two types of position effects may be distinguished: the "cis-trans type" (or "stable type") and the "variegated type":

1. The cis-trans position effect consists (as shown by the cis-transtest) of different phenotypic expressions of two mutated sites within one gene or cistron; they are present in the heterozygous state and separable by genetic recombination. The cis-configuration (both mutated sites on one chromosome, the normal sites on the other) results in a normal phenotype, the trans-configuration (one normal and one mutated site on each chromosome) in a mutant phenotype. This type of position effects can also occur over an interval greater than one cistron (intergenic cistrans p. e..), between mutations occurring in the cistrons of one operon controlled by a single operator and between mutations of the operator itself.

The intergenic cis-trans p. e.. may result if

a) the product of one or both of the wild-type genes involved is unstable (except in combination with the other) and the products decay unless the normal alleles of both genes are translated in the immediate neighorhood of both, or

b) the translation and/or transcription of one or the other of two complementing (genetic complementation) wild-type genes is partially suppressed by a nonsense mutation in the cis-position to it.

2. The variegated (V-type) position effect (Muller 1930) involves chromosomal structural changes whereby the chromosomal "environment" of the wild-type genes under consideration is changed (by translocation or inversion). Classically, genes localized in euchromatin are juxtaposed to heterochromatin and vice versa. V-type position effects are confined to genes present in the heterozygous state and result in a partial repression (by heterochromatinization) of the wild-type gene when it is brought into contact with heterochromatin.

Precocity theory (Darlington 1930) - a theory which regards meiosis as a precocious mitosis. Meiotic (as compared with mitotic) prophase is assumed to start precociously with chromosomes which have not yet split and this is held responsible for chromosome pairing, chiasma formation, nondivision of the centromeres and insertion of a second division, together with the reduction of chromosome number and the segregation of genes (Darlington & Mather 1949). Since both DNA and protein synthesis have been shown to be completed before meiotic chromosome pairing occurs, the p. t. appears to be no longer tenable.

Proband - in human genetics, the "affected" person with whom the study of a particular character in a family (pedigree) usually begins (= propositus if male, proposita if male, proposita if female, or index case). A method of comparing the proportion of progeny in families in which an "index case" or p. shows a specific character with the proportion expected if this character fulfills Mendelian inheritance is referred to as the "proband method".

Promoter (Jacob et al. 1964) - in an operon, a nucleotide sequence located at the operator end (adjacent but external to the operator) which contains all the signals for correct initiation of genetic transcription by the RNA polymerase holoenzyme and determines the maximal rate of RNA synthesis. Different promoters probably have different rates of initiation with RNA polymerae and some need additional factors for highlevel expression of their operons (catabolite repression). Genetically, the p. is defined by mutations that coordinately alter the level of expression of all the structural genes in the operon. These mutations lie outside the structural genes and do not affect induction of the operon.

The primary nucleotide sequence of the promoters analysed up to now differs widely indicating that the DNA-dependent RNA polymerase recognizes a tertiary structure in the DNA. The p. sequences contain symmetric elements which may be significant for the attainment of three-dimensional structures.

Prophage (Lwoff & Gutman 1950) - a temperate bacteriophage integrated into the genome of lysogenic bacteria which replicates in coordination with the replication of the genetic material of the host cell. The p. contains the genetic information for phage production (induction) and, thus, constitutes a potentially lethal factor for the lysogenic bacterium. It behaves like a gene or gene group that exercises only a small part of its potential for phenotypic expression. The presence of a p. in a bacterial cell is recognized by the immunity conferred on the lysogenic cells to further infection by an extrinsic, homologous phage. Furthermore, the p. may endow the host cell with other characters whose phage mediated acquisition is referred to as lysogenic conversion or phage conversion.

Pseudoallele (Morgan, Bridges & Sturtevant 1938; Lewis 1948) - any of two (or more) mutations which are allelic (similar or related) in a functional but not in a structural sense. Pseudoalleles map at different positions (genetic map) and show a low order of genetic recombination via crossing-over. Such mutations (m1 and m2) are phenotypically mutant in trans-heterozygotes (m1+/+m2), but wild-type in cis-heterozygotes (m1m2/++), that is, exhibit a cis-trans position effect ("position pseudoalleles").

Pseudoallelism is evidence that the gene cannot be defined as a unit out structure and function (cistron) and may be interpreted in two ways:

1. Each of the separable elements of a (pseudoallelic or complex) locus is a discrete gene (operon) with a distinct function and the cis-transposition effects is the result of a gene product acting as the substrate (not able to diffuse from one chromosome to its homologue) for the activity of the adjacent gene.
2. The separable elements represent mutations at different recombinable sites within one and the same gene (mutational site). According to this interpretation, the gene functions only when all its parts are assembled together in one sequence, and its function can be impaired by mutation at any of numerous separable points along its length. Today, pseudoalleles are generally interpreted as closely linked and functionally related genes. A cluster of pseudoalleles is referred to as a "pseudoallelic series", a "complex locus" (Dunn 1954), or a "region" (Benzer 1957).

Pseudovirion (Michel et al. 1967) - any viral capsid containing linear fragments of cellular DNA, instead of the viral genome or fragments thereof. Pseudovirions are not infections.

Psi factor (Travers et al. 1970) - any of a class of bacterial positive control elements assumed to act as secondary specificity determinants allowing the RNA polymerase holoenzyme (not the core enzyme) to transcribe several classes of transcription units. Each element is assumed to be specific for a given class. There are indications that psi factors are regulated by low molecular weight effectors. This provides for a positive control which is readily reversible and would allow the bacterial cell to respond readily to environmental changes.

Puff (Bridges 1937) - any of the structurally modified regions in polytene chromosomes involved in RNA synthesis. Puffs originate from single bands and involve the despiralization of the deoxyribonucleoprotein (DNP). The despiralized DNP, the accumulated RNA (due to site-specific synthesis), and the accumulation of acidic proteins (synthesized elsewhere) contribute to the morphology of the p. site.

Pure line (Johannsen 1903) - the descendants, through self-fertilization, of a single homozygous parent, or a highly inbred line of plants or animals obtained by long continued inbreeding. All members of a p. l. are of the same genotype and are homozygous for all allele pairs within the limits set by the probability of new gene mutations.

Q

Quadrivalent - a multivalent consisting of four chromosomes which are either completely (as in autopolyploids) or partially homologous (as in translocation heterozygotes).

R

Race - an intraspecific category, primarily a population or aggregate of populations, with characteristic gene frequencies or features of chromosome structure that distinguish a particular group of individuals from other groups of the same kind within formally recognizable subspecies or within species. Race difference are relative, not absolute. The term subspecies is frequently used in the same sense as race. Any race is able to interbreed freely with any other r. of the same species. Whenever different races of a cross-fertilizing species occupy geographically separate territories, they are said to be allopatric; those occupying the same territory are sympatric. Races may become distinct species (speciation) by the production of reproduction isolation (with respect to the other races of the same species) and thus the formation of isolated gene pools.

Radiomimetic (Dustin 1947) - of chemical agents (typically represented by alkylating agents) which mimic the most important end effects induced by ionizing radiations in living systems. These include gene and chromosome mutations, carcinostatic and carcinogenic effects and, in vitro, depolymerized nuclei acids. In a broader sense, the term r. is used for all chemicals ("radiomimetics") which are capable of inducing gene and chromosome mutations.

Random fixation (Wright 1931) - the complete loss of one of two alleles in a population, the other allele reaching a frequency of 100% under certain calculable circumstances (gene frequency; genetic drift). Whether or not r.f. will occur in a population depends on the interrelationship between the number of breeding individuals, the selective value of the allele (s), the mutation pressure, and the gene flow,

Reactivation - r. or recovery is the term used to describe the phenomenon in which a given physical (radiation) or chemical treatment is rendered nonlethal or nonmutagenic when in the same biological system under different circumstances it would be lethal or mutagenic (restoration; photo-reactivation; host-cell reactivation; UV-reactivation; catalase reacti-vation; multiplicity reactivation; cross reactivation).

Reactor - a group of cells which react on the activity of an inductor (or evocator) during embryonic induction.

Reading frame - the mechanism which moves the ribosome three nucleotides at a time during genetic translation. A shift in the r.f.. by any number of nucleotides other than three or a multiple of three will create an entirely new sequence of codons (frameshift mutation).

Recessive (Mendel 1865) - of genetically controlled characters and the alleles responsible for their production when the phenotypic expression is associated with alleles in the homozygous state, as opposed to dominant and recessive character difference (AA x aa) the recessive character is not evident, but reappears in one quarter of the second filial generation (F2).

Reciprocal (Mendel 1865) - 1. of a second cross (A ♁B ♂) which is similar to a first one (B ♂ A ♁) but with the sexes of the parents (the sources of male and female gametes) interchanged;

Recombinant - any of the individuals or cells arising as a result of interchromosomal and intrachromosomal (via crossing-over or conversion) genetic recombination. The number of recombinants in relation to the number of offspring in appropriate crosses is referred to as the "recombinant proportion, or fraction".

Recombination frequency - the total number of recombinants (arising by reciprocal genetic recombination via crossing-over) divided by the total number of progeny individuals (or gametes) scored. R.f. (= recombination fraction) equals the crossover value (and represents a measure of map distance) for those chromosomal regions so short that multiple crossovers constitute an insignificant class (complete interference). As intervals between the genetic markers under observation increase, this equality no longer holds.

Recon (Benzer 1957) - the smallest unit (corresponding to a single DNA nucleotide) which is exchangeable but not divisible by intragenic recombination (genetic recombination) between homologous linkage structures (chromosomes) and which may be detected by the techniques of genetic fine structure analysis of the linear array of mutational sites within the genetic material (cistron; muton; polaron hybrid DNA hypothesis).

Regulator gene (Jacob & Monod 1961) -

1. in the broadest sense, any gene which regulates or circumscribes the activity of other genes.

2. A gene which codes for a (allosteric) protein (the repressor) which functions alone (inducible systems) or in combination with a corepressor (repressible system) in regulating the genetic transcription of the structural genes in an operon by binding to the operator. Mutations of r. genes may result either in a complete shutdown in repressor production or may give rise to a repressor whose ability to

interact with the operator of the relevant operon is greatly reduced (constitutive enzyme synthesis determined by the structural genes of the operon). A mutationally altered repressor may be unable to bind the corepressor readily or may fail to undergo the proper conformational change (allosteric) which activates the repressor after its interaction with the corepressor. Hyperrepressed mutants may act by producing a repressor with very high affinity for the sensitive operator or which more readily assumes the active repressor configuration in the presence of metabolites representing normal cell constituents.

Release factor (Ganoza 1966) - any of the special protein factors (= termination factors) which recognize the terminator codons UAA, UAG, and UGA in messenger RNA and stimulate the codonspecific release of polypeptides from ribosomes during genetic translation. R.f. dependent hydrolysis of the peptidyl-tRNA ester bond requires location of the peptidyl-tRNA in the ribosomal P-site, NH4+ or K + ions, and is inhibited by several antibiotics that also inhibit peptide bond formation.

Renner complex - in complex heterozygous species (Oenothera), a group of chromosomes (and the genes which they contain) which are distributed as one unit from generation to generation.

Renner effect (Darlington 1932 - competition (= megaspore competition) among the four genetically different spores formed by one meiosis in regard to which shall form the embryo (Darlington & Mather 1949).

Repetitious DNA (Britten & Kohne 1968) - in eukaryotes, nucleotide sequences (abbreviation rep-DNA) occurring repeatedly in the chromosomal DNA and characterized by 1. the number of nucleotide pairs per sequence, 2. the specific order of base pairs per sequence, and 3. the number of copies of a sequence per genome. A particular family of repeated sequences can exist either dispersed between unique DNA sequences or as clusters of repeating units limited to one or a few areas (called regional repetitions by Lee & Thomas 1973) in the genome.

Two categories of rep-DNA may be distinguished on the basis of denaturation-renaturation kinetics (Cot value): 1. a highly repetitive or "fast fraction" repeated about 106 times and consisting of simple sequence DNA; 2. a middle repetitive or intermediate fraction made up of sequences repeated 102 to 104 times each and 300 ± 200 base pairs long, interspersed in the majority of instances between regions of unique of non-repetitive DNA. In each of these fractions, several families of repeated sequences are found. The evolution of such families is assumed to involve divergence through mutation of highly repetitive sequences which arose through saltatory multiplication events (Thomas circle).

Highly repetitive or simple-sequence DNA tends to be associated with constitutive hetero-chromatin and is often concentrated close to the centromere or chromosome ends. It appears to be transcribed to a very small extent, if at all, within the cell.

Replication - in genetics, the nondestructive read-out and storage of genetic information giving rise to a new carrier of information by copying a preexisting unit of the same kind. Subcellular biological systems in which the process of r. occurs are the nucleic acids (in most cases DNA, in some RNA).

Replicon (Jacob & Brenner 1963) - any genetic element (bacterial chromosome, virus genome, plasmid) that behaves as an autonomous unit of DNA replication (deoxyribonucleic acid), i.e., capable of replication under its own control. The r. possesses a functional group which controls its replication and is called the replicator. The diffusible product of a particular gene controlling (positive control) the replication of DNA is termed the initiator and is assumed to interact specifically with the replicator. Interaction results in the initiation of a new round of replication at specific sites (the origin). Replication ends at the terminus and is bidirectional.

Repressor (Pardee et al. 1959) - any of the specific allosteric protein molecules (products of regulator genes) which bind to the operator of operons and prevent RNA polymerase (which binds at the preceding promoter) from proceeding into the operon to transcribe messenger RNA from DNA (negative control). Each set of independently regulated genes is controlled by a different repressor produced by a different regulator gene, For each r, -operator system there exists an effectr molecule (the inducer or co-repressor) that interacts with r. affecting its affinity for the operator. The co-repressor (an end product or related compound of anabolic systems) increases the repressor's operator binding activity; the inducer acts in the reverse manner. A fully functional r. must, therefore, contain two sites: one (A) must be specific for the recognition of, and association with, the operator on the DNA; the second (B) must specifically recognize and interact with an effector or antirepressor.

Restriction endonuclease (Meselson & Yuan 1968) - any of the strain-specific restriction enzymes that enable bacteria to recognize and rapidly degrade DNA introduced from foreign sources (such as bacteria, phages, plasmids) which has not been modified by replication in the appropriate bacterial host cell (DNA modification-restriction enzyme).

Restriction enzyme - = restriction endonuclease.

Reverse transcriptase (Baltimore 1970; Temin & Mizitany 1970) - RNA - dependent DNA polymerase (originally isolated from single-stranded RNA viruses but not unique to these, and virus-infected or tumor cells) that transcribes synthetic or natural RNAs into DNA. The enzyme requires Mg2+ or Mn2+ as a metal activator, deoxyribonucleioside triphosphates, an RNA template, and an RNA primer to initiate the synthesis is initiated by adding a deoxynucleotide to the 3'-hydroxyl group of the primer, i.e., the mechanism of DNA synthesis by r. t.. seems to be essentially identical to that of the Kornberg DNA polymerase I of E. coli, except that the r. t.. can use an RNA primer and an RNA template.

Revertant - a mutant which has regained, partially or completely, the wildtype phenotype by either genetic or nongenetic mechanisms of reversion. Genotypic revertants which have regained the enzyme or function that was lost by gene mutation may arise by several mechanisms and fall into two classes:

Rho factor (Roberts 1969) - a protein factor in *E. coli* (= termination factor) which binds to the RNA polymerase holoenzme and catalyzes specific (correct) termination of genetic transcription and release or RNA in vitro. The rho factor (mol. weight about 200,000) is a tetramer or hexamer causing the release of RNA species of discrete sizes, i.e., termination of transcription at DNA sites not recoginized by RNA polymerase alone (M factor; psi factor). The rho factor is complementary to the sigma factor and plays an essential role in the mechanism which results in RNA polymerase recognizing termination signals (terminator codons). Rho is antagonized by the antiterminator proteins synthesized during phage infection of bacterial cells. Rho and sigma determine the specificity of transcription of RNA off DNA templates.

Ribonuclease - any of a class of enzymes that inactivate and degrade ribonucleic acid (RNA).

Ribonucleic acid - a polynucleotide (symbol RNA) consisting essentially of chains with a repeating backbone of phsophate and sugar units to which nitrogenous bases are attached. RNA governs all protein manufacture and exists in a variety of different species. Most sequenced RNAs have the potential to form secondary structures by means of local base pairing and loop formation in otherwise single-stranded molecules. These structures (pre-messenger RNA; transfer RNA metabolism of eukaryotes is more complex than that of prokaryotes and many species of RNA have been found to have functions other than those classically assigned to RNA (heterogeneous nuclear RNA; low molecular weight nuclear RNA; chromosomal RNA).

Three main differences distingusih deoxyribonucleic acid (DNA) and RNA:

The chemical structure of uracil ribonucleoside. Ribonucleic acid

(RNA) and deoxyribonucleic acid (DNA) differ in that RNA contains uracil instead of thymine, and ribose instead of deoxyribose.

1. The D-ribose sugar of RNA differs from the 2-deoxy - D - ribose of DNA in having a total of four rather than three hydroxyl groups. The extra OH group is at the C 2 position.

2. Both DNA and RNA contain four principal bases but only three of these (adenine, guanine, and cytosine) are common to both nucleic acids. The RNA counterpart of thymine is uracil, the former being the 5-methyl derivative of the latter.

3. The secondary and tertiary structures of RNA show structural and functional heterogeneities which allow one distinguish two main classes of RNA:

3.1 Genetic RNA: The carrier of primary genetic information in most plant viruses, some animal viruses and certain bacteriophages (reverse transcriptase; replicative form) which may be single- or double-stranded.

3.2 Non-genetic RNA: RNA specified by DNA in the course of genetic transcription. It is heterogeneous and three principal types can be distinguished: ribosomal RNA (80 % of the total cell RNA), transfer RNA (10 to 15% of the total), and messenger RNA (5 to 10 % of the total). These types of RNA are usually synthesized via the sequential processing of primary transcription units (transcription) which are longer in nucleotide sequence rather than the mature products (precursor molecules).

Regulation of RNA synthesis (transcription) in vivo is of two types, generalized and specific. Generalized regulation as a function of growth rate or of the presence of essential amino acids is coordinate, i.e., applies simultaneously to all species RNA (relaxed; stringent). Regulation of the synthesis of specific RNA species occurs at the level of genetic transcription (transcriptional control; genetic regulation). Two types of reactions leading to the formation of functional RNA molecules are known which require either DNA (DNA-dependent synthesis of cellular RNA) or RNA (RNA-dependent synthesis of cellular RNA) as the primer and are catalyzed by distinct classes of enzymes.

1. DNA-dependent RNA synthesis: DNA acts as the template, i.e. its nucleotide sequence determines by the formation of complementary base pairs the nucleotide sequence of the synthesized RNA molecules. The process is genetic transcription and is catalyzed by DNA-dependent RNA polymerase which links together the monomeric precursors (the ribonucleoside triphosphates ATP, GTP, UTP and CTP).

2. RNA-dependent RNA synthesis: RNA acts as a template and the process corresponds to the replication of viral RNA. It is catalyzed by RNA-dependent RNA

polymerase (RNA replicase, RNA synthetase) capable of forming new RNA strands upon parental RNA templates. When viral RNA enters the cell and single strands attach to host ribosomes this enzyme is formed and ribonucleoside-5'-triphosphates are polymerized to form a long chain of polynucleotides (with elimination of inorganic phosphate) by transfer of the activated 5'-phosphate end of a nucleotide to the free 3'-hydroxyl group of a second, adjacent nucleotide in the growing polynucleotide (replicative form).

Chloroplast and mitochondrial rRNAS in eukaryotes have lower molecular weights than their cytoplasmic counterparts, and different average nucleotide composition. The genes coding for these RNAs are contained in chloroplast DNA and mitochondrial DNA, respectively.

Ribosome (Roberts 1958; Dintzis et al. 1958) - an exceedingly complex cell organelle in pro- and eukaryotes which is made up of about 60 unique proteins and three (in prokaryotes) or four (eukaryotes) species of ribosomal RNA (rRNA) organized into two dissimilar ribosomal subunits (30 S and 50 S in prokaryotes; 40 S and 60 S in eukaryotes). Ribosomes are more or less spherical in shape and measure approximately 20 nm across. Those of bacteria are slightly smaller than those of eukaryotes. A growing bacterial cell contains 10,000 to 30,000 ribosomes, most or perhaps all of which are involved in protein synthesis during the exponential growth phase.

Ring bivalent - a bivalent at the first meiotic division consisting of two paired chromosomes (chromosome pairing) with terminalized chiasmata in both arms (chiasma terminalization; ring multivalent).

Ring chromosome - a physically circular chromosome (genetic circularity). Ring chromosomes are a normal feature of chromosome organization in some prokaryotes (e.g., *Escherichia coli* and some viruses). In eukaryotes, ring chromosomes arise as a result of chromosomal structural changes (chromosome mutation) and are usually meiotically unstable due to the production, by meiotic crossing-over, of double-sized dicentric chromosomes which normally are eliminated.

RNA polymerase - any of a number of enzymes (ribonucleoside triphosphate: RNA nucleotidyltransferase) capable of polymerizing ribonucleoside triphosphates into polyribonucleotides (RNA). DNA (DNA-dependent RNA p.) or RNA (RNA-dependent RNA polymerase or RNA replicase) is used as the template. DNA-dependent RNA p. is involved in genetic transcription, RNA-dependent RNA P. in the replications of RNA viruses with double-stranded RNA molecules.

S

S_1. S_2, S_3, etc. - shorthands for generations of self-fertilization in plants: S^1 denotes the generation obtained by selfing the parent plants, S^2 the generations obtained by selfing the S^1 plant, etc.

S-allele - any allele of those genes which control incompatibility in plants. Alleles present in both style and pollen are referred to as "matched S-alleles", whereas those present only in either style or pollen are called "unmatched S-alleles" (Lewis 1943). SF alleles are alleles belonging to the same series of multiple alleles as S-alleles but which control self-fertility.

SAT - chromosome (Heitz 1931) - a chromosome with a secondary constriction which separates a satellite from the rest of the chromosome. SAT - chromosomes whose secondary constriction is associated with the formation of the nucleolus are referred to as "nucleolar" SAT - chromosomes and the constriction is a "nucleolar" secondary constriction.

Satellite (Navashin 1912) - a chromosomal segment separated from the main body of the chromosome by one secondary constriction if located terminally (terminal s.). Together, the s. and the secondary constriction are termed "satellite region".

Satellite DNA (Sueoka 1961; Kit 1961; Britten & Kohne 1968) - in eukaryotes, any DNA fraction sufficiently different in base composition from the major DNA component to separate as distinct bands during isopycnic CsCl gradient centrifugation. Sat. DNAs may have a buoyant density in neutral CsCl lower or heavier than the main band DNA. Nuclear and cytoplasmic sat. DNAs may be distinguished: the former are of solely nuclear origin, the latter are composed of DNA from cytoplasmic sources including DNA from chloroplasts (in plants), kinetoplasts, and mitochondria.

Self-incompatibility (Stout 1917) - the inability of a plant producing functional gametes to set seed when self-pollinated (incompatibility).

Semiconservative (Delbriick & Stent 1957) - of the normal mode of replication of deoxyribonucleic acid (Fig. 31) in which each of the two parental strands is conserved in a different daughter molecule (nonconservative; DNA polymerase).

Sex determination - the determination of sex differentiation is eusexual (parasexual) organisms by process which shift the socalled "bisexual potency") (Hartmann 1923) or the balance between male (M) and female (F) determining genes (sex realizators) of cells, organs and individuals in either the male or female direction.

Sex duction (Jacob & Wollman 1960) - the incorporation of bacterial genes by sex factors and their subsequent transfer by conjugation (F-duction). The sex factors act as conjugons and are plasmids.

Sex factor - in bacteria, transmissible plasmids that can promote the transfer (by conjugation) of genetic material of the host chromosome or a coexisting plasmid from the bacterial donor to the recipient cell. The fertility properties of most s. factors (some, but not all, of the colicinogenic factors and resistance factors) are generally repressed in the majority of cells (the F-plasmid being an exception to the rule). S. factors determine specific cellular traits (sex pilus) that render a male bacterial cell capable of mating with a (female) recipient.

Sex fimbrium (Diguid & Anderson 1967) - = sex pilus.

Sex-influenced - sex-controlled.

Sex-linkage (Morgan 1914) - linkage of genes located in the sex-chromosomes (X- and Y-chromosomes) of eukaryotes. Such genes and their mode of inheritance, as well as the phenotypic characters whose manifestion these control, are all referred to as "sex-linked" (sex-limited)

Sigma factor (Burgess et al. 1969) - a positively acting transcription factor (mol. weight 85, 000 to 90, 000; 4.5 to 5 S) which is the primary determinant for promotor recognition by bacterial DNA-dependent RNA polymerase. Sigma is one of the subunits of the RNA polymerase holoenzyme and restricts the RNA polymerase core enzyme (without s. f.) to initiating transcription at certain genes. Transcription of some bacteriophages during infection is controlled by production of successive phage coded sigma-like factors which replace the host s. f. and direct the host core RNA polymerase to transcribe specific phage genes (psi factor; rho factor; catabolite activating protein; M factor).

Sister chromatid exchange (Taylor 1958) - a deviation from the regular semiconservative replication pattern of the chromosomes of eukaryotes consisting in the exchange of segments between sister chromatids of a chromosome. After incorporation of ^3H - thymidine into chromosomal DNA, s. c. e. is evident from segmental switches of radioactive label from one chromatid to the other in the second post-labeling mitosis (sister label exchange). S. c. e. may be of two types: single exchanges which effect only one of the two pairs of granddaughter chromosomes (prevention of anaphase separation of chromatids results in four homologues per tetraploid cell), and twin exchanges in which both chromosomes show an exchange at the same point.

Somatic reduction - reduction of (normally polyploid) somatic chromosome complements (reductional grouping; cell hybridization) by formation of aneuploid intermediates. S. r. in premeiotic cells suggests a possible origin for some existing diploid and tetraploid taxa from polyploids through a process of depolyploidization. S. r. may give rise to haploid cells (haploidization) during the parasexual cycle of certain fungi.

Spacer DNA (Brown & Weber 1968; Birnstiel et al. 1968) - "nongenic" DNA one another. S. DNA is assumed to account for a large fraction of the DNA within a chromosome. Spacer gene arrangement of cistrons is generally established for ribosomal DNA where the spacer separates the blocks of the 28 S and 18 S ribosomal RNA cistrons. In this case, the untranscribed spacer may have an average molecular weight as high as 6.6 X 10^6 daltons for each repeating unit and seems to vary in length by as much as a factor of 10 even within one species. Its sequence is very different even in two closely related species. Where amplification of ribosomal DNA occurs (gene amplification), the spacer takes part in this process.

Speciation (Simpson 1944) - the evolutionary formation of species which, in essence, occurs by the production of barriers to gene exchange (reproductive isolation) between genetically diverging populations. Isolating mechanisms are promoted by natural selection if gene exchange lowers the fitness of its products.

The main modes of s. are (Mayr 1963):

1. Sympatric speciation: s. without geographic isolation, i.e., the acquisition of isolating mechanisms within a deme.

2. Allopatric speciation: s. with geographic isolation, i.e., the acquisition of isolating mechanisms by a population during a period of geographic isolation ("geographic s."), or the partitioning of species along lines of secondary intergradations, or along lines of strong ecological contrast ("semigeographic s.").

3. Stasipatric speciation (White et al. 1967(: s. whereby an ancestral species gives rise to a number of descendent species through chromosomal rearrangements, arising within its range, spreading to occupy limited areas, and acting as incipient isolating mechanisms in narrow zones of hybridization (White 1973).

4. Instaneous or cytocatalytic speciation: s. by the origination of a single (e.g., alloploid) individual which is reproductively isolated from the species to which its parents belong. It is reproductively isolated from the species to which

its parents belong. It is reproductively and ecologically capable of establishing a new population. An analogous result may be obtained by multiple structural changes of the karyotype (saltational s. of Lewis 1954) which, after hybridization between the ancestral species and the resultant derivatives with structurally changed chromosomes, give rise to a high degree of hybrid sterility.

Species (Ray 1670) - groups of actually or potentially interbreeding natural populations which are reproductively isolated from other such groups. S. are the largest and most inclusive reproductive communities of sexual and cross-fertilizing individuals that share in a common gene pool (Mayr 1940; Dobzhansky 1950). Species containing only a single subspecies are called "monotypic", while those containing more than one subspecies are termed "polytypic" (semispecies, sibling species, superspecies).

Spheroplast (Tulasne et al. 1959) - a bacterial cello (stable protoplast) the wall of which is incompletely removed or reduced to a template moiety. Frequently L form is used synonymously with s.

Spherosome (Dangeard 1919; Perner 1952) - in plants, organelles that correspond to the lysosomes of animal cells.

Spindle - in eukaryotes, an ellipsoidal (bipolar) collection of fibers (s. fibers) visible during nuclear division (mitosis and meiosis) and cytokinesis (Fig. 90). Some spindles are long, thin and pointed at the poles, others are much shorter and may appear barrel-shaped. In the meiosis of some animals, each bivalent has s. of its own and the mitotic apparatus as a whole consists of a bundle of these individual spindles. In some ratus as a whole consists of a bundle of these individual spindles. In some protozoa, the centrioles are intranuclear and the whole mitotic apparatus is formed within the nuclear envelope (intranuclear s.), which does not break down (premitosis). In flagellates and all higher organisms, the centrioles have become extranuclear and division is by eumitosis. The spindles of higher plants are anastral, i.e., they lack asters and centrioles.

Stickiness (Beadle 1931) - a sort of chromosomal "agglutination" of unknown nature which results in a pycnotic or sticky appearance of chromosomes. S. may give rise to sticky adhesions between two or more chromosomes and to the formation of "sticky bridges" at anaphase when the chromosomes or chromatids associated in such adhesions move to opposite spindle poles. Sticky effects result from certain gene mutations and may be induced by the application of a wide variety of chemical and physical agents during the condensed state of chromosomes in mitotis and meiosis.

Subspecies (v. Wettstein 1898; du Rietz 1930) an aggregate of (local) breeding populations of a given species occupying a geographic subdivision of the species range (= geographic race). A s. usually differs from other similar breeding groups of the same species both taxonomically and with respect to certain gene pool characteristics (such as the frequency or prevalence of certain genes: gene frequency). Subspecies like semi species and species may be either allo - or sympatric. The small local populations of which a. s. consists are called "microgeographic races" or "microsubspecies (Huxley 1940). A s. represented by widely separated, phenotypically identical populations is said to be "polytopic".

Substitution line - a line in which one chromosome pair is transferred from the donor variety into the recipient variety (alien addition line, alien substitution line). S. lines are being used in determining the number and dominance relationships of the genes governing a character, for the genetic analysis of quantitative characters, for determining the number of genes controlling a quantitative character on a single chromosome, and for mapping these genes (Khush 1973).

Supergene (Darlington & Mather 1949) - a group of linked genes mechanically held together on a chromosome and usually inherited as a unit. If co-adapted, such genes (which are not necessarily functionally related) may cooperatively produce some adaptive characteristic. Supergenes are preserved intact either by heterozygous structural changes in the chromosomes concerned (inversions or reciprocal translocation suppressing or reducing the amount of genetic recombination via meiotic crossing-over in the relevant segments) or by localized chiasmata formation in the organism under observation.

Suppression (Sturtevant 1920) - the restoration (partial or complete phenotypic repair) of a mutationally altered function of a genotype which may be due to:

Suppressor mutation (Bridges 1932) - a secondary mutation ("second-site mutation") that totally or partially restores a function lost due to a primary mutation which is located at a mutational site different from the primary mutation. In contrast to back mutation, s. m. does not result in a reversal of the original mutation. Frequently suppressor mutations do not lead to a complete suppression or phenotypic repair in the mutants under observation, but only to a partial reversal toward wild-type activity of the double mutants, the s. m. can be separated from the original mutation by genetic recombination.

Sympatric (Poulton 1903; Mayr 1942) - of two or more populations occupying the same geographic area, or of a population existing in a breeding condition within the cruising range of individuals of another population, as opposed to allopatric. S. population existing in a breeding condition within the cruising range of individuals of another population, as opposed to allopatric. S. populations are not necessarily in physical contact, or at least reproductive contact; they may remain completely separated by different habitat preferences, or by different habits.

Synaptonemal complex (Moses 1958) - as seen in the electron microscope a regularly occurring, coplanar set of proteinaceous parallel strands (usually three), coaxial to chromosomes involved in meiotic chromosome pairing.

The presence of the s.c. is restricted to meiotic cells and prerequisite, but not alone sufficient for chiasma formation (and, hence, crossing over). The s.c. shows a basic structural similarity in very diverse eukaryotic organisms: two dense lateral elements flank a relatively unstructured area (the central space), containing a central element of variable prominence. The dimensions and spacing of the various components may vary according to species. The lateral elements vary in diameter from 30 to 65 mm, the central element between 12 and 50 mm, and the central space between 65 and 120 mm. The total width of the s.c. ranges between 160 and 240 mm,

T

Tandem duplication - duplication.

Taxon (Rickett 1958) - a taxonomic group of individuals (population) reconized as a formal unit at any level of a hierarchical classification. The individuals possess certain common characters assumed to be controlled by identical hereditary determinants (genes).

Telocentric (Darlington 1939) - of chromosomes or chromatids with a terminal centromere. Such chromosomes may arise by centromere misdivision or breakage induced within the centromere region. Usually telocentrics are unstable and are either eliminated in the course of a few cell divisions or transformed into so-called isochromosomes. This instability is assumed to be the reason for the absence or rarity of telocentrics in the chromosome complements of wild species.

Telomere (Muller 1940) - a term used to denote the natural unipolar chromosome ends in eukaryotes. They cannot normally be transposed (by chromosomal structural changes) to intercalary positions in the chromosomes. Their absence (deletion) usually impairs normal chromosome behavior (bridge-breakage-fusion-bridge cycle).

Any product of a chromosomal structural change usually requires two telomeres and one centromere in order to survive.

Template - a macromolecular mold for the synthesis of another macromolecule.Through template process a limited number of building blocks (e.g., four nucleotides, twenty amino acids, etc) are polymerized into macromolecular structures whereby the actual arrangement of the building blocks is uniquely determined by a preexisting one, which is identical, complementary, or otherwise related to the newly synthesized macromolecule. The product of a biosynthetic template. Replication of deoxyribonucleic acid (DNA), genetic transcription of DNA into a number of ribonucleic acid (RNA) species, and genetic translation of RNA into polypeptides are the best known template processes (Pontecorvo 1966).

Teratogen - any (teratogenic) agent that produces or increases the incidence of congenital malformations in a population.The best known types of teratogens in man are viruses, radiation, and drugs. To act as a t., an agent must affect some specific metabolic process in the developing embryo. Most teratogens act during a rather limited period at the time the embryo is undergoing rapid organogenesis. The extent of damage produced varies with the timing of teratogen application.

Terminator codon - in prokaryotes and eukaryotes, any of the specific codons (UAA, UAG, and UGA in mRNA) which direct polypeptide chain termination during genetic translation. T. codons are recognized by release factors. UAA (ochre) is probably the chief t. c. while UAG (amber) and UGA (opal) provide accessory fail-safe signals to ensure that termination never fails. Both UAA and UAG are t. codons that lead to 100% efficient translational termination in bacteria. UGA is seldom more than 98% efficient and in a "leaky" t. c., i.e., a terminator apparently misread as an amino acid codon at a low but significant rate (even in non-suppressing strains of bacteria: suppressor mutation). This results in polypeptide chain elongation beyond the t. c. (read-through); the amount of misreading appears to vary with the bacterial strain.

Tetrad (Nemec 1910) - 1. the four chromatids of any bivalent in the first division of meiosis (Nemec).
2. A quartet of cells (each with reduced chromosome number) resulting from a single meiotic event (meiosis) which takes place in a meiocyte or mother cell. In some organisms (certain fungi, bryophytes and algae), all four meiotic products may be recovered and analyzed genetically (tetrad analysis). Among such organisms the immediate products of meiosis (referred to as gones) undergo 1 or 2 subsequent mitoses, giving rise to S or 16 products which represent exact replicates of the original four.

Tetrad analysis - the genetic analysis of all the products of a single meiotic event in organisms with (ordinarily) cytologically observable chromosomes (eukaryotes) and a standard meiosis. T. a. is possible in those organisms in which the four products (or their derivatives) of a single nucleus which has undergone meiosis are grouped together (tetrad).

Tetraploid (Nemec 1910) - of auto or alloploid cells, tissues, and individuals with four chromosome sets in their nuclei (symbol 4n).

Tetrasomic (Blakeslee 1921) - of polysomic cells, tissues, or individuals with one chromosome represented four times instead of twice in an otherwise diploid (disomic) chromosome complement (symbol 2n + 2). Tetrasomics may result from intercrossing or selfing of trisomic individuals (2n + 1). The frequency of tetrasomic individuals among the offspring of such matings depends upon the frequency of transmission (in the male and female line) of the extra chromosome present in trisomics.

Tetrasomics rarely arise spontaneously from diploids.

Transcription (Hayashi et al. 1964; Szybalski et al. 1970) - the general unit of genetic transcription (= scripton) principially defined by the presence of a single autonomous promotor (i.e., a promotor not under the cis dominant control of another promotor). It represents a cluster of genes and may contain a variety of positive and negative control elements (transcriptional control). Its various segments need not be expressed in a co-ordinate manner and may even become physically separated during certain stages of development. Several messenger RNA molecules can be synthesized at varying rates by one t., but the transcription of promotor distal regions must always depend on initiation at the single autonomous promoter. Transcriptons may overlap in divergent manner and, thus, some regions of the (prokaryotic) genome may be part of several transcriptons. The operon is a special, rather simple kind of it.

Transduction (Zinder & Lederberg 1952) - The transfer of genes from one cell to another by means of a viral (in the case of bacteria, a bacteriophage) vector or a vector, which, in its external characters at least, is similar to a virus particle (pseudovirion). The genes picked up by the virus particles are incorporated at the expense of all or part of the virus genome. This deletion of viral genes leads to defects which vary in kind and extent depending on the nature and the length of the host genome segment. (genetic transformation; transfection).

two types of phage mediated t. of bacterial cells may be distinguished:

1. generalized t.: the transducing vector consists of a random fragment of the bacterial chromosome (the exogenote or merogenote) and a phage coat after some occurs during lytic growth. Upon subsequent infection of a bacterial strain by such a particle the bacterial genes are injected and can recombine with the host chromosome. Generalized t. requires homology between the transduced material and the recipient chromosome as well as the host-recombination system (rec). The recombinants arising by this type of t. are stable.

2. Specialized or restricted t.: the transducing genome is a defective prophage which, as a consequence of inproper prophage excision, carries a specific fragment of the bacterial chromosome incorporated into the phage genome. Lysogeny and prophage induction are generally prerequisites for the formation of transducing phages. Only bacterial genes located near the prophage attachment site are subject to this type of t. Since the transduced genes are part of the phage genome, they can insert into the recipient chromosome by the phage integration systems.

Specialized t., therefore, does not require the rec system. Like a prophage, the transducing phage genome can be lost from the host chromosome (curing) and his makes the transducing clones unstable in that inherited genes are subject to loss. The transducing phage genomes are usually deleted in phage genes on the side opposite to the incorporated bacterial DNA. A generalized transducing phage may be capable of specialized t. When its genome is exceptionally linked to a definite bacterial marker.

Transfer RNA (Hoagland et al. 1957) - any of a class of ribonucleic acid molecules (= amino acid acceptor RNA; adaptor RNA; soluble RNA. Symbols tRNA or sRNA) each with a molecular weight of about 25,000 (4 S) consisting of 73 to 90 nculeotides (Fig. 94). Transfer RNA molecules are involved in genetic translation of messenger RNA into the amino acid sequences of protein molecules (aminoacyl tRNA), in nonsense and frameshift suppression (suppressor tRNA), and (in bacteria) in the regulation of a number of amino acid biosynthetic operons.

In genetic *translation*, tRNAs are able to recognize specific amino acids and corresponding genetic code words in messenger RNA. In doing so, they translate the information encoded in nucleic acid messages into protein products. Multiple steps in this process include amino acid binding (aminoacyl-tRNA-synthetase), transport to the ribosome, binding to one ribosomal site (the A site), peptide bond formation, translocation to another ribosomal site (the P site), and release from the ribosome.

transition (Freese 1959) - the substitution of one pyrimidine for another (e.g., replacement of deoxycytosine by thymidine) or of one purine for another at any base location (site) in DNA (or RNA if this is the primary carrier of genetic information, as in some viruses). At the next replication of the altered region, a similar t. will be induced in the daughter strand, i.e., pyrimidine t. will result in a purine t. in the newly synthesized complementary strand. With this type of base pair substitution the purine-pyrimidine orientation is preserved. By means of spontaneous or induced transitions gene mutations (t. mutations) may arise. Any transition mutation can revert spontaneously or by any transitional agent.

transversion (Freese 1959) - the substitution of a purine for a pyrimidine or of a pyrimidine for a purine at any base location (site) in DNA (or RNA if this is the carrier of primary genetic information). T. results in a distortion of the nucleic acid backbone in addition to base pairing difficulties, since two purines hydrogen-bonded in the helix would increase the diameter of the molecule while hydrogen-bonded pyrimidines would decrease it. Transversions may give rise to gene mutations (t. mutations) which can be detected by means of amino acid analysis. No chemical or physical mutagen has been shown to induce transversion exclusively (tran-sition). Transversions may come about by, e.g., the mistaken pairing of one purine with either another purine or with a gap produced by depurination, or the mistaken pairing of one pyrimidine with another pyrimidine. Transition agents are unable to revert transversion mutants and transversion agents do not revert transition mutants.

trisomic (Blakeslee 1921) - of polysomic cells, tissues, or individuals with one or more (doubly t., etc.) extra chromosomes in an otherwise diploid (disomic) chromosome complement. This phenomenon is referred to as trisomy and is symbolized by 2n + 1, 2n + 1 + 1, etc. The extra chromosome is either completely (primary trisomics) or partially homologous (secondary and tertiary trisomics) to certain chromosomes of the standard complement.

1. *Primary trisomy* (Blakeslee 1924; Rhoades 1933; Goodspeed & Avery 1939): The extra chromosome is completely homologous to one pair of homologues in the chromosome complement and, therefore, one chromosome is represented three times in an otherwise diploid complement. Such homology permits trivalent formation at the first meiotic division. The trivalents may assume various configurations (multivalent). Instead of one trivalent, a bivalent plus a rodshaped univalent may be produced (Fig. 100). The number of different primary trisomics possible for a given karyotype corresponds to the number of chromosomes per chromosome set.

The principal sources of p. trisomics are (Khush 1973): 1. normal disomics (spontaneous occurrence or by treatment with physical and chemical agents); 2. asynaptic and desynaptic disomics; 3. polyploids (e.g. haploids, triploids, and tetraploids); 4. translocation heterozygotes; 5. primary trisomics, tetrasomics, multiple trisomics, and secondary, tertiary, and compensating trisomics; 6. monosomics; and 7. other chromosomal abnormalities. Frequently, primary trisomics differ from normal diploids and from each other in a number of phenotypic characters due to changes in the genic balance brought about by the extra chromosome. Their viability and fertility is usually lower than that of the diploids from which they arose.

2. **Secondary trisomy** (Blakeslee 1924; Rhoades 1933; Goodspeed & Avery 1939): the extra chromosome has two identical arms and is an isochromosome. For each primary two types of secondaries are possible. Thus, in secondary trisomics, one chromosome segment (arm) is represented four times in an otherwise diploid chromosome complement.

The diagnostic pairing configuration at the first meiotic division is either a ring trivalent (which only rarely occurs in primaries), or a bivalent and one U-shaped ring resulting from the internal pairing of the univalent with identical arms. Secondary trisomics may occur among the progeny of normal diploid individuals, but their best source is a among the progeny of individuals with one or more univalent chromosomes in which centromere misdivision takes place.

3. Telosomic trisomy (Burnham 1962): the extra chromosome is a telocentric fragment chromosome homologous to one arm of a chromosome pair in the standard chromosomes. Its ends are homologous with the ends of two different chromosomes. Tertiary trisomics occur regularly (and possibly exclusively) in the progenies of translocation heterozygotes by 3:1 segregation of the four chromosomes which are partly homologous in such heterozygotes and may pair meiotically as a quadrivalent.

The meiotic pairing configuration diagnostic for a tertiary trisomy is a chain of five chromosomes or two pairs of chromosomes connected by a single chromosome. A tertiary trisomic individual has two chromosome segments represented three times each in an otherwise diploid complement. The potential number of different tertiaries for a given karyotype is almost infinite.

5. Compensating trisomy (Blakeslee 1927): one chromosome of the diploid standard complement is missing but is compensated for by the presence of two other chromosomes which together are equivalent to the missing one. The missing chromosome in a compensating trisomic may be compensated for by: 1. two tertiary chromosomes, 2. two isochromosomes, 3. one tertiary and one telocentric chromosome, 4. one iso- and one telocentric chromosome, and 5. one iso- and one tertiary chromosome (Khush 1973). Compensating trisomics may occur among the progeny of individuals with a ring multivalent consisting of six chromosomes due to the presence of two heterozygous reciprocal translocations. They may also be derived from a secondary trisomic crossed with the proper interchange. In the case of compensating trisomy, a five-membered ring pairing configuration may occur at the first meiotic division.

U

Univalent - an unpaired chromosome (chromosome pairing) at the first meiotic division which corresponds to a particular chromosome of the mitotic complement (bivalent; multivalent). Univalents either lack a homologue or result from asynapsis (absence of meiotic pairing is due either to asynaptic genes or to the influence of environmental factors) and desynapsis (premature dissociation of pairing configurations in diplotene due to the absence of chiasma formation).

Univalent chromosomes usually lie at random on the spindle at metaphase I of meiosis and any follow two main courses at anaphase (Darlington 1957):

1. Those lying near the equator are randomly distributed (without division) to the spindle poles ("postdivision");
2. those lying near the equator move onto the plate where they orientate themselves axially and divide into their two chromatids which in turn pass to opposite poles ("predivision").

Some univalents may not be included in one of the daughter nuclei and are then lost in the cytoplasm (formation of micronuclei). Further more, the presence of several univalents in one meiocyte may completely hinder the separation of the chromosomes into two daughter groups and result in a restitution nucleus. Univalents frequently show centromere misdivision at first or second meiotic division.

V

Variance - a numerical calculation describing the extent of dispersion of data around a mean (symbol s^2).

Variation - the occurrence of heritable or nonheritable differences in the permanent structures of cells ("intraindividual v."), among individuals of one population ("individual v."), or among populations ("group v.).

The chief sources of v. among the characteristics of related organisms are genic differences (due to mutation or new assortment of genes) or environmentally induced differences which only cause temporary changes of the phenotype.

Primarily, biological v. may be subdivided into three categories:

1. phenotypic variation: The total biological v. of a given character (symbol V^P). V. without natural discontinuities is called continuous v. and characters that exhibit it are quantitative or metric characters, as opposed to discontinuous v. and qualitative characters.

The components of continuous and discontinuous phenotypic v. in a segregating population may be grouped into two major classes referred to as genetic (symbol V^G) and nongenetic or environmental (symbol V^E4) variation. By definition, phenotypic v. is then $V^P = V^G + V^E$.

2. Environmental variation: V. due to all intra- and extracellular factors which influence the expression of the genotype. This part of v. is expressed in terms of environmental variance (symbol V^E) and consists of two main components. One is represented by the intangible statistical residue known as "error" and certain genotype-environment interactions; the other is the controllable environment.

3. Genetic variation: V. due to the contribution of segregating genes and gene interactions (symbol V^G) represented as the proportion of the total phenotypic variance that is exclusively genetic. This production is called the heritability (symbol H) and is commonly expressed as a percentage (H = 100% if all the variation is due to genetic causes or there is no environmental variance; as the environmental component of variance increases, H decreases):

$$H = \frac{V_G}{V_P} \text{ or } H = \frac{V_G}{V_G + V_E}$$

The main components of genetic variance are "additive genetic variance" (due to any two alleles that have different quantitative effects contributing to this component), "dominance devi-ations" (deviations from additive variance due to dominance) and "interaction" or "epistatic deviations" (deviations from additive variance due to epistasis and other nonallelic interactions).

Genetic v. in a population is controlled by three main factors (Mayr 1963): The input of new genetic information due to mutation and gene flow; the erosion of this variation be selection and sampling errors; and the protection of the stored variability by cytophysiological devices and ecological factors (genetic variability).

Viroid (Diener 1971) - any of a class of plant pathogenic agents consisting of and infectious, single-stranded free RNA molecule with unusually low molecular weight (about 50,000). The RNA has transfer RNA-like conformation, i.e., the single-strand is folded in such a way that it has a partial double-stranded character.

Virosome (Dahl & Kates 1970) - any of a class of ribonucleoprotein particles (heterogeneous in size and density) which contain (vaccinia) virus DNA and can be found in the cytoplasm of cells infected with this virus. They contain the newly replicated viral DNA and an endogeneous RNA polymerase activity.

Virulent - of bacteriophages which are incapable of establishing the lysogenic response (temperate).

Virus - any of the infectious, subcellular and ultramicroscopic particles (usually nucleoprotein, more exclusively DNA or RNA in nature) representing potentially pathogenic agents, whose replication in the cell and transmission by infection results in characteristic reactions of the host cells and host individuals. Viruses are independent genetic systems (endowed with genetic continuity and mutability) which possess and evolutionary history of their own. They use the synthetic machinery of living cells to direct the synthesis of specialized particles, the virions, which contain the viral genome and transfer it to other cells. major groups of viruses are bacterial viruses (bacteriophages), animal viruses, and plant viruses.

V-type position effect (Lewis 1950) - position effect

W

Wobble (Crick 1966) - the ability of the third base in the transfer RNA anticodon (5' end) to hydrogen bond with any of two or three bases at the 3' end of the codon in mRNA, i.e., to undergo unusual base pairing. Due to w. a single tRNA species can recognize several different codons. Although w. allows some limited ambiguity in the codon - anticodon recognition, this is compensated by the degeneracy of the genetic code, so that no mistranslation follows. The rules of w. are:

X

X-chromosome (Henking 1891; McClung 1902) - in organisms with male heterogamety and diploid sex differentiation ("diplogenotypic sex determination"), the sex chromosome which is present in both sexes (sex linkage). One sex (the homogametic, female sex) is homozygous (XX) with respect to this chromosome, the other (the heterogametic, male sex) is heterozygous (XO or XY). In organisms with haploid sex differentiation ("haplogenotypic sex determination"), the sex-chromosome of the female.

Xenia (Focke 1881) - seeds whose hybrid nature is recognizable by particular phenotypic characters (color, shape, size, etc.) owing to direct influences exerted by the genotype of the pollen on the embryo or on the maternal tissue (endosperm) of the fruit.

X-inactivation - dosage compensation.

Y

Y-chromosome (Wilson 1909) - in organisms with male heterogamety and diploid sex differentiation ("diplogenotypic sex determination"), the sex chromosome limited to one sex (the heterogametic, male sex) and which pairs with the X-chromosome (sex linkage). In organisms with haploid sex differentiation ("haplogenotypic sex determination"), the sex chromosome of the male sex.

Part - 3

Specialized Genetics: Bio-Technology

A to V

Adventitious - Developing from unusual points of origin, such as shoots or root tissues, developing from callus or embyros, developing from sources other than zygotes.

Adventive embryony - Embryo formation and development resulting from asexual cells as occurs in vivo in certain members of the Rutaceae.

Androgenesis - Development of plants from male gametophytes.

Aneuploid - When the nucleus of a cell does not contain an exact multiple of the haploid number of chromosomes; one or more chromosomes being present in a greater or lesser number than the rest. The chromosomes may not show rearangements.

Antibiotics - Substances secreted by fungi or some bacteria which are capable of inhibiting the growth or killing various bacteria. They can be made naturaly using appropriate microrganisms, or synthetically in the laboratory.

Antibody - Specific protein produced by the immune system of higher animals and humans as part of the immune response to the presence of a specific antigen.

Antigen - Substance or well defined part of a substance which is recognized and bound by a matching antibody.

Aseptic cuture - Raising cultures from a tissue or an organ after freeing it of bacteria, fungi, and other mieroorganisms.

Assay - Technique for measuring a biological process.

Attenuated - Weakened. Applied to vaccine formation, a pathogen used to induce antibody formation has been treated to render it incapable of causing a disease.

Autoimmune Disease - An immunological disorder in which the body produces antibodies against its own tissues.

Auxins - A class of growth hormones which cause cell elongation, apical dominance, root initiation, Indole acetic acid (IAA), naphthalene acetic acid (NAA), indolebutyric acid (IBA), and 2,4-dichlorophenoxyacetic acid (2, 4-d) are some of the auxins commonly used in tissue culture.

Axenic-Totally free form association with other organisms.

Axillary - Developing in the axil of leaves, e.g., axillary bud.

B

Bacillus substilis - A bacterium used in genetic engineering experiments because of its importance in secreting proteins it makes.

Bacillus thuringiensis - A common soil bacterium which produces a protin toxic to insects.

Base - A component of the nucleotides which make up the DNA Four different organic bases are involved in the structure of the DNA: adenine, guanine, cytosine and thymine. Their sequence is responsible for the genetic information in the DNA molecule.

Base Pair - A unit of the DNA double helix consisting of a base of each DNA strand loosely connected to each other. Only two pair combinations are possible: adenine-thymine and guanine-cytosine.

Batch Culture - A suspension culture in which cells grow in a finite volume of nutrient medium and follow a sigmoid pattern of growth.

Batch Processing - An industrial fermentation technology in which defined amounts of inorganic and living material are joined in a bioreactor. The desired product is selectively removed upon completion.

Biocatalyst - An enzyme that accelerates a biochemical reaction.

Biochips - Biological molecules that can replace semicondutors in electronic circuits.

Bioconversion - Capable of being broken down into simpler molecular components by living organisms such bacteria or fungi.

Bioelectro catalysis - Incorporation of enzymes into fuctioning electrodes and their use to catalyze electrochemical and energy transfer reactions.

Biohazard - Term used by scientists to denote that genetically engineered organisms are being experimented on; intended as a warning to be cautions.

Biomass - The total mass of living matter in a given area in biotechnologically produced organismes or substances.

Bioprocess - Use cells or cellular components to produce a desired end product.

Bioreactor - Apparatus used for bioprocessing.

Biosynthesis - Formation of complex biological molecules from simpler ones by living organisms.

Biotechnology - Development of products by exploting biological processes or substances. Production may be carried out by using intact original or modified organisms, e.g. bacteria or yeasts, or by using acive cell components such as enzymes from organisms.

B Lymphocytes (B Cells) - Lymphocytes developed in bone marrow and involved in the production of antibody.

C

Callus - Cluster of undifferentiated plant cells that may be induced to form whole plants by suitable treatments.

Carcinogen - Agent that can induce cancer.

Cell culture - A technique for growing cells outside the body of n organism.

Cell Fusion - Formation of a hybrid cell by fusing two different cells.

Cell Line - Cells grown by culture methods having the same genetic makeup.

Chemostat - A growth vessel that maintains constant volume of cells or microbes by adding fresh nutrient medium and removing spent culture.

Clone - A group of individual organisms or cells all derived from a single progenitor by asexual reproduction and genetically identical to it.

Cloning - Multiplication method viaclones.

Co-Metabolism - Process in which a substrate is modified but is not utilized for growth by an organism that is grown on or metabolizing another substrate.

Complementary DNA (cDNA) - DNA synthesized from a messenger RNA rather than from a DNA template. This type of DNA is used for cloing a gene or as a DNA probe to locate specific genes.

Continuous Processing - A bioprocessing method in which new material is added and the products removed continuously at a rate that maintains the volume at a constant level.

Cosmid - A vector for carrying large DNA fragments into host cells; made in the laboratory. It is formed from a plasmid by introducing "cos" (insertion) sites from lambda phage DNA at two exposed ends of the plasmid.

Cybrid - The viable cell resulting from the fusion of a cytoplast with a whole cell, thus creating cytoplasmic hybrid.

Cytokinins - A class of growth hormones which cause cell division, cell diferentiation, shoot differentiation, breaking apical dominance, etc. Some of the cytokinins commonly used in tissue culture are kinetin, benzylaminopurine (BPA), 2- isopentenyladenine (2-ip), and zeatin.

Cytoplast - The intact cytoplasm remaining following the enucleation of cell.

Cytotoxic - Capable of killing cells.

D

Diagnostics - Agents used as a help to diagnose disease or disorders, i.e., identify a disease or disorder and distinguish one from another.

Differentiation - The development of different physiological and/or morphological characteristics; the formation of different cell types, roots, shoots, embryos, or any other organ in the callus or cell culture.

DNA Probe - Usually a nucleic acid (or some other molecule) which has been labelled in some way and used to locate a particular base sequence or gene or a DNA molecule.

Downstream Processing - Stages in industrial processing following the bioconversion step (e.g., fermentation). Includes separation, purification and packaging of the product.

E

Electrophoresis - Separation of molecules based on their differing electrical charges.

Embryo - A very young plant developing inside the female gametophyte with or without fertilization.

Embryogenesis - The process of embryo initiation and development.

Embryoid - Non-zygotic embryo formed in culture.

Encapsulated Embryos - Plant embryos derived from somatic cells (somatic embryogenesis) artificially encapsulated together with nutrients and possibly also growth enhancers and pesticides. Intended for replacing seeds.

Endonuclease - Enzyme that breaks nucleic acids at specific internal sites, producing fragments of various lengths.

Epigenetic Variation - Phenotypic variability which has a non genetic basis.

Escherichia coli: (E.coli) - A bacterium that normally inhabits the intestine of mammals. commonly used in genetics experiments.

Excise - A segment of DNA that is transcribed into mRNA and translated into protein.

Exon - A segment of DNA that is transcribed into mRNA and translated into protein.

Explant - Tissue taken from its origins state and transferred to an artifical medium for growth or maintenance.

F

Feedstock - Raw material used in chemical or biological processing.

Fermentation - An anaerobic process that releases energy from a sugar of other fermentable substance. Used to synthesize various products.

Friability - A term indicating the tendency for plant cells to separate from one another.

G

Gene - A segment of DNA carrying, in its base sequence, a very specific information. Some genes carry the information for the synthesis of proteins (structural genes), others carry information for regulatory functions (regulatory genes).

Gene Therapy - Therapy for congenital diseases involving the replacement of a deficient gene.

Geneic Engineering - A technology used to alter the genetic make up of living cells through direct interferece with the genome in order to make them capable of producing substances or performing functions alien to the unmanipulated cells.

Genome - The total set of hereditary elements in a cell or organism.

Genomic Library - Fragments of cloned DNA from a single species of organism obtained by restriction enzyme digests; fragments are used to locate specific genes using the hybridizaton technique.

Genotyope - The genetic make-up of a individual as determined by the set of genes carried in the chromosomes.

Germplasm - The sum total of genetic variability available to a particular population of organisms.

H

Habitution - Ability of cells to grow in the absence of phytohormones after prolonged cultivation in the presence of hormones.

Heterokaryon - A cell possessing two or more genetically different nuclei, in a common cytoplasm usualy derived as a result of cell-to cell fusion.

Heteroploid - The term given to a cell culture when the cell comprising the culture possess nuclei containing chromosome numbers other than the diploid number. This is a term used only to describe a culture and is not used to describe individual cells. Thus, a heteroploid culture would be one which contains aneuploid cells.

High Fructose Corn Syrup (HFCS) - Sweeteners made of maize, rich in fructose.

Homokaryon - A cell possessing two or more genetically identical nuclei, in a common cytoplasm, derived as a result of cell-to-cell fusion.

Host-Vector system - Combination of DNA-transporting unit (vector) and DNA receiving cell (host); used for introducing foreign DNA into a cell or organism.

Huymoral Response - An immune response involving the production of antibodies by B lymphocytes.

Hybrid cell - The term used to describe the mononucleate cell which results from the fusion of two different cells, leading to the formation of a synkaryon.

Hybridoma - A hybrid cell that produces monoclonal antibodies in culture; formed by the fusion of a myeloma (cancer) cell with a normal antibody producing lymphocyte.

I

Immune System - In vertabrates, the surveillance system that recognizes and acts against alien invaders or foreign cells.

Immunoassay - Technique for identifying substances by using antibodies that combinbe with them.

Immunodeficiency - Deficiency in the normal defence reaction of higher organisms against a foreign substance especially against disease causing agents.

Immunoglobulins - Circulating proteins comprising the antibodies.

Immunosuppressant - A substance which suppresses the immune system of an animal to lessen the body's rejection of, for example, transplanted organ.

Immunotoxin - A molecule that kills cells attached to an anthibody.

Induction - Intitiation of a plant structure, organ, or process in vitro.

Interferon - A chemical messenger of the immune system that inhibits viral replication and may have anticancer properties; there major types (alpha, beta and gamma) are known.

Interleukin - A class of lymphokines important in the function of the immune system.

Intron - 'A part of eucaryotic gene that does not code for protein.

In vitro - Literally "in glass", now applied to any process carried out in sterile cultures.

In Vitro Propagation - Propagation of plants in a controlled, artifical environment, using plastic or glass culture vessels, aseptic techniques, and a defined growing mdium.

In vivo - Literally "in life", applied to any process occurring in a whole organism.

J

Juvenile - A phase in the sexual cycle of a plant characterized by differences in appearance from the adult and which lacks the ability to respond to flower-inducing stimuli.

K

Karyoplast - A cell nucleus surrounded by a narrow rim of cytoplasm and a plasma mem-brance which is obtained form the cell by enucleation.

L

Ligase - Enzyme used to join nucleic acid fragments together.

Lymphokine - A class of soluble proteins that play a role in the immune response whose mechanism is not yet understood.

Lymphoma - Cancer of the lymph tissue.

M

Macrophages - White blood cells that ingest dead cells and other debris in tissues and are involved in the production of interleukin; may also kill tumour cells when exposed to lymphokine "macrophage activating factor."

Mericloning - A popular term, not in scientific usage, referring to the in vitro vegatative propagation of orchids from excised shoot tip, axillary buds or floral organs.

Meristem Culture - In vitro culture of a generally shiny dome like structure measuring less than 0.1 mm in length when excised, most often excised from the shoot apex.

Microprogation - This term is synonymous with in vitro propagation.

Monoclonal Antibodies - Highly specific antibodies derived from only one clone of a specific hybridoma cell and hence exactly of the same type. They recognize only one specific site of an antigen.

Morphogenesis - The porocess of growth and development of differentiated structures; the evolution of a structure from an undifferentiated to differentiated state.

Mutagenesis - A process by which changes in genetic constitution of a cell through alterations in its DNA are brought about.

Mutant - A phenotypic variant resulting from a changed or new gene.

Myeloma - A type of cancer cell used in the monoclonal antibody technique to form hybridomas.

O

Oligonucleotide - Short segments (up to about 10 nucleotides) of DNA or RNA.

Oncogene - Cancer-causing gene.

Organogenesis - The evolution, from dissociated cells, of structure which show natural organ form or function or both.

P

Parasexual Hybridization - Hybridization by non-sexual methods, e.g., by protoplast fusion.

Phagocyte - The name adopted by Tulecke et al., (1965) for an apparatus designed for the semi-continuous chemostat culture or plant cells.

Plantlet - A small rooted shoot or germinated embryo.

Plasmid - A small rooted shoot or germinated embryo.

Plasmid - A small bacterial DNA circle occurring in the bacterial cell separated from the bacterial chromosome. A plasmid carries additional genes not essential for bacterial growth. Plasmids can be replicatedindependently from the bacterial chromosome and can be passed on to another bacterial cell upon contact. They are efficient tools for genetic engineering.

Promoter - A DNA site where RNA polymerase initiates transcription.

Protoplast - A plant cell without its outer retaining cell wall.

Pseudodiploid - This describes the condition where the number of chromosomes in a cell is diploid but, as a result of chromosomal rearrangements, the karyotype is abnormal and linkage relationships my be disrupted.

Pseudogene - A silent gene; a copy of a gene that is not transcribed.

R

Radioimmunoassay - A technique that uses a radioactively labelled antibody to identify a molecule or measure a process.

Recombinant DNA (rDNA) - Segments of DNA from two different organisms spliced together in the laboratory into a single molecule.

Regulatory Gene - A gene that acts to control the protein-synthesizing activity of other genes.

Replicon - Any genetic element that can reproduce independently, e.g., chromosome or a plasmid.

Restriction Enzyme - An enzyme that catalyzes the cleavage of DNA at a highly specific state. Each one recognizes a specific site. Each one recognizes a specific DNA sequence where it catalyzes cleaveage.

Retroviruses - A virus belonging to a group of extremely small viruses whose carrier of genetic information is not DNA but RNA. In these retroviruses, RNA is first reverse transcribed into DNA which is then inserted into the genome of the infected host cell, thus changing its genetic composition. This can be one of the reasons for cancerous diseases.

S

Somatic Cell Hybridization -The in vitro fusion of animal cells or plant protoplasts derived from somatic cells which differ genetically.

Somatic Embryogenesis - Induction via hormones of the development of an embryo out of a somatic cell cluster, in plants.

Subculture - With plant cultures, this is the process by which the tissue or explant is first subdivided, then transferred into fresh culture medium.

Super ovulation - The maturation and release of more than the usual number of ova in an animal induced by the additional application of hormones.

Splicing - A stage in RNA processing in which introns are removed and exons are joined to form a continuous coding sequence of RNA.

Suspension culture - A type of culture in which cells, or aggregates of cells, multiply while suspended in liquid medium.

Synchronous Culture - A culture in which the cell cycles (or a specific phase of the cycle) of a proportion of the cells (often a majority) are synchronous.

Synkaryon - A hybrid cell which results from the fusion of the nuclei it carries.

T

Tissue Culture - Commonly used as a blanket term to refer to all types of aseptic plant or animal cultures. Strictly speaking. however, this term should include only the culture of unorganized tissue or callus.

Tisssue Plasminogen Activator - A protein involved in the process of dissolving blood clots. It activates the enzyme plsinogen which is important in dissolving blood clots.

T-lymphocytes (T CELLS) - Lymphocytes which recognize, engulf, and destroy specific foreign cells.

Totipotency - A cell characteristic in which the potenital for forming all the cell types in the adult organism is retained.

Transfection - The transfer, to cells in culture, of a gene(s) from another cell.

Transgenic Organism - A genetically manipulated organism containing in its genome one of more inserted genes of another species.

Transposon - A segment of DNA carrying one or more genes that can move from one DNA molecule to another to result in a change of the altered DNA.

Turbidostat - An open contiuous culture into which fresh medium flows in response to an increase in the turibidity of the culture. A preselected biomass density is uniformly maintained by wash-out of excess cells.

V

Vaccine - A preparation containing an antigen made up of whole disease causing organisms (attenuated or parts of such organism; used ot confer immumity against the antigen.

Variant - A culture exhibiting a stable phenotypic change whether genetic or epigenetic in origin.

Vector - Agent used to transfer DNA into a host cell, e.g., plasmid or bacteriophage.

Vegetative Propagation - Reproduction of plants using a non-sexual proces involving the culture of plant parts such as stem and leaf cuttings.

Part - 4
Gene Cloning

A to Y

2 um circle - A plasmid found in the yeast *Saccharomyces cerevisiae* and used as the basis for a series of cloning vectors.

Adaptor - A synthetic, double-stranded oligonucleotide used to attach sticky ends to a blunt-ended molecule.

Agrobacterium tumefaciens - The soil bacterium which, when containing the T1 plasmid, is able top form crown galls on a number of dicotyledonous plant species.

Autoradiography - A method of detecting radioactively labelled molecules through exposure of an X-ray sensitive photographic film.

Auxotroph - A mutant micro-organism that will grow only if supplies with a nutrient not required by the wild-type.

Avidin - A protein that has a high affinity for biotin and is used in the detection system for biotinylated probes.

B

Bacteriophage or Phage - A virus whose host is a bacterium. Bacteriophge DNA molecules are often used as cloning vectors.

Baculovirus - A virus that has been used as a cloning vector for the production of recombinant protein in insect cells.

Batch culture - Growth of bacteria in a fixed volume of liquid medium in a closed vessel, with no additions or removals made during the period of incubation.

Biolistics - A means of introducing DNA into cells that involves bombardment with high-velocity microprojectiles coated with DNA.

Biological containment - One of the prcautionary measures taken to prevent the replication of recombinant DNA molecules in microorganisms in the natural environment. Biological containment involment. Biological containment involves the use of vectors and host organisms that have been modified so that they will not survive outside of the laboaratory.

Biotechnology - The use of living organisms, often but not always microbes, in industrial processes.

Biotin - A molecule that can be incorporated into dUTP and used as a non-radioactive label for a DNA probe.

Blunt end or Flush end - An end of a DNA molecule at which both strands terminate at the same nucleotide position with no single stranded extension.

Bovine papilloma virus (BPV) - A group of mammalian viruses, derivatives of which have been used as cloning vectors.

Broth culture - Growth of microorganisms in a liquid medium.

Buoyant density - The density possessed by a molecule or particle when suspended in a aqueous salt or sugar solution.

G

Capsid - The protein coat that encloses the DNA or RNA molecule of a bacterio-phage or virus.

Cassette - A DNA sequence consisting of promoter-ribosome binding site -unique restriction site-terminator carried by certain types of expression vector. A foreign gene inserted into the unique restriction site will be placed under control of the expression signals.

Cauliflower mosaic virus (CaMV) - The best studied of the caulimoviruses, used as a cloning vector for some species of higher plant.

Caulimoviruses - One of the two groups of DNA viruses to infect plants, the members of which have potential as cloning vectors for some species of higher plants.

Cell extract - A preparation consisting of a large number of broken cells and their released contents.

Cell-free translation system - A cell extract containing all the components required for protein synthesis (i.e. ribosomal subunits, tRNAs, amino acids, enzymes and cofactors) and able to translate added mRNA molecules.

Chimaera - A recombinant DNA molecule made up of DNA fragments from more than one organism, named after the mythological beast.

Chromosome - A self-replicating nuc-leic acid molecule carrying a number of genes.

Chromosome walking - A technique used to identify a series of overlapping restriction fragments, often to determine the relative positions of genes on large DNA molecules.

Cleared lysate - A cell extract that has been centrifuged to remove cell debris, subcellular particles and possibly chromosomal DNA.

Clone - A population of identical cells, generally those containing identical recombinant DNA molecules.

Compatibility - Refers to the ability of two different types of plasmid to coexist in the same cell.

Competent - Refers to a culture of bacteria that have been treated to enhance their ability to take up DNA molecules.

Complementary - Refers to the ability of two different types of plasmid to coexist in the same cell.

Competent - Refers to a culture of bacteria that have been treated to enhance their ability to take up DNA molecules.

Complementary - Refers to two polynucleotides that can base-pair to form a double-stranded molecule.

Complementary DNA (cDNA) cloning - A cloning technique involving conversion of purified mRNA to DNA before insertion into a vector.

Conformation - The spatial organization of a molecule. Linear and circular are two possible conformations of a polynucleotide.

Conjugation - The process whereby two bacteria exchange genetic material via a transient intercellular connection called a pilius.

Consensus sequence - A nucleotide sequence used to describe a large number of related though non-identical sequences. Each position of the consensus sequence represents the nucleotide most often found at that position in the real sequences.

Continuous culture - The culture of microorganisms in liquid medium under controlled conditions, with additions to and removals form the medium ovar a lenghty period of time.

Copy number - The number of molecules of a plasmid contained in a single cell.

Cos site - One of the cohesive, single-stranded extensions present at the ends of the DNA molecules of certain strains of 1 phage.

Cosmid - A cloning vector consisting of the 1 cos site inserted into a plasmid, used to clone DNA fragments up to 40 kb in size.

Covalently closed-circular (CCC) - A completely double-stranded circular DNA molecule, with no nicks or discontinuities, usually with a supercoiled conformation.

D

Defined medium - A bacterial growth medium in which all the components are known.

Deletion analysis - The identification of control sequences for a gene by determining the effects on gene expression of specific deletions in the upstream region.

Denaturation - Of nucleic acid molecules; breakdown by chemical or physical means of the hydrogen bonds in base pairing.

Density-gradient centrifgation - Separation of molecules and particles on the basis of buoyant density, by centrifugation is a concentrated sucrose or caesium chloride solution.

Dideoxynucleotide - A modified nucleotide that lacks the 3'hydroxyl group and so prevents further chain elongation when incorporated intoa growing polynucleotide.

Disarmed plasmid - A Ti plasmid that has had some or all of the T-DNA genes removed, so it is no longer able to promote cancerous growth of plant cells.

DNA sequencing - Determination of the order nucleotides in a DNA molecule.

Double digestion - Cleavage of a DNA molecule with two different restriction endonucleases, either concurrently or consecutively.

E

Electrophoresis - Separation of molecules on the basis of their net electric charge.

Electroporation - A method for increasing DNA uptake by protoplasts through prior exposure to a high voltage which results in the temporary formation of small pores in the cell membrance.

End-filling - Conversion of a sticky end to a blunt end by end by enzymatic synthesis of the complement to the single-stranded extenstion.

Endonuclease - An enzyme that breaks phosphodiester bonds within a nucleic acid molecule.

Episome - A plasmid capable of integration into the host cell's chromosome.

Ethanol precipitation - Precipitation of nucleic acid molecules by ethanol plus salt, used primarily as a means of concentrating DNA.

Exonuclease - An enzyme that sequentially removes nucleotides from the ends of a nucleic acid molecule.

Expression vector - A cloning vector designed so that a foreign gene inserted into the vector will be expressed in the host organism.

F

Fermenter - A vessel used for the large - scale culture of micro-organisms.

Footprinting - The identification of protein-binding site on a DNA molecule by determinging which phosphodiester bonds are protected from cleavage by DNase 1.

G

Gel electrophoresis - Electrophoresis performed in a gel matrix so that molecules of similar electric charge can be separated on the basis of size.

Gel retardation - A technique that identifies a DNA fragment that has a bound protein by virtue of its decreased mobility during gel electrophoresis.

Geminivirus - One of the two groups of DNA viruses that infect plants, the members of which have potenitial as cloning vectors for some species of higher plants.

Gene - A segment of DNA that codes for an RNA and/or polypeptide molecule.

Gene cloning - Insertion of a fragment of DNA, carrying a gene, into a cloning vector, and subsequent propagation of the recombinant DNA molecule in a host organism.

Gene mapping - Determination of the relative positions of different genes on a DNA molecule.

Genetic engineering - The use of experimental techniques to produce DNA molecules containing new genes or new combinations of genes.

Genetics - The branch of biology devoted to the study of genes.

Genome - The complete set of genes of an organism.

Genomic library - A collection of clones sufficient in number to include all the genes of a particular organism.

H

Harvesting - The removal of microorganisms from a culture, usually by centrifugation.

Helper phage - A phage that is introduced into a host cell in conjunction with a related cloning vector, in order to provide enzymes required for replication of the cloning vector.

Heterologous probing - The use of a labelled nucleic acid molecule to identify related molecules by hybridization probing.

Homology - The degree of identity displayed by the nucleotide sequencs of two related but not complementary polynucleotides. 85% homology means that 85 nucleotide positions out of 100 are identical in the two polynyucleotides (e.g. AAAAA) to the end of a nucleic acid molecule, usually referring to the synthesis of single-stranded homopolymer extensions on the ends of a double-stranded DNA molecule.

Horseradish peroxidase - An enzyme that can be complexed to DNA and with is used in a nonradioactive procedure for DNA labelling.

Host-controlled restriction - A mechanism by which some bacteria prevent phage attach through the synthesis of a restriction endonuclease that cleaves the non-bacterial DNA.

Hybrid-arrest translation (HART) - A method used to identify the polypeptide coded by a cloned gene.

Hybrid-release translation (HRT)-A method used to identify the polypeptide coded by a cloned gene.

Hybridization probe - A labelled nucleic acid molecule that can be used to identify complementary or homologous molecules through the formation of stable base-paired hybrids.

I

Immunological screening - The use of an antibody to detect a polypeptide synthesized by a cloned gene.

Incompatibility group-Comprises a number of different types of plasmid, often related to each other, that are unable to coexist in the same cell.

Induction - (1) Of a gene; the switching on of an expression of a gene or group of genes in response to a chemical or other stimulus. (2) Of I, phage: excision of the integrated form of induction, and switch to the lytic mode of infection, in response to a chemical or other stimulus.

Insertional inactivaton - The cloning strategy whereby insertion of a new piece of DNA into a vector inactivates a gene carried by the vector.

Insertion vector - A insertion vector constructed by deleting a segment of non-essential DNA.

In situ hybridization - A technique for gene mapping involving hybridization of a labelled sample of a cloned gene to a large DNA molecule, usually a chromosome.

In vitro mutagenesis - Any one of several techniques used to produce a specified mutation at a predetermined position in a DNA molecule.

In vitro packaging - Synthesis of infective particles form a preparation of a capsid proteins and a concatmer of DNA molecules separated by cos sites.

K

Klenow fragment - (of DNA polymerase 1) The enzyme that sythesizes a new DNA strand on an existing template, used primarily in DNA sequencing.

L

Labelling - The incorporation of a radioactive nucleotide into a nucleic acid molecule.

Lambda (z) - A bacteriophage that infects E. coli, derivatives of which are extensively used as cloning vectors.

Ligase (DNA ligase) - An enzyme that repairs single-stranded discontinuities in double-stranded DNA molecules in the cell. Purified DNA ligase is used in gene cloning to join DNA molecules together.

Linker - A synthetic, double-stranded oligonucleotide used to attach sticky ends to a blunt-ended molecule.

Lysogen - A bacterium that harbours a prophage.

Lysogenic infection cycle - The pattern of phage infection that involves integration of the phage DNA into the host chromosome.

Lysozyme - An enzyme that weakens the cell walls of certain types of bacteria.

Lytic infection cycle - The pattern of infection displayed by a phage that replicates and lyses the host cell immediately after the initial infection. Integration of the phage DNA molecule into the bacterial chromosome does not occur.

M

M13-A bacteriophage that infects E.coli, derivatives of which are extensively used as cloning vectors.

Microinjection - A method of introducing new DNA into a cell by injecting it directly into the nucleus.

Minimal medium - A defined medium that provides only the minimum number of different nutrients needed for growth of a particular bacterium.

Multicopy plasmid - A plasmid with a high copy number.

Multigene family - A number of identical or related genes present in the same organism, usually coding for a family of related polypeptides.

N

Nick - A single-strand break, involving the absence of one or more nucleotides, in a double-stranded DNA molecule.

Nick translation - The repair of a nick with DNA polymerase I, usually to introduce labelled nucleotides into a DNA molecule.

Northern transfer - A technique for transfering bands of RNA from an agarose gel to a nitrocellulose or similar membrane.

Nucleic acid hybridization - Formation of a double-stranded molecule by base pairing between complementary or homologous polynucleotides.

O

Oligonucleotide-directed mutagenesis - An in vitro mutagenesis technique that involves the use of a synthetic oligonucleotide to introduce the predetermined nucleotide alteration into the gene to be mutated.

Open-circular - The non-supercoiled conformation taken up by a circular double-straned DNA molecule when one or both polynucleotides carry nicks.

Origin of replication - The specific position on a DNA molecule where DNA replication begins.

Orthogonal field alternation gel electrophoresis (OFAGE) - A gel electrophoresis technique that employs a pulsed electric field to achieve separation of very large molecules of DNA.

P

Partial digestion - Treatment of a DNA molecule with a restriction endonuclease under such conditions that only a fraction of all the recognition sites are cleaved.

Phenotype expression - A technique designed to maximize the transformation frequency obtained when using a plasmid vector.

Pilus - One of the structures present on the surface of a bacterium containing a conjugative plasmid, through which DNA transfer occurs during conjugation.

Plaque - A zone of clearing on a lawn of bacteria caused by lysis of the cells by infecting phage particles.

Plasmid - A usually circular piece of DNA, primarily independent of the host chromosome, often found in bacterial and some other types of cell.

Plasmid amplification - A method involving incubation with an inhibitor of protein synthesis aimed at increasing the copy number of certain types of plasmid in a bacterial culture.

Polyethylene glycol - A polymeric compound used to precipitate macromolecules and molecular aggregates.

Polylinker - A synthetic double-stranded oligounucleotide carrying a number of restriction sites.

Primer - A short single-stranded oligonucleotide which, when attached by base pairing to a single-stranded template molecule, acts as the start point for complementary strand synthesis directed by a DNA polymerase enzyme.

Promoter - The nucleotide sequence, upstream of a gene, that acts as a signal for RNA polymerase binding.

Prophage - The integrated form of the DNA molecule of a lysogenic phage.

Protease - An enzyme that degrades protein.

Protoplast - A cell from which the cell wall has been completely removed.

R

Radioactive marker - A radio-active atom used in the detection of a larger molecule in which it is incorporated.

Random priming - A method for DNA labelling that utilizes random DNA hexamers which will anneal to single-stranded DNA and act as primers for complementary-strand synthesis by a suitable enzyme.

Recombinant - A transformed cell that contains a recombinant DNA molecule.

Recombinant DNA molecule - A DNA molecule created in the test-tube by ligating together pieces of DNA that are not normally contiguous.

Recombinant DNA technology - All the techniques involved in the construction, study and use of recombinant DNA molecules.

Recombinant protein - A polypeptide that is synthesized in a recombinant cell as the result of expression of a cloned gene.

Recombination -The exchange of DNA sequences between different molecules, occurring either naturally or as a result of DNA manipulation.

Relaxed - Refers to the non-supercoiled conformation of open-circular DNA.

Replacement vector - A vector designed so that insertion of new DNA is by replacement of part of the non-essential region of the DNA molecule.

Replica plating - A technique where by the colonies on an agar plate are transferrred en masse to a new plate, on which the colonies will grow in the same relative positions as before.

Replicative form of M13 - The double-stranded form of the M13 DNA molecule found within infected *E. coli* cells.

Reporter gene - A gene whose phenotype can be assayed in a transformed organism, and which is used in, for example, deletion analyses of regulatory regions.

Repression - The switchig off of expression of a gene or a group of genes in response to a chemical or other stimulus.

Restriction analysis - Determination of the number and sizes of the DNA molecules only at a limited number of specific nucleotide sequences.

Restriction fragment length polymorphism (RFLP) - A mutation that results in a detectable change in the pattern of fragments obtained when a DNA molecule is cut with a restriction endonuclease.

Restriction map - A map showing the positions of different restriction sites in a DNA molecule.

RFLP linkage analysis - A technique that uses a closely linked **RFLP** as a marker for the presence of a particular allele in a DNA sample, usually as a means of screening individuals for defective genes responsible for genetic diseases.

Ribonuclease - An enzyme that degardes DNA.

Ribosome binding site - The short nucleotide sequence upstream of a gene, which after transcription forms the site on the mRNA molecule to which the ribosome binds.

S

Selectable marker - A gene carried by a vector and conferring a recognizable characteristic on a cell containing the vector or a recombinant DNA molecule derived from it.

Selection- a means of obtaining a clone containing a desired recombinant DNA molecule.

Shotgun cloning - A cloning strategy that involves the insertion of random fragments of a large DNA molecule into a vector, resulting in a large number of different recombinant DNA molecules.

Shuttle vector - A vector that can replicate in the cells of more than one organism (e.g. in E. coli and in yeast).

Simian virus 40 (SV40) - A mammalian virus used as the basis for a series of cloning vectors.

Southern transfer - A technique for transferring bands of DNA from an agarose gel to a nitrocellulose or similar membrane.

Sphaeroplast - A cell with a partially degraded cell wall.

Stem-loup - A hairpin structure, consisting of a base paired stem and a non-base paired loop, that may form in a polynucleotide.

Sticky end - An efficient promoter that can direct synthesis of RNA transcripts at a relatively fast rate.

Stuffer fragment - The part of a fragment replacement vector that is removed during insertion of new DNA.

Supercoiled -The conformation of a covalently closed circular DNA molecule, which is coiled by torsional strain into the shape taken by a woun-up elastic band.

T

T-DNA -The portion of theTi plasid transferred to the plant DNA.

Temperature - sensitive mutation - A mutation that results in a gene product that is functional within a certain tempertature range (e.g. at less than 30° C), but non-functional at different temparatures (e.g. above 30° C).

Template - A single-stranded polynucleotide (or region of a polynucleotide) able to direct synthesis of a complementary polynucleotide.

Terminator - The short nucleotide sequence downstream of a gene that acts as a signal for termination of transcription.

5'-terminus - One of the two ends of a polynucleotide that which carries the phosphate group attached to the 5' position of the sugar.

3'- terminus - One of the two ends of a polynucleotide; that which carries the hydroxyl group attached to the 3' position of the sugar.

Ti plasmid - The large plasmid found in those Agrobacterium tumefaciens cells able to direct crown gall formation on certain species of plants.

Total cell DNA - Consists of all the DNA present in a single cell or group of cells.

Transcript analysis - Experiment aimed at determining which portions of DNA molecule are transcribed into RNA.

Transfection - The introduction of any DNA molecule into any living cell.

Transformation fequency - A measure of the proportion of cells in a population that are transformed in a single experiment.

U

Undefined medium - A growth medium in which not all the components have been identified.

UV absorbance spectroscopy - A method for measuring the concentration of a compound by determining the amount of ultraviolet radiation absorbed by a sample.

V

Vector - A DNA molecule, capable of replication in a host organism, into which a gene is inserted to construct a recombinant DNA molecule.

Vehicle - Often used as a substitute for the word 'vector', emphasizing that the vector transports the inserted gene through the cloning experiment.

W

Watson-Crick rules - The base pairing rules that underline gene structure and expression. A pairs with T, G with C.

Western transfer - A technique for transfering bands of protein from an electro-phoresis gel to a membrane support.

Y

Yeast artificial chromosome (YAC) - A cloning vector comprising the structural components of a yeast chromosome and able to clone very large pieces of DNA.

Yeast episomal plasmid (YEp) - A yeast vector carrying the 2 xxm circle origin of replication.

Yeast integrative plasmid (YIp) - A yeast vector that relies on integration into the host chromosome for replication.

Yeast replicative plasmid (YRp) - A yeast vector that carries a chromosomal origin of replication.